D1000711

THE WORK ENVIRONMENT

VOLUME ONE

Occupational Health Fundamentals

Doan J. Hansen
EDITOR

Library of Congress Cataloging-in-Publication Data

The work environment / Doan J. Hansen, editor.
p. cm.
Includes bibliographical references and index.
Contents: V. 1. Occupational health fundamentals.
ISBN 0-87371-303-6
1. Industrial hygiene. I. Hansen, Doan J.
RC963.W67 1991
613.6'2—dc20 91-8213
CIP

LEWIS PUBLISHERS, INC.
121 South Main Street, P.O. Drawer 519, Chelsea, Michigan 48118

PRINTED IN THE UNITED STATES OF AMERICA

Preface

The Work Environment: Volume I, Occupational Health Fundamentals introduces workplace health and safety to those who are not themselves occupational health professionals. As defined here, "occupational health" refers to workers' health (or perhaps to their ill health) as it is affected by the chemical and physical stressors present in the workplace.

The professionals who deal with the recognition, evaluation and control of workplace hazards are collectively known as *occupational health professionals.* These professionals work to prevent dangerous exposures in the workplace, or, to mitigate the results of such exposures. This very broad title includes a wide spectrum of specialists: industrial hygienists, occupational health physicians or nurses, safety engineers, toxicologists, radiation specialists, ergonomists and others. An explanation of "industrial hygienist" and "safety engineer" seems appropriate. In greatly simplified terms, an *industrial hygienist* is a professional trained to recognize environmental factors (and their effects on humans), to evaluate the magnitude of those factors and to devise and apply methods for the control and alleviation of adverse health effects. A *safety engineer* is a professional similarly trained in the techniques and means of controlling and preventing accidents.

The need for this volume had become evident during my work as an industrial hygienist when I frequently seemed to be answering many of the same basic questions about occupational health. I looked, unsuccessfully, for a text which appropriately covered the topics I felt needed to be discussed. Consequently, this book is designed to appeal to "*nonoccupational*" health professionals — those who must deal with occupational health issues on the job but find themselves unprepared to respond to some (or all) portions of the work environment.

The full range of occupational health topics is broad enough to require many different chapters with each written by a different expert. All the chapters in this book were written by well-qualified professionals with broad experience and expertise in the fields about which they write.

As the Table of Contents shows, the book itself is divided into sections, with each section made up of topically related chapters. Each section is introduced by a brief preamble which explains (in general) the section's contents. The reading of any chapter (written to introduce the reader to its specific occupational health topic) is not intended to make the reader an expert, however, it should provide readers with a basic knowledge of health and safety hazards to be encountered, and how they should be dealt with, in the work environment.

Doan Hansen

Introduction to the Work Environment

Today we recognize that there often is "something" about many occupations which causes an illness or adverse health effect specific to that trade. The cause of an *occupational disease* can be a chemical agent (coal dust, asbestos, benzene, etc.) or a physical agent (noise, heat, radiation, etc.). Likewise, most people know that employers and employees frequently must take certain precautions to protect health and safety while on the job. For example, coal miners must avoid breathing coal dust because chronic inhalation of coal dust has been shown to lead to a disease known as black lung. Many of us are aware of similar examples in our own vocations or avocations, and just as certainly we are aware of occupationally caused illness and injury from reports in the contemporary media.

Unfortunately, there are other less well-known diseases associated with other trades and/or with other types of exposures. For example, asbestos workers who do not use the proper controls are more likely to get certain lung cancers than are the rest of the population. People who work around sources of infrared radiation, such as glassblowers or furnacemen, can develop a characteristic cataract known as glassblower's cataract. Working around moldy hay can cause farmer's lung; working around bird droppings can lead to bird fancier's disease. While the names and circumstances associated with some of these maladies can seem a little ridiculous, it is important to understand that each represents an occupational disease which has been caused by an occupational (or, on-the-job) exposure to a chemical or physical agent.

Injury as a byproduct of work has been around for about as long as there has been work to do, but in olden times the connection between work and illness or death was not often made. Many occupational diseases, with or without a specific name like those mentioned above, have been identified only relatively recently, that is, now that science and society are sophisticated enough to note excess illnesses within afflicted trades, industries, and professions. In some cases, science is only now identifying occupational illnesses that occurred long ago. One example of this is in the chemical exposures received by painters such as Rubens and Renoir (who used brightly colored paints containing a number of heavy metals). Exposure to these heavy metals can lead to rheumatic diseases (from which a number of painters, Rubens and Renoir included, suffered).

The History of Occupational Health

Because it may be useful, in an introductory examination of the general topic of occupational health, to look at man's past awareness of occupational diseases, certain events related to occupational illness and injury are shown in a chronology presented in Table 1. The list contains a selection of events from around the world which were of significance with respect to the work environment; use the glossary if some of the terminology or abbreviations in Table 1 are unfamiliar to you; also, remember that each chapter of this book introduces and discusses these topics as well.

Table 1. A Selected Occupational Health Chronology

BC	
400	First occupational disease (colic) was observed in slaves working on lead mines by Hippocrates (460-377 B.C.).
23-79	Lead, mercury, sulfur , and zinc were written about as occupational hazards by Pliny the Elder. He also noted that workers in dusty trades tied bladders over their mouths (to breathe), and that workers who used asbestos to make textiles became ill.
AD	
250	Danger of acid mist in copper mines was identified by Galen.
1473	Dangers of metal fumes were identified by Ellenbog in a pamphlet about gold miners (which also discussed carbon monoxide, lead, mercury, and nitric acid). He also provided instruction in preventative measures.
1500s	Paracelsus (1493-1541) wrote a book on occupational diseases. The book identified diseases of miners, which he attributed to vapors from the metals. Paracelsus advised that they be avoided.
1500s	*De Re Metallica,* a series of books about the hazards of mining, was published by Agricola (1494-1555). Agricola recommended the use of ventilation and protective masks (to limit exposures to dusts). As a physician in a mining town, he noticed that some women had been married to (and outlived) many husbands (all of whom were miners and all of whom died from asthma caused by their jobs). Trivia: in 1912 Herbert Hoover and his wife, Lou Henry, translated *De Re Metallica* into English.
1600s	Mercuric nitrate, used by French hatters to make felt, brought about chronic mercury poisoning in members of that trade. Exposure to mercury can lead to symptoms of excitability, etc.; this lead to the phrase "mad as a hatter".

1700s — *De Morbis Artificium* (The Diseases of Workmen), a book dealing with mercury, lead posioning, the pathology of silicosis, etc., was published by Bernardo Ramazzini (1633-1714). Ramazzini is known as the father of occupational medicine. A physician, he advised other physicians to ask patients what their trade was, and, to study work environments to learn about occupational diseases.

1780(ca.) — Devonshire colic shown to have been caused by lead in cider (perhaps the result of acid cider leaching lead from containers).

1788 — The Chimney Sweeps Acts passes (England). Percival Pott (1714-1788) was a London surgeon who had identified cancer of the scrotum in chimney sweeps; this discovery precipitated the legislation.

1800(ca.) — First miner's safety lamp was invented (Sir Humphrey Davey, 1778-1829).

1802 — The Health and Morals of Apprentices Act was passed in England.

1833 — The English Child Labor Laws were passed. This act limited children's hours of work. The act provided a mechanism for inspections of certain industries.

1837 — First U.S. article on occupational disease.

1881 — Switzerland passed the first worker's compensation law. Formerly, workers, if injured (1) had to sue to collect and (2) could not collect any damages if it was shown that the injury was a part of the ordinary risks of the job, or, was due to the negligence of the worker himself or another worker.

1895 — First industrial nurse in U.S., Ada Mayo Stewart, worked as a visiting nurse for the Vermont Marble Company.

1905 — First state (Massachusetts) employed health inspectors to investigate occupational hazards.

1906-07 — The "Pittsburgh Survey" was conducted. This was an exhaustive study of Pittsburgh for one year which revealed that 526 occupational fatalities took place in one county in one year. In addition, according to the survey, another 500 "human wrecks" were maimed.

1908 — The first Federal Compensation Act, covering civil service employees, was passed in the United States. This act gave the right to compensation for occupational injuries to certain employees.

1910 — U.S. Bureau of Mines became an agency in the Department of the Interior.

1911 — First Workmen's Compensation Laws passed (New York and New Jersey).

1911 — Labor unions entered the health field - the International Ladies Garment Workers Union Health Center established.

1913 — U.S. Department of Labor was established. Bureau of Labor Statistics organized to manage data and information about occupational accidents and health problems.

1913 — National Council for Industrial Safety (eventually the National Safety Council) was established.

1914	U.S. Office of Industrial Hygiene and Sanitation was formed (as a part of the U.S. Public Health Service). This agency was a precursor for NIOSH (1979).
1915	American Society of Safety Engineers was organized.
1917	First United States occupational disease legislation was enacted by Hawaii.
1918	Harvard Medical School established the first program leading to advanced degrees in industrial hygiene.
1919	Alice Hamilton, M.D., was appointed first woman faculty member at Harvard.
1923	"Dusty Trades" studies were begun. These studies showed that the incidence of occupational diseases was very high for some trades. For example, there was a 100% incidence of silicosis for granite industry workers with more than 15 years on the job.
1923(ca.)	Radium, used on luminous portions of watch faces, was found to cause fatal cancers of the jaw.
1930-36	The excavation of a tunnel in Gauley Bridge, West Virginia, resulted in many deaths from silicosis in what is known as the "Gauley Bridge Disaster". Estimates of the number of fatalities range from a few hundred to a few thousand (out of a work force of approximately 4000). Contractors did not provide any controls in spite of the fact that they evidently knew that the rock was almost pure silica.
1930s	U.S. Public Health Service initiated a study of mercury exposures for workers in the hat-making industry.
1935	The Social Security Act was passed, providing funds to expand state and local programs.
1936	Walsh Healy Public Contracts Act passed in U.S. This act was a predecessor to OSHA (1970) and outlined safety and health requirements for government contractors.
1938	The ACGIH (American Conference of Governmental Industrial Hygienists) was formed.
1939	The first list of standards for chemical exposures in industry, known as Maximum Allowable Concentrations (MAC), were agreed on by consensus of the American Standard Association and members of the ACGIH.
1946	The American Industrial Hygiene Association Quarterly (a precursor of the *AIHA Journal)* was published for the first time.
1948	All states had worker's compensation laws.
1948	A serious air pollution episode took place in Donora, Pennsylvania, causing approximately 6000 illnesses and 20 deaths.
1961	First issuance of the TLV booklet by ACGIH (TLVs grew out of the MACs which had first been issued in 1939).
1966	Metal and Nonmetallic Mine Safety Act passed, defining health and safety standards for miners (U.S.).

1969	Coal Mine Health and Safety Act (U.S.).
1970	Occupational Safety and Health Act (U.S.). This act created NIOSH, OSHA, and the OSH Review Commission.
1986	Hazard Communication Standard (U.S.)

Note the large number of years separating the early events, which were often just discoveries of what would seem, in retrospect, to have been obvious cause and effect relationships. In more recent years the discoveries appear to have been followed, overshadowed, and in some cases, preempted, by events of a regulatory nature. In historical times, science had not advanced to the point where it was possible to make observations about the cause and effect relationship between work and illness; as a result, such discoveries at first were rare. This has changed as technologies have continued to improve. In more recent times, the increasing regulation regarding occupational health issues could be interpreted as reflecting both an awareness of the costs to society, of occupational injury, in suffering and in dollars, and, less willingness to accept those conditions which were once commonplace.

As science in general has progressed, so too has the general science of occupational health progressed. Table 1 also shows (in later years) an increasing frequency of discoveries and actions by societies in response to those discoveries. Some changes were brought about in response to catastrophic events, and some evolved in more subtle and reasoned manners; the result is today's regulated, relatively socially aware (but often confusing) manner of trying to prevent occupationally caused diseases.

The Work Environment

Every workplace can be thought of as an environment in much the same way that we broadly consider the woods, streams, and trees out of doors to be an environment. Certainly for the people employed there, each workplace is really a *work environment* where there are interactions between people and the chemical and physical demands involved with performing the job. There are also interactions on the work environment from without, such as the regulatory impacts which occupational health legislation has. This book introduces the work environment to those who know little or nothing about it, at least in the occupational health sense, and hopefully can clarify portions of it for those who already do (but who desire an updating in certain areas).

Summary

It is not easy to fully characterize any work environment; it is also not fair or honest to unquestioningly attribute illnesses to exclusively occupational causes, either, since there is ample evidence that many diseases are caused by personal choice such as smoking of tobacco, by genetics, and by nonoccupational accidents (to name but a few). Not all occupational illnesses have cute names; there are many chemicals and physical stressors capable of producing anonymous adverse health effects. This volume should help promote an understanding of exposures to chemicals, relevant laws and exposure limits, how we protect workers, and how we study them to reduce future injury.

Doan J. Hansen, PhD, CIH, is an Industrial Hygienist employed as a Project Engineer in Chemical Emergency Response at Brookhaven National Laboratory, Upton, NY. He is also engaged as the Acting Director of the Occupational Safety and Health Program at New York University, as a Clinical Assistant Professor in Allied Health Sciences for SUNY at Stony Brook and as a private consultant. He has an MPH in Environmental and Industrial Health and a PhD in Industrial Health, both from the University of Michigan, and a BA from Albion College. He is a Certified Industrial Hygienist (CIH) and is an active participant in the AIHA Committees on Emergency Response Planning Guidelines (ERPGs), Exposure Assessment and Computer Applications. He is an avid, but ineffective, golfer.

Acknowledgments

In addition to extending a thank-you to the chapter authors, the editor would like to identify several other individuals and to recognize the assistance they provided:

Thanks to Stan Alter, for coming through with a last minute photograph which saved the day

Thanks to PT, for being on the spot when needed

Thanks to David Birney, my TA, for his conscientous work throughout this project

Thanks to Therese for putting up with me.

DJH

To Jerri, for always transcending the obstacles

Contributors

Andrea Bailey, MS, CIH, is a Corporate Industrial Hygienist for The Upjohn Company, a worldwide manufacturer of pharmaceutical products based in Kalamazoo, MI. She received her Bachelor of Arts degree from Wells College in New York and a Master of Science in Environmental and Industrial Health from The University of Michigan. As an industrial hygienist at Upjohn, her interests lie in methods of exposure control during manufacturing operations, reproductive hazards in the workplace, and the development of workplace exposure limits for pharmaceutical agents.

James E. Brower, PhD, is Group Leader, Center for Assessment of Chemical and Physical Hazards, Brookhaven National Laboratory. Dr. Brower is an environmental health scientist involved in assessing health risks and impacts of toxic chemicals. He manages chemical information resources and databases for worker and community right-to-know programs. He is the editor of the ASTM Special Technical Publication, *Hazard Communication: Issues and Implementation.* He has also served as an environmental health consultant on hazardous waste and solid waste management projects.

Leo J. DeBobes, MA, CSP, is Assistant Director of Environmental Health and Safety with the State University of New York at Stony Brook. He previously held positions in occupational safety and health at Brookhaven National Laboratory, United Technologies Corporation/Norden Systems, Estee Lauder, Inc., and Stone and Webster Engineering Corporation. He holds a BS from the New York Institute of Technology and an MA in Occupational Safety and Health from New York University. He is a Certified Safety Professional and has served as President of the Long Island Chapter of the American Society of Safety Engineers.

David Fulton, BSc, DIH, CIH, ROH, has held senior occupational hygiene positions with Imperial Oil Limited, Ontario Hydro and Clayton Environmental Consultants. He is a Certified Industrial Hygienist and is registered with the Canadian Board of Registered Occupational Hygienists. He is a chemistry graduate from the University of Windsor and has a Diploma in Industrial Health from the Faculty of Medicine, University of Toronto. His research includes the application of sensor technologies to personal industrial hygiene monitoring and he developed the Quest Electronics Questemp I personal heat stress monitor.

Richard P. Garrison, PhD, CIH, CSP, is an Associate Professor of Industrial Health in the School of Public Health at The University of Michigan, Ann Arbor. He has worked in industry for eight years as a consultant and as a corporate manager of occupational health and safety before joining the university faculty. He teaches courses in industrial ventilation and industrial hygiene. His research interests and publications focus on engineering controls, emphasizing ventilation, for occupational health hazards. He is a Certified Industrial Hygienist and a Certified Safety Professional.

Joe Levesque is a graduate from the University of Maryland's Fire Protection Engineering Program. He has three years experience with Factory Mutual Engineering Association as a Loss Prevention Specialist in the Highly Protected Risk Category of property insurance. His most recent experience is with Brookhaven National Laboratory, a 170 building complex employing 3500 people working on basic research programs for the Federal Government.

Lisa D. Lieberman, PhD, is currently Assistant Professor in the Department of Health Education, New York University. Among her teaching, advising, and research activities, she is currently Academic Advisor for the Occupational Safety and Health Education Program. She earned her PhD in Health Behavior and Health Education from The University of Michigan (1986) and a Master of Arts in Health Education from Kent State University (1979). She is a former member of the Health Education faculty at the University of Toledo (Ohio) from 1979 to 1982 and was a Senior Research Scientist in the Division of Health Promotion of the American Health Foundation from 1983 to 1988. Her work has included federally funded school and community-based cigarette, alcohol, and drug prevention research projects. In addition, she has been a consultant to the American Lung Association, the Biological Sciences Curriculum Study, the Women's Action Alliance, and the Federal Employees Health Unit.

Mary Beth Love, PhD, is Chair and Associate Professor of Health Education at San Francisco State University. She works as a consultant to businesses interested in the development and implentation of worksite health promotion programs. Dr. Love is an editor for the American Journal of Health Promotion and has published research in the area of worksite health promotion and individual beliefs and behavior.

William Marletta, MA, CSP, CHCM, is a self-employed safety consultant in Copiague, NY, specializing in the prevention of pedestrian and machine accidents (particularly with respect to trip and slip hazards on stairs, ramps, walks, and floor surface accidents). He has an MA in Occupational Safety and Health from New York University (1985) and a BA in Biology from Indiana University (1977); he is a doctoral candidate in the Occupational Safety and Health Program at New York University. He is a professional member of the ASSE and represented the Long Island chapter as a National Delegate from 1984 to 1989. He is presently the Chairman of the LI-ASSE Education Committee and is also a member of the AIHA, ASTM, NFPA, and HCM. He is Secretary of ASTM Committee F13 (Safety and Traction for Footwear) and is a member of ASTM F06 (Resilient Floor Coverings). He previously was Senior Director for Ogden Risk Management Control Services (a subsidiary of Ogden Allied Services) and also was Regional Casualty Technical Consultant for Loss Control at Commercial Union Insurance Companies.

Joseph L. McGuire, PhD, MS, MPH, founded Acoustical Research International in 1974 and has been actively engaged in hearing conservation and industrial noise abatement activities worldwide since that time. He has served as a principal investigator for EPA and serves on OSHA and AIHA noise advisory committees. He taught at The University of Michigan and the University of Tennessee, Knoxville. He has worked as a research scientist and environmental research engineer for Bendix Research Laboratories, Ford Motor Scientific Laboratory, and the General Motors Co. His last position at Ford Motor was as a consultant to the Vice President of the Engineering and Environmental Activities and Ford World Headquarters Corporate Management. His consulting group specializes in building new chemical plants and industrial plant noise abatement engineering and setting up and maintaining hearing conservation programs and audiometric records. He obtained a Master's degree in engineering, a Master's in industrial hygiene, and a doctorate in occupational health from The University of Michigan, specializing in occupational hearing loss and industrial noise abatement engineering.

Braxton D. Mitchell, MPH, PhD, received his undergraduate degree from Princeton University in 1978 and his MPH and PhD from The University of Michigan (1982, 1987, respectively). He is currently an Assistant Professor in the Division of Clinical Epidemiology, Department of Medicine, at the University of Texas Health Science Center at San Antonio. His current research interests are in the epidemiology of noninsulin-dependent diabetes mellitus, cardiovascular disease, and Baltimore Orioles baseball.

Paul D. Moss, MPH, CIH, is currently Manager of Safety, Health, and Environmental Affairs for the NutraSweet Company in Mt. Prospect, IL. Prior to this position, he had been a corporate manager of health and safety for Ecology and Environment, Inc., particularly for investigations at hazardous waste sites, for 6 years. He received his BS in Environmental Toxicology from the University of California-Davis in 1980 and his MPH from The University of Michigan in 1982.

His special interests include safety and health training, respiratory protection, and personal protective equipment. He is currently Chairman of the American Industrial Hygiene Association (AIHA) Toxicology Committee, and is also a member of the AIHA Gas & Vapor Detection Systems Committee. His credentials include Certification in Industrial Hygiene, and Registration as an Occupational Hygienist by the Canadian Registration Board of Occupational Hygienists.

Mark Puskar, PhD, received his BS in Chemistry from Central Michigan University (1981) and worked as an on-site chemist at hazardous waste clean-up sites in the early 1980s. Dr. Puskar received his MS in 1984 and his PhD in 1986 from The University of Michigan. He is currently Manager of the Corporate Industrial Hygiene Laboratory for Abbott Laboratories in North Chicago, IL, and has numerous publications in the areas of hazardous waste identification and in the development and validation of air monitoring technologies.

Glenn Shor, MPP, PhD, is currently Program Coordinator of the California Occupational Health Program, a joint program of the California State Department of Health Services and the California Public Health Foundation. He is also a lecturer in Public Administration at California State University, Hayward. He has been involved in research and education on workers' compensation issues at the U.S. Department of Labor, the University of California at Berkeley, the California Workers' Compensation Institute, and as a private consultant.

Neil J. Zimmerman, CIH, PE, PhD, is an Assistant Professor of Occupational Safety and Health in Purdue University's School of Health Sciences. He teaches in and coordinates the undergraduate and graduate industrial hygiene programs. His research interests include low-level exposure risk assessment studies for radon and asbestos, indoor air quality, and respiratory protection. He also has worked as a consultant for a number of organizations in comprehensive industrial hygiene, indoor air quality, and asbestos.

Contents

PART I

Evaluating Workplace Hazards

Many years ago man began to realize that working in certain occupations contributed to ill health and death. For example, around 400 BC Hippocrates noted that slaves working in lead mines developed colic (perhaps the first recognized occupational disease). *De Re Metallica,* a series of books written by Agricola in the 1500s (and translated into English in 1912 by Herbert Hoover and his wife Lou Henry), talked about the occupational hazards of mining. Agricola, who was a physician in a mining town, observed that some women were married to (and outlived) many husbands (all of whom were miners and all of whom died prematurely as a result of their profession). *De Moribis Artificium* (The Diseases of Workmen), written in the 1700s by Bernardo Ramazzini (the father of occupational medicine), discussed mercury, lead poisoning, silicosis, etc. Early observations about the adverse effects of occupations often were only made in the form of a body count - that is, after workers had contracted a disease and/or were dying or dead.

Today, the study and prevention of adverse health effects in the workplace is much advanced. The three chapters in the section *Evaluating Workplace Hazard* illustrate some of the methodologies available to *recognize* and *evaluate* hazards in the workplace:

Recognition and Evaluation of Chemical Hazards describes the different types of chemical hazards which might be found in the workplace. This chapter discusses how airborne chemicals in the workplace can be detected and measured.

Toxicological Terms and the Material Saftety Data Sheet (MSDS) addresses the effects which chemicals can have on biological systems such as laboratory animals and the human body. This chapter also introduces one way that our knowledge of chemical hazards can be communicated, namely, the Material Saftey Data Sheet (MSDS).

Occupational Epidemiology: Investigating Health Hazards in the Workplace describes the field of epidemiology. Epidemiology is the study of diseases in populations, including the measurement of diseases, incidence and prevalence rates, etc.

CHAPTER 1

Recognition and Evaluation of Chemical Hazards

Mark A. Puskar

INTRODUCTION

Many adverse health effects, i.e., occupational diseases, affecting workers are caused by the inhalation of airborne chemical contaminants. The inhalation of certain chemicals can lead to long-term health effects, such as lung cancer, or to short-term effects such as dizziness, unconsciousness, etc. Often, multiple exposures, and possibly multiple adverse health effects, can result. For example, a worker spraying paint onto car bodies can create paint mist and solvent vapors, and the inhalation of either of these can be harmful to the worker's health.

Unfortunately, many examples of airborne workplace contaminants exist in daily life. It is the measurement of these workplace air contaminants, and the comparison of the worker's measured exposure level(s) to occupational exposure limits (either legislated by the government or recommended by a professional association or agency) that permits us to evaluate the relative hazard to the worker and to establish proper controls. As a simplistic example, if a worker's exposure to Chemical X was 100 ppm and the standard was 50, the conclusion is that the worker was overexposed to X and that steps should be taken to reduce the worker's exposure. Workplace exposure standards are discussed in another chapter of this book as well as in the Glossary. This chapter addresses chemical hazards found in workplace air: what they are, how they are classified, and how they are collected (or sampled for).

TYPES OF CHEMICAL HAZARDS

Chemical hazards can be classified into three general categories: gases, vapors, and aerosols. In the occupational health professions, a substance is considered to be a gas if at normal temperature and pressure (70°F and 1 atm) its normal physical state is gaseous.[1] A vapor is the gaseous form of a substance normally in a solid or liquid state at normal temperature and pressure. Many organic compounds exist as a gas and a vapor concurrently, therefore, the usage distinction between gas and vapor is not precise. Aerosols are liquid droplets or solid particles dispersed in a stable aerial suspension. Liquid droplet aerosols are typically referred to as mists or fogs. The difference between mists and fogs is the droplet size. Mists are large liquid droplets, usually large enough to be individually visible, while fogs have small particle sizes. Solid aerosols exist as dusts, fumes, smoke, or fibers. Dusts are particles formed from solid organic or inorganic materials by reducing their size through mechanical processes such as grinding, crushing, or drilling. Fumes are generally metal oxides (welding fume) and are formed by processes such as combustion, sublimation, and condensation. Smoke is airborne particulates formed from the incomplete combustion of organic materials (wood, coal, tobacco). Fibers are particles having parallel sides and an aspect ratio (length to width) of 3 or more. Airborne fiber length is generally under 50 μm.

IDENTIFYING AIRBORNE HAZARDS

To control hazards in the workplace, the health professional's first step is to identify the specific hazard. If an airborne chemical hazard is suspected to be present in a workplace, the specific chemical must be identified. Once identified, the next step is to measure personal exposures to the chemical hazard in the workplace atmosphere, and then to compare those exposures to recognized exposure standards. If the concentration of the chemical exposure is near or above the standard, appropriate control steps must be taken to lower the exposure below the standard. These control steps may be in an administrative (rotating employees), a personal (protective equipment respirators) or preferably an engineering (ventilation) form. Engineering controls are always considered the best form of control for chemical exposures; different controls are discussed elsewhere in this book.

In certain situations this sequence of recognition, evaluation, and control process is straightforward, especially when the specific manufacturing process is understood, the chemicals and chemical reactions are known, and the toxicological properties of the raw materials, the intermediates and the products, are completely characterized. In these cases, a series of specific industrial hygiene sampling and analytical methods, which have been individually developed to monitor a specific chemical over a specific concentration range, can be used to estimate the human exposure.

Unfortunately, knowledge of the chemicals used or their toxicology is frequently incomplete. For example, when a potential health hazard situation arises in the workplace, initially it may not be known if the hazard is a chemical contaminant, a physical agent, a biological agent, a psychological effect, or the result of poor employee relations. If the cause is assumed to be an unknown airborne chemical agent, identifying it may not be straightforward and in certain cases, it may be impossible. Certain technologies exist for the identification of unknown chemical hazards in the atmosphere. These analytical technologies can be very useful in specific applications, however, they are limited. Once the chemical has been identified, an exposure guideline may or may not be available for the estimation of a safe exposure level. If the chemical can be identified and an exposure guideline does exist, a 50% possibility exists that an air sampling method may be available to measure its concentration in the atmosphere. If an air monitoring method exists, its accuracy should not automatically be assumed. The method may never have been validated at the concentration, air temperature, or humidity levels of the atmosphere of interest. Also, other compounds in the unknown atmosphere may interfere with the accurate measurement of the compound of interest.

The results from industrial hygiene samples have serious health, economic, legal, and regulatory implications. Possible consequences include whether to allow workers to continue a potentially hazardous operation, to spend millions of dollars on engineering controls to lower exposures, and/or to levy a fine against a company for overexposing its work force. Unfortunately, these decisions may be made often on the basis of a small amount of data that have been generated with an air sampling method, perhaps of questionable accuracy. Therefore, the reliability of air sampling methods and their potential errors must be measured and understood if the data generated from these methods are to be used properly.

For the vast majority of occupational health professionals, the only contact with chemical qualification (identifying the compound) and quantification (measuring the amount present) in the workplace is in the interpretation and use of the monitoring results. For this reason, the limitations of the processes associated with the recognition and evaluation of airborne chemical hazards are highlighted in this chapter and may give the reader a better ability to recognize the advantages and disadvantages of different sampling and analysis technologies. This will better prepare the reader for the interpretation of his/her exposure data.

QUANTIFICATION OF KNOWN CHEMICAL HAZARDS

The quantification of known chemical hazards in the workplace is one of the fundamental aspects of industrial hygiene. To accurately determine the concentration of any compound in air, a validated sampling and analytical method is needed. Developing and validating air sampling methods is difficult, time consuming, and expensive. Industrial hygiene air monitoring methods are available from governmental agencies,[2,3] professional organizations,[4-9] and manufacturers of air sampling products.[10] Accredited industrial hygiene laboratories are also a good source of sampling and analytical methods. A complete list of these laboratories is printed in each March, June, September, and December issue of the *American Industrial Hygiene Association Journal*. The accuracy of any air sampling method should always be thoroughly investigated prior to beginning any monitoring program. The use of two independent sampling and analysis methods is recommended, but may not be possible.

Proper sampling is essential for representative assessments of workplace exposures. If the sample collected is not representative of the actual exposure, all future work performed to analyze the sample is wasted. More importantly, all work toward correcting the potential problem is wasted. The effects of humidity, temperature, coexisting compounds in the atmosphere, sample preservation, and sample analysis quality control must be considered and understood.[11] Samples should be collected by trained personnel and should only be analyzed by accredited laboratories showing proficiency with the specific analysis, such as the American Industrial Hygiene Association's Proficiency Analytical Testing (PAT) program.

MONITORING TECHNOLOGIES

Sampling and analytical methods for industrial hygiene air monitoring can be divided into two main monitor groups, personal and area. Generally, personal monitoring is a two-step process with sample collection in the field followed by sample analysis in the laboratory; however, this is not always true. There are certain personal monitoring systems that give real-time or near real-time results in the field. (Real-time means the user can take an instrument into the field and obtain a reading.[12]) Area monitoring methods are generally real-time or near real-time. This is currently the largest area of new research and product development in the industrial hygiene sampling field.

Personal monitors are used to estimate the actual exposure of the individual in the workplace. The monitor is affixed to the worker in his/her breathing zone and is worn throughout the worker's shift. This is the type of monitoring that is required by the Occupational Safety and Health Act (OSHA) of 1970. Results from personal monitors are usually expressed in time-weighted averages (TWAs), meaning the average concentration over the sampling period. For OSHA permissible exposure limit (PEL) monitoring, this length is 8 hr. For short-term exposure limit (STEL) monitoring, the length is 15 min. For ceiling (C) monitoring, the sampling length is defined as instantaneous; however, direct reading techniques are not available for all compounds and 15-min STEL sampling may be needed to measure ceiling exposures.

Active methods and chemical dosimeters are thc two main sampling technologies for sampling using personal monitors. The majority of published industrial hygiene air sampling methods are active methods, which use a calibrated sampling pump to draw air through the collection media. The pumps are calibrated against a primary standard (generally a bubble meter) prior to and after sampling. More specifically, a small mechanical pump is fastened to the worker's belt. A piece of tubing attaches to the sampling media from the pump. When the pump is started, workplace air (and its airborne contaminants, if present) is drawn through the sampling media. If the proper sampling media has been chosen, the airborne contaminants will be collected, depending on the type of sampling media and contaminant. After a known period of time, the pump is turned off and the pump and sampling media are removed from the worker. The sampling media is then sent to a laboratory for analysis. After chemical analysis of the sampling media, it is possible to calculate the worker's exposure. In short, the mass of chemical contaminant found is divided by the volume of air pumped to express the airborne contaminant as a concentration (usually either in ppm, mg/m^3, f/cc, or f/cm^3).

SAMPLING MEDIA

Vapors and gases are collected using either liquid impingers (bubblers), solid sorbents, or whole air samples collected in either gas samplings bags, cylinders, or evacuated bulbs. The type of collecting media is prescribed in the published air sampling method. Impingers are liquid holders, generally made out of either glass or plastic, designed to allow the sampled air to be drawn through the liquid. The compounds in the air sampled are dissolved in the solvent. Solid sorbents are usually packed into glass tubes that are attached to a sampling pump. The pump draws the air through the tube, across the sorbent surface where the compounds of interest are either adsorbed or derivatized onto the sorbent. Activated charcoal is the most widely used sorbent, while silica gel, alumina, and porous polymers (Tenax) are used for specific applications. Generally, solid sorbents are preferred over impingers and whole air samples because they are simple to use, lightweight, easy to transport, and concentrate the sample. By concentrating the sample prior to analysis, the limit of detection of the air sampling method will be lower and precision error will be smaller.

Aerosols (dusts, fibers, smoke, mist, fumes, or fogs) are usually collected using either impingers, filters housed in cassettes, sorbent tubes, or impactors. Common filter media used include cellulose acetate, polyvinyl chloride, polycarbonate, glass fiber, and Teflon®.* The filter cassettes may contain cyclones, which are used to separate respirable dust from nonrespirable dust by creating a vortex within the collector, propelling nonrespirable particles to locations where they may be removed from the collector.

Due to the limited capacity of sorbents, most industrial hygiene sampling trains

---

* Registered trademark of E. I. du Pont de Nemours and Company, Inc., Wilmington, DE.

(the sampling media plus pumps) have duplicate samplers inline (i.e., one sampling media followed by another). The second sampler is a backup for the collection of any of the compounds of interest that may have escaped a loaded primary collector. Most solid sorbents have backup sections of sorbent for this reason and after sample collection, both sections are analyzed separately. If 5% or more of the total amount of the sample collected is found on the backup section, the capacity of the collector has been exceeded and breakthrough is said to have occurred. The sampling result is questionable and resampling is necessary.

Impinger sampling is generally avoided, if possible, due to the complexity of the sampling train, the possibility of the worker spilling the impinger sample, and in some cases, the collection solution may be as harmful to the worker as the compound being monitored. At times there are few alternatives, and impinger sampling must be used.

The rate at which the air sample is drawn through the collection media (sampling rate) usually ranges from 20 mL/min to 2.5 L/min; however, each method has a stated optimum sampling rate range that was experimentally determined during its development and validation. Many times two different types of collection media may be used in series (two filters or a filter/sorbent tube) to collect both the particulate and vapor of a compound of interest in the work environment.

CHEMICAL DOSIMETERS

Chemical dosimeters (passive/diffusional dosimeters) are an alternative sampling method for the collection of gases and vapors. With passive monitoring, no sampling pump is needed to collect the sample. The compound is collected by the principle of mass transport across a diffusion layer or by permeation through a membrane as the rate-limiting step.[13,14] The rate at which a given gas permeates a given membrane is a simple function of concentration in the air and the time of exposure. The major advantages of passive monitoring are the simplicity of performing a monitoring survey. No sampling pumps need to be purchased, calibrated, or maintained. Numerous passive monitoring methods have been developed for monitoring organic compounds and are also available for inorganic compounds such as NO_2, ammonia, and Hg. Most chemical dosimeters must be returned to an industrial hygiene laboratory for analysis; however, recently, many new products such as direct reading or on-site analysis-capable chemical dosimeters have been introduced into the marketplace. The accuracy of many of these new techniques have not been verified independently.

Fundamentally, passive monitoring should work as well, if not better, than active monitoring methods. However, recent studies performed on some of these products have identified manufacturing quality control problems that will affect the ability of the monitor to meet the manufacturers' stated accuracy requirements.[15,16]

STORAGE OF SAMPLES

A time when industrial hygiene samples are commonly compromised (unknowingly ruined) is during shipping and storage by elevating the sample temperature. Most common organic solvents collected on charcoal will migrate during storage and must be kept at -4°C until analysis. Although the rate of sample loss is compound- and method-specific, most industrial hygiene samples must be analyzed within a few weeks of collection.

ANALYSIS

The most common form of analysis for industrial hygiene air samples is desorbtion of the compound of interest from the sorbent into a solvent followed by analysis of the solution. When sampling for organic compounds, carbon disulfide (CS_2) is typically used as the desorption solvent for two reasons: (1) a wide range of compounds are soluble in CS_2, and (2) more importantly, flame- and photoionization detectors do not respond highly to CS_2, making separation and analysis easier. A secondary method used to remove the compounds of interest from solid sorbent collection devices is thermal desorption. Instead of liquid extraction of the solid sorbent, the sampling tube is heated and the adsorbable compounds are purged directly into either the analytical instrument or a secondary cryogenic trap (liquid nitrogen) for focusing prior to analysis. Thermal desorption eliminates the use of solvents and is more sensitive than solvent desorption techniques; however, this technique does not lend itself well to the automated analysis of samples that is necessary for the economical analysis of industrial hygiene samples. The high sensitivity of thermal desorption is generally not needed for typical PEL monitoring; however, thermal desorption has become a valuable technique in indoor air quality monitoring.

Organic samples are generally analyzed by chromatography separation (gas, liquid, thin-layer, or ion) with detection by flame ionization, photoionization, spectrophotometry, or mass spectrometry. Inorganic samples are generally analyzed by atomic adsorption, X-ray fluorescence, inductively coupled plasma emission spectroscopy, neutron activation, electrochemical detection, or chemical selective electrodes.

Dust samples may be dissolved in a solvent and analyzed using one of the techniques listed above to determine the amount of the active ingredient in the sample or are analyzed with an analytical balance to determine the total mass of dust. Note that total dust samples must be weighed prior to sampling.

Industrial hygiene monitoring results are reported in mass of analyte per a specific air volume sampled, in mg/m^3 or $\mu g/m^3$. Gases and vapors can also be reported in ppm. The relationship between these two methods of reporting results is discussed in detail in Reference 17 and in the Glossary; however, it is worth noting that ppm in air is not a direct mass per volume relationship, as is the case with liquid samples. Molecular weight, temperature, and pressure must also be considered.

AREA MONITORS

Area monitoring is useful in the identification of exposure sources and is sometimes useful in the estimation of human exposure; however, it cannot be used as an alternative to personal monitoring to meet OSHA standards. The most common use is as a real-time alarm after accidental chemical releases. All personal monitoring techniques can be used as area monitoring techniques, but because samples collected with these methods generally need to be sent to an accredited industrial hygiene laboratory for analysis, these methods are not as useful as having direct real-time alarms. The following discussion of area monitoring techniques focuses on direct reading techniques, whose primary advantage is the ability to generate air concentration data in real-time (within seconds of sampling). The main disadvantage is their lack of specificity — the ability to differentiate between similar compounds. Successful monitoring of atmospheres with area monitors requires knowledge of the other compounds in the air in conjunction with the compound of interest.

Area monitors can be separated into two systems: mobile and fixed. Mobile direct reading systems allow the individual collecting the data to move from source to source, while fixed systems generally have an active sampling system that draws air samples from different locations into a fixed detector system for analysis. The major advantages of mobile systems are their low cost and portability. The major advantages of fixed systems are that they are microprocessor based and therefore monitor continuously without constant human interaction. Prior to use, quantitative area monitors must be calibrated across the concentration range of interest. Semiquantitative area monitors (detector tubes) should be checked at a specific standard concentration to verify that they are functioning properly.

MOBILE AREA AIR MONITORS

If the sampling atmosphere has already been characterized, a mobile area air monitoring technique can be identified to measure the concentration of the compound(s) of interest. These monitors are available for most gases, vapors, and aerosols. The most common techniques for gases and vapors are based on colorimetry (detector tubes), ionization of carbon (flame ionization detectors [FID] and photoionization detectors [PID]), electrochemical cells, the absorption of infrared (IR) radiation (mobile IR units), and chemiluminescence. The most common techniques for aerosols are based on gravimetric, electrostatic, and light-scattering-based analyses. The selection of the method of choice is usually based on sensitivity, specificity, portability, speed of analysis, and cost.

These techniques have distinct advantages and disadvantages depending on the monitoring situation. The major disadvantage in monitoring the concentration of a specific compound is the positive and negative interferences in the measurement caused by other compounds in the atmosphere being monitored. Direct detector-based mobile area monitors (detector tubes, FIDs, and PIDs) are very useful when the organic compound of interest is the only major compound present. For example, if that compound is benzene, any of the technologies previously mentioned can be used to quantify its concentration; however, if other similar organic compounds (toluene and xylene) are present, interference problems may occur. It must also be stressed that these technologies do not differentiate between the compound of interest and any interferences. Again, using the example of benzene, a detector tube will change color in the presence of most aromatic hydrocarbons. Therefore, simply because the benzene detector tube changed color, the presence of benzene cannot be confirmed since toluene will also change the color of the tube.

Mobile predetection separation techniques, such as gas chromatography (GC), can separate the compound(s) of interest from others in the environment for carbon ionization detectors. Using the benzene example, a set of chromatographic conditions could be set up on a mobile gas chromatogram to separate benzene, toluene, and xylene in an air sample and quantify each; however, known air standards of each compound must also be prepared in order to calibrate the instrument. Chromatographic separation techniques are limited in that numerous compounds may be retained by the same chromatography column for the same amount of time, e.g., three peaks on a gas chromatogram do not ensure that three compounds are present in the air sample; however, recent improvements in mobile gas chromatograms, including column ovens, column backflushing, and microprocessors have greatly increased their usefulness as mobile air monitors. Generally, these techniques will work well if it is known which compounds are present and if a set of standards can be prepared containing each of the major components.

FIXED POINT MONITORS

Fixed point air monitoring systems are typically used as evacuation warning systems in processes where the sudden release of a hazardous chemical may threaten life. They are also used to monitor the concentration of specific chemical compounds at or near OSHA PEL levels as indicators of the effectiveness of engineering controls. The major advantage of these systems is that they are designed to work continuously.

Fixed point air monitoring systems have four main components: an air sampling system, a separation and/or detector system, an instrument calibration system, and a microprocessor controller. The air sampling system may be a series of Teflon® or stainless steel air sampling lines connected to a sampling pump, in the case of a GC-based system, or a series of optical sources and mirrors, in the case of a laser- or IR-based system. The technology of the separation and detector systems is the same as in the mobile area monitors except that they are completely computer controlled in fixed point air monitoring systems. Certain fixed point systems also have built-in calibration systems. For example, a GC-based fixed point system for monitoring ethylene oxide (EtO) in a hospital products sterilization facility will have an automated calibration system for the calibration of the detector to EtO. This calibration system will contain a permeation device inside an oven set to a specific temperature. The known permeation rate of this device at the oven temperature will have been previously determined gravimetrically. By diluting the permeated EtO in the oven with a known amount of air, an exact EtO standard concentration airstream can be produced. In the case of EtO, this will probably be at the PEL of 1.0 ppm. After a series of samples have been processed through the instrument, the calibration airstream is sampled and the detector response is checked with the the new standard stream. An adjustment of the compound response factor (peak concentration ÷ peak area) will be made if the detector response has changed since the last calibration. Because fixed point systems are microprocessor controlled, automated reports can be generated from the data of each sampling point in a timely manner. The major disadvantages of fixed point monitoring systems are their complexity and cost. These instruments need a large amount of maintenance and initial costs may exceed $100,000.

IDENTIFICATION AND QUANTIFICATION OF UNKNOWN CHEMICAL HAZARDS

The identification of unknown chemical hazards in the workplace has received much attention over the past 10 years. The reasons for this are directly related to increased energy costs and the problem of hazardous waste handling and disposal. With the energy crises of the 1970s, emphasis was placed on reducing heating and cooling costs in buildings. This was accomplished by increasing the amount of insulation and reducing the amount of fresh air changes distributed throughout buildings. These changes also increased the amount of worker complaints associated with increased levels of unknown chemical contaminants in the air. These higher levels are typically referred to as sick building syndrome.[18]

The handling of hazardous wastes and the clean up of abandoned hazardous waste sites have directly exposed a large number of workers to unknown chemical hazards in the workplace. Sampling and analysis methods were needed to identify the major constituents of the waste so that proper worker and community protection programs could be developed.[19] Currently, only screening techniques exist for real-time identification of unknown chemical hazards; for example, a series of detector tube screening techniques exist for the classification of unknown chemical hazards.[20] Unfortunately, the "unknown" is classified only into general categories such as hydrocarbon, amine, or acid.

Gas Chromatography/Mass Spectroscopy

The majority of organic "unknown" identification is performed by collecting the air sample on a solid sorbent, usually charcoal or Tenax, followed by either liquid or thermal desorption and then analysis by gas chromatography/mass spectrometry (GC/MS). GC/MS is the same technique used to confirm traces of illegal drugs in blood and urine samples. The mass spectrometer is another detector that is connected to a gas chromatograph (separation tool). Unlike the previously discussed FID and PID, the mass spectrometer is a truly qualitative tool (i.e., it can identify unknowns).

The mass spectrometer converts the compounds of a sample into rapidly moving ions and resolves them on the basis of their mass-to-charge ratios, which produces a chart called a mass spectrum.[21] The mass spectrum produced is characteristic of the compound, and each organic compound has a unique mass spectrum. Large numbers of mass spectra of known compounds have been assembled, digitized, and stored in libraries on computer systems used to operate mass spectrometers. After generating the mass spectrum of an unknown compound, the unknown spectrum is compared to the known spectra in the libraries. The closest matches are usually returned for inspection by the chemist. Final confirmation is made by a trained mass spectroscopist.

Because mass spectra are easier to interpret than nuclear magnetic resonance (NMR) or IR absorption spectra, the GC/MS technique has become the definitive tool for the identification of unknowns; however, the GC/MS air monitoring technique does have major shortcomings when it is used to identify unknowns:

1. The GC/MS technique can only identify compounds that have been collected by the collection sorbent selected. For example, if Tenax was selected for sampling and the unknown of interest was not readily collected by Tenax, the technique would not identify the unknown. These types of problems are limited by collecting a single air sample with an array of different sampling media, each designed to collect a specific class of organic compound.
2. The mass spectrometer spectral searching algorithms can only match unknown spectra to known spectra of pure compounds in the libraries. Not only must the unknown of interest be included in the library, but the chromatography column selected to perform the chemical separation must possess the ability to separate the unknown from other compounds in the sample. The selection of a proper column is difficult when dealing with unknowns; however, this problem has been limited by both the advent of fused-silica capillary columns, some with lengths <100 m for increased separation, and new computer searching algorithms for the interpretation of the mass spectrum of mixtures.

3. The majority of organic compounds known to man are not chromatographable with GC. This means that a gas chromatograph cannot be used as a predetection separation technique prior to identification with a mass spectrometer. These compounds are thermally labile and will break down at typical GC temperatures. For example, most pharmaceuticals are nonchromatographable by GC. Separation of these compounds is usually performed by high performance liquid chromatography (HPLC). Liquid chromatography/mass spectroscopy (LC/MS) systems are available for the identification of unknowns. These are extremely complicated and expensive pieces of equipment.

It is best to keep in mind the limitations of the sampling and analysis method when interpreting data generated from an air sample collected to identify unknowns by GC/MS. If the GC/MS technique has identified a series of compounds, the compounds were probably there (if the blanks were clean). If the GC/MS technique has not identified the hazard, it does not automatically mean that the unknown hazard is not an organic compound.

Development and Validation of Industrial Hygiene Methods

Because of its limited resources, the National Institute for Occupational Safety and Health (NIOSH) has been unable to keep pace with demands to generate and validate industrial hygiene sampling methods. The validation procedure followed during the joint NIOSH/OSHA Standards Completion Program[22] was considered brief validation at best. Due to the need for measuring employee exposure to a variety of chemical agents, many corporations have developed industrial hygiene method generation laboratories to meet their own demands for industrial hygiene methods. Although validation criteria and guidelines have been developed by NIOSH[23] and others,[24] a generally accepted procedure for the development and validation of industrial hygiene methods does not exist.

The criterion for validated sampling and analytical methods for workplace air monitoring is defined in terms of total method accuracy,[23] a function of both the precision and the bias of the method. Precision is a measurement of the reproducibility of the method. If six samples of the same atmosphere were collected, how close to each other would the six results be? The precision of a method is estimated by the pooled coefficient of variation (CV_t) of multiple samples over the validation range of concentrations. Bias (B_i) is a measurement of the closeness between the concentration determined by the method and the "actual" concentration. Bias is calculated as:

$$B_i = \frac{\text{Amount Found} - \text{Amount Known}}{\text{Amount Known}} \quad (1)$$

Having estimated both the precision and the bias of a method, the total method accuracy can be calculated:

$$\text{Accuracy } (\pm\%) = [1.96\,(CV_t) + |B_i|] \times 100 \qquad (2)$$

The NIOSH criterion for industrial hygiene method accuracy states that the total method accuracy (sampling plus analysis) should be less than 25% in at least 95% of the samples analyzed between 0.5 and 2.0 times the level of the standard. It appears that the ±25% was picked arbitrarily, based on previously set analytical chemistry guidelines.[25]

At first glance, a total method accuracy of ±25% seems quite large; however, when experiments are designed to estimate this accuracy at the 95% confidence interval, the number of validation samples needed to estimate the accuracy make the validation project both complicated and expensive. It is also worth noting that although many NIOSH/OSHA methods met the ±25% accuracy criteria during the Standards Completion Program, results of the interlaboratory PAT program suggest that a large number of these methods do *not* currently meet the criteria.[26] The reason for this decrease in accuracy is that the exposure standards have been lowered over the past 15 years, while the sampling and analysis methodologies have remained constant. As the concentrations measured by a method approach the method's limit of detection, total error increases.

When one considers that trace chemical analyses are being conducted for hundreds of occupationally hazardous substances for which no collaborative testing, complete method validation, or reference sample programs have been established, the magnitude of this "suspect" quality problem becomes evident. Also, obviously, the method is only as good as the person collecting the sample. If the sample collector makes random and/or systematic errors during the collection process, the sample's results are flawed even with an accurate method.

The initial step in the development of a new air sampling method takes place in the library, where two questions must be asked: Does a validated method already exist? If not, does a validated method exist for a similar compound? One must then obtain an exposure guideline: Over what concentration range is this new method expected to accurately monitor? Setting exposure guidelines is best left to trained occupational toxicologists.[27]

The new method should be validated over a range of concentrations bracketing the exposure guideline. Although the NIOSH criteria states 0.5 to 2 times the exposure guideline as the initial concentration range for validation, 0.1 to 10 times the exposure guideline is recommended as a better starting point. This wider validation range may limit further work if an exposure guideline is lowered.

In the development and validation of a new air sampling method, there are three major steps: analytical validation, laboratory validation, and field validation. Figure 1 is a flow diagram used by an industrial hygiene laboratory within Abbott Laboratories during the development and validation of industrial hygiene air sampling methods.

Once it has been verified that a method has not already been developed and validated for the compound of interest, it becomes important to identify analytical techniques that are suitable for that class of compounds. For example, if a method

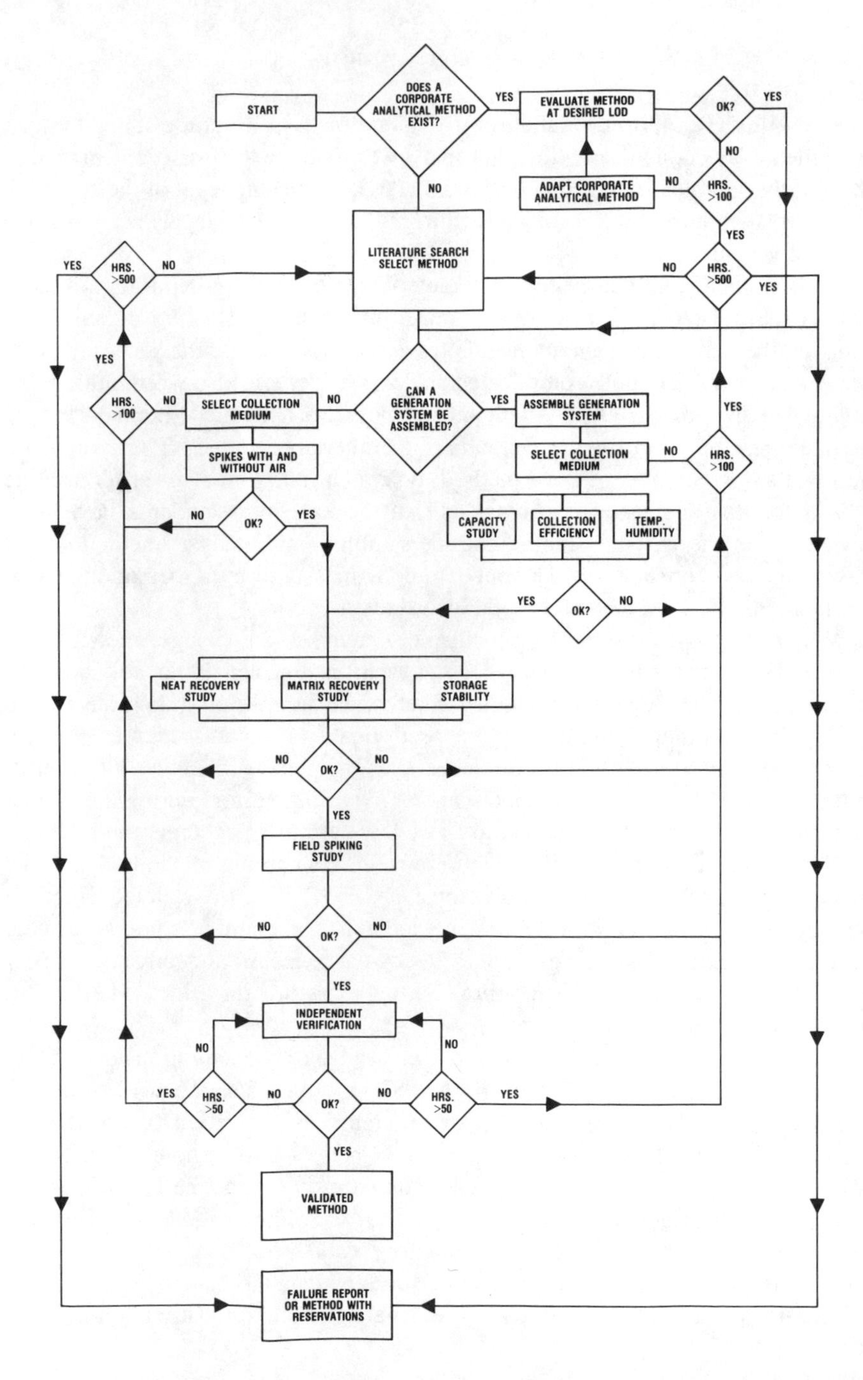

Figure 1. Development and validation of industrial hygiene air sampling methods: flow chart of process. (From Corporate Industrial Hygiene Laboratory, Abbott Laboratories, Chicago, 1989. With permission.)

is under development for a new phenothiazine pharmaceutical, analytical techniques for the identification of similar phenothiazines are useful.

Based on the exposure guideline of the compound of interest, the expected sample desorption amount (5 mL is typical for a 37-mm filter), and the limit of detection of the analytical technique, an analysis technique is identified that can detect a fraction of the exposure guideline (usually 10%). Analytical standards stability is also studied and documented.

Once an analytical technique with sufficient sensitivity has been identified, the next step is the selection of an appropriate collection media. Most of this selection process is based on past experience with similar compounds as well as trial and error. One must first investigate whether the compound of interest can be spiked onto the collection media at the lowest concentration expected and recovered. If these experiments are successful, known concentration dynamic atmospheres are prepared for further laboratory studies.[17] These studies include analyte capacity on the collection media, sample recovery, and sample stability over the expected storage time, temperature, and humidity ranges.

Laboratory chamber studies are an important factor in the validation of an industrial hygiene monitoring methods; however, these controlled experiments generally lack the capability to evaluate the ability of a new method to:

1. Accurately monitor under transient conditions, including variations in concentration, temperature, and humidity over the course of one sampling period. For example, a typical laboratory chamber exposure of a group of EtO monitors is made at 1.0 ppm, at 25°C, and 60% relative humidity for 8 hr. In an actual exposure setting, the concentration, temperature, and humidity may change dramatically over the course of an 8-hr shift, but still average the chamber readings.
2. Study the effects of chemical interferences common in specific workplaces. What affect these additional components have on measuring the correct concentration of the compound of interest must be understood.
3. Study the effects of shipping samples over long distances over extended periods.

For these reasons, field validation of a method is necessary after laboratory validation is completed. The major problem with field validation studies is that the "actual concentration present" is generally not known in the field atmosphere, therefore, method accuracy cannot be calculated. Thus, the new method is generally compared side by side to a reference method and closeness of the results is reported. In some cases the new method is compared during a field validation to other sampling methods that may be "suspect" themselves.

When reference methods are not available, the field validation study usually involves collecting samples that have been previously spiked with a known amount of the compound of interest in both suspected clean and contaminated areas. The recovery of these spike samples are compared to normal samples collected side by side. Using these data along with trip spikes (spiked samples sent into the field and returned to the laboratory without additional sampling), a level of confidence regarding the ability of the method to accurately collect the compound of interest can be obtained.

SUMMARY

This chapter summarizes the subject of industrial hygiene air sampling of chemical hazards. Major sampling and analytical techniques have been described. Major advantages and disadvantages of sampling and analysis technologies are summarized in Table 1. The subject of air sampling method development and validation has been briefly discussed. The collection and analysis of industrial hygiene samples is best left to trained professionals.

Table 1. Advantages and Disadvantages of Common Industrial Hygiene Sampling and Analysis Techniques

Method	Advantages	Disadvantages
A. Active sampling methods	Numerous validated methods exist. Can be used for personal monitoring.	Pumps must be purchased, calibrated, and maintained.
1. Impingers	May be only collection method available.	Complex sampling train for personal monitoring. May be dangerous.
2. Solid sorbents	Lightweight, easy to transport. Concentrates samples.	Need pumps. Sorbent compound-specific.
3. Whole air samples (bags, cylinders, and bulbs)	Easy to collect.	Sample stability problems. Wall losses during storage. Do not concentrate sample.
4. Filter cassettes	Lightweight, easy to transport. Ideal for personal monitoring.	Difficult to remove filter without disturbing sample. A large portion of the sample may end up on the inside surface of the cassette.
B. Chemical dosimeters	No pumps needed. Simplifies sampling. Can use for personal monitoring.	Production problems will effect monitor's accuracy.

Table 1 (continued). Advantages and Disadvantages of Common Industrial Hygiene Sampling and Analysis Techniques

Method	Advantages	Disadvantages
C. Personal monitors analysis		
1. Liquid desorption	Enough sample for repeat analysis. Better sample stability.	Less sensitivity. Solvent and solvent's contaminants may interfere with compounds of interest.
2. Thermal desorption	Increased sensitivity; sub-ppb levels possible.	Not easily automated. No repeat analysis.
D. Area monitors		
1. Mobile systems	Portable. Real-time analysis possible.	Not specific. Major interference problems possible. Cannot be used for personal monitoring.
2. Fixed systems	Specific. Designed to work continuously. Good alarm systems.	Complex. Require high levels of maintenance. High initial costs.
E. Unknown identification		
1. Screening techniques	Real-time analysis.	Not specific.
2. GC/MS	True identification possible.	Not real-time. Identification is sampling media, GC column, and searching library-specific.

REFERENCES

1. Clayton, G. D., and F. E. Clayton, Eds. *Patty's Industrial Hygiene and Toxicology: Volume I, General Principles,* 3rd rev. ed. (New York: John Wiley & Sons, 1978).
2. *OSHA Analytical Methods Manual, Vols. 1 and 2* (Salt Lake City, UT: Occupational Safety and Health Administration, 1985).
3. *NIOSH Manual of Analytical Methods*, 3rd ed. (Cincinnati: National Institute for Occupational Safety and Health, 1987).
4. *Fundamentals of Analytical Procedures in Industrial Hygiene* (Akron, OH: American Industrial Hygiene Association, 1987).
5. Advances in Air Sampling (Cincinnati: American Conference of Governmental Industrial Hygienists, 1988).
6. Melcher, R. G. "Industrial Hygiene," *Anal. Chem. Appl. Rev.* 55(5):40R-56R (1983).
7. Melcher, R. G., and M. L. Langhorst. "Industrial Hygiene," *Anal. Chem. Appl. Rev.* 57(5):238R-254R (1985).

8. Langhorst, M. L., and L. B. Coyne. "Industrial Hygiene," *Anal. Chem. Appl. Rev.* 61(12):128R-142R (1989).
9. Lodge, J. P., Ed. *Methods of Air Sampling and Analysis,* 3rd ed. (Chelsea, MI: Lewis Publishers, 1988).
10. *The 1989 SKC Comprehensive Catalog and Guide* (Eighty Four, PA: SKC. Inc., 1988).
11. *Quality Assurance Manual for Industrial Hygiene Chemistry,* (Akron, OH: American Industrial Hygiene Association, 1988).
12. Swift, D. L. "Direct Reading Instruments for Analyzing Airborne Particles," in Air Sampling Instruments, (Cincinnati: American Conference of Governmental Industrial Hygienists, 1978), pp. 1-34.
13. Harper, M., and C. J. Purnell. "Diffusive Sampling — A Review," *Am. Ind. Hyg. Assoc. J.* 48(3):214-218 (1987).
14. Bartley, D. L., L. S. Doemeny, and D. G. Taylor. "Diffusive Monitoring of Fluctuating Concentrations," *Am. Ind. Hyg. Assoc. J.* 44(4):241-247 (1983).
15. Puskar, M. A., and L. H. Hecker. "Field Validation of Passive Dosimeters for the Determination of Employee Exposures to Ethylene Oxide in Hospital Product Sterilization Facilities," *Am. Ind. Hyg. Assoc. J.* 50(1):30-36 (1989).
16. Bishop, E. C., and M. A. Hossain. "Field Comparison between Two Nitrous Oxide Passive Monitors and Conventional Sampling Methods," *Am. Ind. Hyg. Assoc. J.* 45(12):812-816 (1984).
17. Nelson, G. O. *Controlled Test Atmospheres: Principles and Techniques* (Ann Arbor, MI: Ann Arbor Science Publishers, 1979).
18. Montgomery, D. D., and D. A. Kalman. "Indoor/Outdoor Air Quality: Reference Pollutant Concentrations in Complaint-Free Residences," *Appl. Ind. Hyg.* 4(1):17-20 (1989).
19. Levine, S. P., and W. F. Martin, Eds. *Protecting Personnel at Hazardous Waste Sites* (Stoneham, MA: Butterworths, 1985).
20. *Drager Review 46* (Germany: Dragerwerk AG, December 1980), pp. 8-9.
21. Skoog, D. A., and D. M. West. *Principles of Instrumental Analysis,* 3rd ed.(Philadelphia: Saunders College/Holt, Rinehart and Winston, 1985).
22. Taylor, D. G., R. E. Kupel, and J. M. Bryant. "Documentation of the NIOSH Validation Tests," DHEW (NIOSH) Publication No. 77-185, (1977).
23. "Development and Validation of Methods for Sampling and Analysis of Workplace Toxic Substances," DHHS (NIOSH) Publication No. 80-133 (September 1980).
24. Melcher, R. G., R. R. Langner, and R. O. Kagel. "Criteria for the Evaluation of Methods for the Collection of Organic Pollutants in Air Using Solid Sorbents," *Am. Ind. Hyg. Assoc. J.* 39(5):349-361 (1978).
25. McFarren, E. F., R. J. Lishka, and J. H Parker. "Criterion for Judging Acceptability of Analytical Methods," *Anal. Chem.* 42(3):358-365 (1970).
26. Saltzman, B. E. "Variability and Bias in the Analysis of Industrial Hygiene Samples," *Am. Ind. Hyg. Assoc. J.* 46(3):134-141 (1985).
27. Paustenbach, D., and R. Langner. "Corporate Occupational Exposure Limits: The Current State of Affairs," *Am. Ind. Hyg. Assoc. J.* 47(12):809-818 (1986).

CHAPTER 2

Exposures to Chemicals

Paul D. Moss

TOXICOLOGICAL TERMS AND THE MATERIAL SAFETY DATA SHEET (MSDS)

What is Toxicology? One might think that this short question might have a simple answer; however, toxicology is a relatively young and evolving field compared to those sciences with great, long histories such as physics and astronomy. The toxicologist must have a strong understanding of the many disciplines that interact within the scope of the science. Figure 1 presents a basic summary of the multidisciplinary building blocks that make up the field of toxicology.

Now that it has been demonstrated just how many extrinsic sciences can affect the field of toxicology, the pure definition of toxicology becomes complicated. James[1] describes toxicology as the study of chemical or physical agents that produce adverse responses in the biologic systems with which they interact. In another definition, Loomis[2] describes toxicology as the study of the harmful actions of chemicals on biologic mechanisms. Casarett and Bruce[3] established that toxicology is more than just the science of poisons; hence, the definition must also include the impact of both intrinsic and extrinsic factors that interact with the toxin. A basic understanding of toxicological concepts must be undertaken before the reader can make a judgment as to whether this science can truly be defined. It is important to understand the role that toxicology plays in the overall interpretation of dose vs effect, and ultimately the outcome. Once understood, the concepts associated with the Material Safety Data Sheet (MSDS) and their subsequent

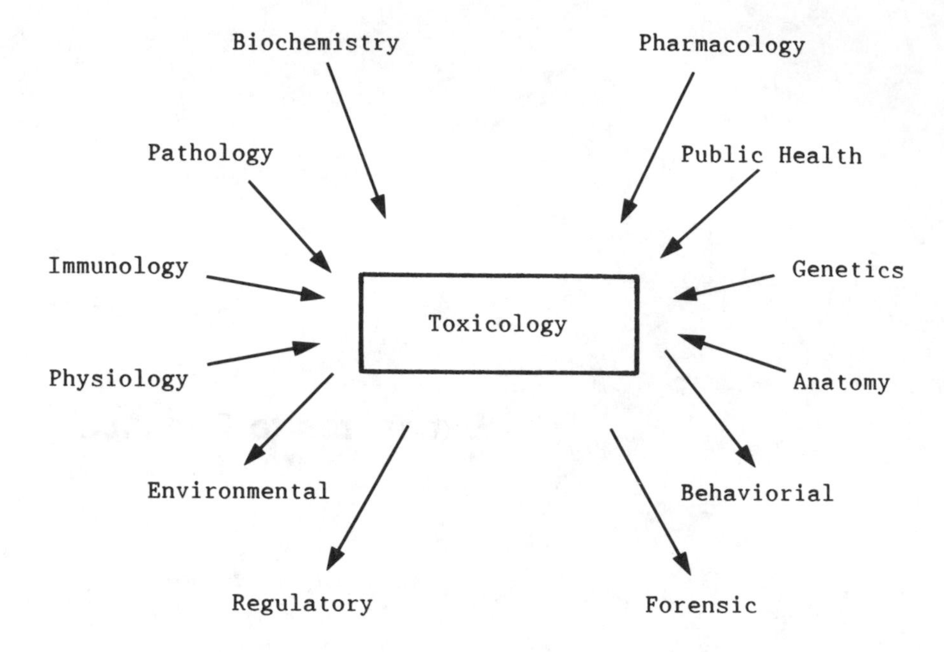

Figure 1. Multidisciplinary building blocks that make up the field of toxicology.

interpretation will become evident, and the reader will be better prepared for decision-making processes with respect to potential exposures in the workplace.

ACUTE TOXICITY VS CHRONIC TOXICITY

In the context of the MSDS, an *acute exposure* is that which occurs once over a short time frame, usually to a high concentration of a chemical or chemical agent. Its effects are commonly reversible, although there certainly are exceptions, based on the dosage, concentration and the sustained damage to biological matter. A *chronic exposure* is one which occurs repeatedly over a long period of time, usually to low level concentrations of an agent, and is commonly irreversible in effect or damage. Both types of exposures can be measured for degree of toxicity, based on local or systemic damage to the biological system. An example of acute toxicity could be the exposure of an individual to a large cloud of ammonia gas following a train derailment of a loaded tank car. The resulting acute health effects might include severe irritation to the respiratory tract, eyes, and skin. Conversely, an example of chronic toxicity to the same chemical might be low level exposures to an individual working around a slow-leaking refrigerant tank in an industrial setting. The chronic health effects might include mild irritation of the respiratory tract, eyes, and skin and also might lead to some residual reduction in pulmonary function, along with possible complications of bronchitis or pneumonia.

Another example of acute vs chronic toxicity is described here for benzene. Acute health effects from benzene include CNS depression, headache and dizziness, convulsions, and possibly death. The chronic health effects of benzene exposure include suppression of the hematopoietic system, possibly leading to aplastic anemia, leukemia or a lymphoma cancer. It is important to remember that acute and chronic toxicity not only differ in duration of exposure as well as potency of effect, but also in symptomatology. For example, a large acute exposure to welding on zinc might result in physical burn or possibly symptoms of metal-fume fever (an influenza-like illness with symptoms including muscle aches, chills, fever, weakness, dry throat, and nausea), while a chronic low-level exposure to zinc fumes might ultimately be reflected by a latent pneumoconiosis or pulmonary fibrosis. The concepts of acute and chronic toxicity can be examined further through a review of the dose-response theory.

THE DOSE-RESPONSE CONCEPT

The basis for determining the toxicology of a chemical compound lies in the relationship between the individual dose and the individual response to the agent. This concept of dose-response is used as the basis for determining the relative safety of a chemical compound in the living organism. The dose-response curve can be presented graphically with the y-axis representing lethal effect or lethal dose, while the x-axis represents the actual dose or concentration administered to the test species. From a dose-response graph, one can detail specific areas of a curve that represent important toxicological concepts. One of these concepts is the LD_{50}, or the dose at which 50% of the test species has died. If one were to do a study on a pharmaceutical where the effect of the drug rather than mortality was being measured, one might call this value an ED_{50}, or an effective dose at which 50% of the test species exhibits the expected effect. In a similar fashion, one can draw two lines at 90° between the ordinate and abscissa and obtain an LD or ED value at any percent of effect (for example, LD_1 or LD_{99}, etc.). Out of these concepts comes the oft-asked question in risk assessment, "How much is too much?"

Figure 2 shows curves for two hypothetical chemical agents that have been administered to biologic specimens, for example, mice. The two hypothetical chemical agents are represented by the curves labeled A and B.

These two curves describe their actual toxicities. Remember that an LD_{50} itself will not provide an idea of what the actual dose-response curve is (i.e., the shape of the curve representing responses at different dosages); therefore, when one compares these values, the range of doses covered by the interval (represented by the width of the curve on the x-axis) is as important as how low or high the dosage is at the beginning of the response interval. If one examines the LD_5 and LD_{50} values for these two curves, it is apparent that the chemical with the most toxic LD

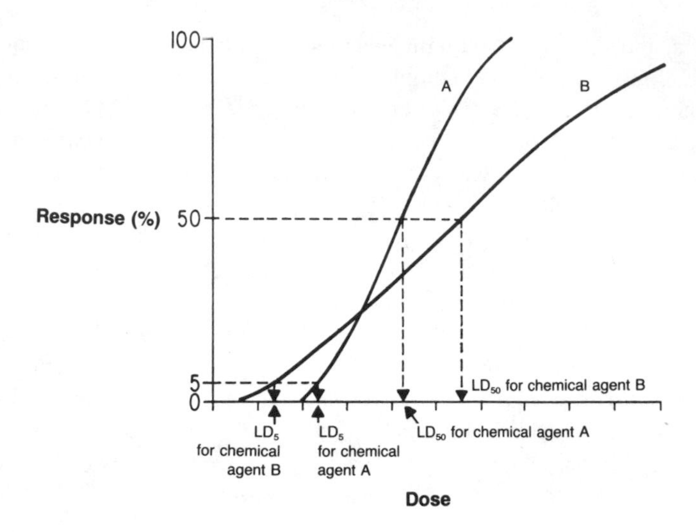

Figure 2. Hypothetical dose-response curves for two different chemical agents (A and B) administered to test mice.[2]

value changes from LD_5 to LD_{50}; if only the LD_{50} values of compounds A and B were compared, A would be more toxic. However, at another point on the curve (LD_5), the opposite is true (i.e., B is more toxic since a lesser dose of B was required to kill 5% of the test animals). Therefore, the shapes and widths of the two curves merit close examination when comparing two compounds. If one does not examine the whole curve instead of a portion of it, misperception of toxicity can occur.

Comparisons among compound LD_{50} values in animal species can provide an approximate level of toxic potency for each compound relative to the other. For example, multiplying the LD_{50} dosage by the assumed weight of the so-called "standard man" (70 kg) will give the researcher a rough extrapolation of the dose required to cause toxicity in the human species. This requires the assumption that man is equivalent in sensitivity to the chemical agent as was the test species. Table 1 shows a toxicity rating chart for humans, based on that concept. The conclusion can be drawn from this table that it is apparent that too much of *any* substance (for example, sugar) can be toxic, but toxicity is relative to the real dose that can be expected to occur. One would not expect an individual to sit down and deliberately eat a quart of salt, yet this serves to illustrate that the dose of a chemical agent is as important as its toxicity in determining risk or hazard. This subject, i.e., what constitutes a realistic dose, has been fodder for many a regulatory argument, including aspects associated with the recent banning of certain food residuals such as ethylene dibromide and Alar.

It must be noted that there are intraspecies differences in dose-response curves, based on gender, age, weight, and numerous other confounding factors. Similarly, there can also be interspecies differences in response, i.e., different species of test

Table 1. Toxicity Rating Chart

Toxicity Rating or Class	Probable Oral Lethal Dose for Humans: Dose	For Average Adult	Example of Substance
1. Practically nontoxic	>15 g/kg	More than 1 quart	Sugar
2. Slightly toxic	5—15 g/kg	Between pint and quart	Salt
3. Moderately toxic	0.5—5 g/kg	Between ounce and pint	2,4-D
4. Very toxic	50—500 mg/kg	Between teaspoonful and ounce	Arsenic acid
5. Extremely toxic	5—50 mg/kg	Between 7 drops and teaspoonful	Nicotine
6. Supertoxic	<5 mg/kg	A taste (<7 drops)	Botulism toxin

animal can show different toxic responses to the same dosage of a particular chemical. This often makes it difficult to interpret the results of toxicity testing (or to attempt extrapolation of the results from chemical exposure tests on animals to humans). These differences in response can be the result of several factors. Each species can have distinct physiological responses or metabolic pathways that makes their response to a chemical different from other species. For example, the surface to volume ratio of the test animal species can play a role in interspecies variations in toxic response.

For purposes of extrapolating animal data to man (the "animal model"), it only makes sense to be as conservative as possible. One often chooses the laboratory species that exhibits the most detrimental response to the test compound, i.e., at the lowest concentration. Although it is this logic that usually dictates which species will be chosen for a certain toxicology testing protocol, other factors also affect the ultimate choice. For example, rabbits are the species of choice for eye irritation and skin testing, since they are known to react more strongly to mild and moderate irritants than humans (thus creating a more conservative extrapolation), while mice and rats are most commonly chosen for oral feeding studies, mostly due to metabolism and physiological likenesses to man, their shorter life span, as well as cost and availability.

The final dose-response concept that deserves attention is the threshold, or the lowest point on the dose-response curve. The threshold or no observed effect level (NOEL), as it is frequently called, is that dose below which an effect by a given agent is nondetectable. This concept is the basis for many useful applications in the legislative framework for government and industrial regulations. Efforts such as the establishment of Occupational Safety and Health Administration (OSHA) Permissible Exposure Limits (PELs) and the American Conference of Governmental Industrial Hygienists (ACGIH) Threshold Limit Values (TLVs) are designed to ultimately strive for threshold values to working populations taking into account conservative safety factors. The American Industrial Hygiene Association (AIHA) Emergency Response Planning Guidelines (ERPGs) estimate acute exposure guidelines for the general public. These terms are discussed elsewhere in this volume. The threshold value has been a point of contention with regard to safety of carcinogens. Current regulatory policy has shown that because of the absence of any *demonstrated* threshold for carcinogens, the rate of cancer should decrease (in probability, linearly) until zero exposure occurs. This has been and will continue to be a great source of controversy among experts in the field. Also, levels at which adverse health effects are seen as a result of chronic exposure are usually lower than those from acute exposure and cannot be predicted on the basis of acute exposure. Extrapolation cannot occur between acute and chronic exposure because the mechanism of effect for the chemical (i.e., bioaccumulation, storage, metabolism) or the target organ can differ for either type of exposure. For example, if an individual receives an acute exposure to a lipophilic compound such as polychlorinated biphenyl (PCB), symptomatology such as chloracne (skin eruptions) might result. An individual's chronic exposure to PCB, however, would cause bioaccumulation of the compound in the fatty tissues and organs of the body; if this individual were to go on a strict diet (and lose these fat cells quickly), a strong exposure to PCB in the bloodstream might result in major organ damage and/or in possible deleterious reproductive effects.

Much literature has been published on the dose-response relationship, offering differing opinions on the strengths and weaknesses of the theory. Toxicology has often been ridiculed for the hierarchy in which dose levels are chosen for laboratory testing. Public perception has been that laboratory subjects are chosen, dosed until termination, counted, and the results are extrapolated to the human species. Fortunately, the science of toxicology goes beyond this simple and false judgment, and explores the actual effects, both beneficial and adverse, of the dose level. Since in most cases it is unethical to use humans as test subjects, we owe thanks to the sacrifice of laboratory animals for the benefits that we share in the form of commercial pharmaceuticals, pesticides, and some foods.

CONCEPT OF THE TARGET ORGAN

As a basic understanding of the science of toxicology is continuously achieved, we realize that a chemical can only exert a toxic effect on an organ system if it can damage the system. Most chemicals that produce systemic toxicity do not cause a similar degree of toxicity in all organ systems in any one individual; rather, they usually manufacture the major toxic effect or damage in one or two organs.[5] These affected organs are termed the *target organs*. For a chemical to produce a toxic effect in a target organ, the agent (after entering the body by inhalation, ingestion, or absorption) must be transported to the organ by a vehicle such as the bloodstream. It must be noted, however, that the organ in which the toxic compound is most highly concentrated may not be the organ in which most of the damage occurs. A possible example is an acute exposure to dichlorodiphenyl-trichloroethane (DDT), a synthetic insecticide widely used in the 1940s and 1950s to reduce outbreaks of mosquito-driven malaria. DDT is highly lipophilic, meaning a majority of the compound will settle in body fat, although the primary toxic effect of DDT is exerted on the human neurologic system. Another example of target organ

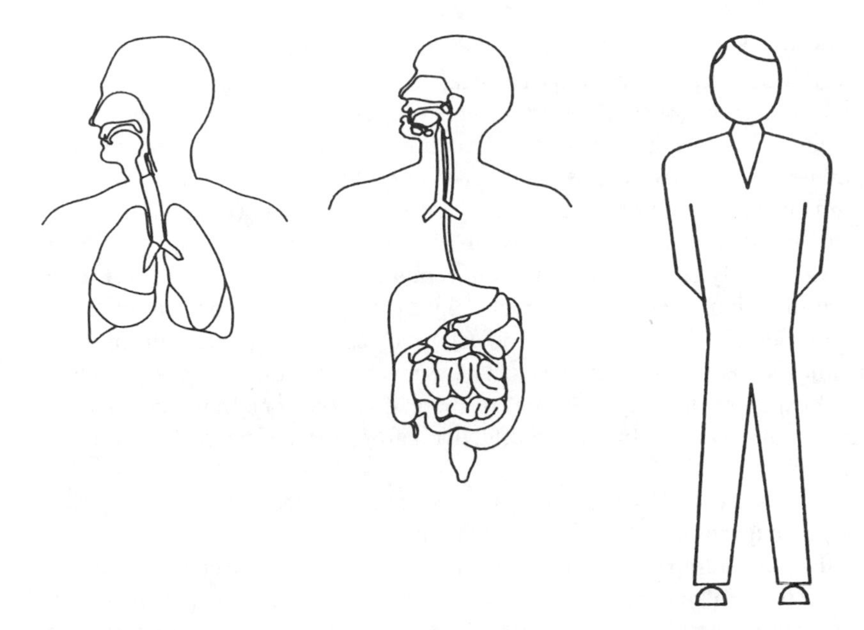

Inhalation
The dose received via the lungs (inhalation) is influenced by the solubility and/or particle size of the substance, also by its airborne concentration.

Oral Ingestion
Caused by eating, smoking, chewing gum, etc., in contaminated areas, also by the inadvertent swallowing of inhaled particles.

Dermal Exposure
Exposure to chemicals through the skin are frequent and usually depend upon the substance's pH, solubility in water and lipids and its molecular size; personal hygiene can also play a role.

Figure 3. The three most common routes of chemical exposure: inhalation, oral ingestion, and dermal exposure. To reduce each of these types of occupational exposures, engineering controls and personal protective equipment are frequently used in the workplace.

toxicity that occurs away from the area of greatest concentration of the agent is that of lead. Since lead mimics calcium once in the body, most lead that enters the human body is transported into the skeletal bone; however, the ultimate biological systems affected by the toxicity of lead are the nervous, hematopoietic, and renal systems. For substances that generally cause a localized effect, the severity and frequency of the toxic reaction depends largely on the route and location of the exposure.

ROUTES OF EXPOSURE

Chemicals can enter the body through a variety of routes. The chemical and physical properties of a given compound will largely determine the route by which intentional or accidental exposure occurs.[6] Figure 3 describes the major routes of exposure of the human body.

Through inhalation via the respiratory system, ingestion into the gastrointestinal tract, and absorption across the dermal (skin) barrier, toxic agents can enter the biologic system to ultimately exert a single or multiple effects. In general terms, the exposure is dependent upon the concentration of the toxic compound and the duration or frequency of exposure(s) that occur. Most industrial exposures occur through the inhalation or dermal routes; hence, most regulatory standards (PELs) or recommendations (TLVs) are created to protect against these types of exposures. Conversely, accidental exposures commonly can be the result of toxicity through ingestion, examples of which might include suicidal poisoning or pica (the behavior of putting accessible objects, such as dirt, in the mouth) in young children. For many toxic compounds, it takes much more dermal exposure to equal a lethal exposure through ingestion. In these situations, one may conclude it is likely that the skin is acting as an effective barrier for the prevention of a toxic dose.[7]

Chemicals can also enter the human system by other routes of exposure, including intravenous (i.v., injection into the vein), subcutaneous (s.c., injection under the skin), intraperitoneal (i.p., injection into the peritoneal cavity), and intramuscular (i.m., injection into the muscle). Since these routes of exposure are only attainable through professional medical treatment or self-inflicted intentions, they are not common or significant for the health and safety setting. They are important, however, for laboratory toxicology testing, medical practice, and unfortunately in current society, drug addiction.

Once exposure has occurred through any of these routes, the pathway of the chemical through the body affects the sequence in which tissues and organs are reached by the chemical, as well as the physiologic and metabolic events that occur along the way.[8]

PATHWAYS IN THE HUMAN BODY

The most common routes of exposure are absorption, inhalation, and ingestion. Once exposure has taken place, the toxicant enters the body and is distributed via any number of different metabolic and transportation pathways to an ultimate fate. Often, the toxicant will exert a localized effect, such as a burn or irritation. An example of this type of agent might be a corrosive chemical or an acid such as nitric acid. If all toxic effects from chemicals were localized, medical treatment would be far easier to perform, based on simple cause and effect relationships. Unfortunately, other pathways also are utilized, and the toxicant can be localized in a target organ, in the bloodstream, or be absorbed across membranes and exert an effect(s). The main purpose of metabolism with regard to the human body and a toxic exposure is to detoxify the chemical into some inert product that can be eliminated from the system. It is fortunate that the liver acts as the detoxification machine with regard to chemical substances; however, there are instances when this pathway can actually be detrimental. An example is the insecticide, parathion; upon entering the body, one of the biotransformation pathways for parathion is its

Table 2. Basic Toxicological Concepts

1. Toxicity (biological)	Ability of a chemical or compound to exert damage on a biological system
2. Dose response	A scientific relationship that correlates dose (concentration) with an observed effect
3. Acute toxicity	Single short-term exposure exerting toxic effect; usually reversible
4. Chronic toxicity	Repeated long-term exposure exerting toxic effect; usually irreversible
5. Target organ	Biologic system that suffers greatest toxic effect
6. Routes of exposure	Inhalation, dermal, and ingestion are the most common routes

conversion to a more toxic and active form, paraoxon. Paraoxon acts as an anticholinesterase inhibitor, and can cause severe damage to the neurological system.

Additional pathways in the body include absorption across membranes such as the skin, lungs, and gastrointestinal tract. The elimination pathway for both inert chemicals and for biotransformed metabolites involves the kidney as the organ of concern. It is obvious that the human body is an intricate and very specialized system that utilizes numerous pathways and organ systems to deal with substances foreign to the body. Thus far in this chapter, these pathways have been touched upon in a very simplistic manner toward the end of directing the reader to the next section: the application of toxicology principles to the interpretation of the MSDS. Table 2 summarizes basic toxicological concepts important in understanding MSDSs.

An MSDS is intended to be a concise source of physical, toxicological, and health and safety data for practically every commercially available chemical product. MSDSs are usually written by the chemical manufacturer, compounder, or importer and must be distributed to the purchaser at the original purchase and/or whenever there has been a change to the data in the MSDS. It is perhaps unfortunate that MSDSs are not standardized, i.e., there is no single form or format that must be used. This can lead to difficulty in finding information on the MSDS or to differences in the degree of detail included in the MSDS. In theory, the MSDS for a chemical or compound should contain everything needed concerning the chemical or chemical product including, but not limited to, the manufacturer; the hazardous ingredients; chemical and physical characteristics; descriptions of hazards for health, fire, explosion, and reactivity; and safe handling precautions and control measures (see Chapter 5 for a complete discussion of MSDS). Figure 4 presents OSHA Form Number 174,[9] a suggested but nonmandatory format for the MSDS.

FUNCTION OF THE MSDS

The MSDS is intended to be an informational source for employers and

Material Safety Data Sheet May be used to comply with OSHA's Hazard Communication Standard, 29 CFR 1910.1200. Standard must be consulted for specific requirements.	U.S. Department of Labor Occupational Safety and Health Administration (Non-Mandatory Form) Form Approved OMB No. 1218-0072
IDENTITY *(As Used on Label and List)*	*Note: Blank spaces are not permitted. If any item is not applicable, or no information is available, the space must be marked to indicate that.*

Section I

Manufacturer's Name	Emergency Telephone Number
Address *(Number, Street, City, State, and ZIP Code)*	Telephone Number for Information
	Date Prepared
	Signature of Preparer *(optional)*

Section II — Hazardous Ingredients/Identity Information

Hazardous Components (Specific Chemical Identity; Common Name(s))	OSHA PEL	ACGIH TLV	Other Limits Recommended	% *(optional)*

Section III — Physical/Chemical Characteristics

Boiling Point		Specific Gravity (H_2O = 1)	
Vapor Pressure (mm Hg.)		Melting Point	
Vapor Density (AIR = 1)		Evaporation Rate (Butyl Acetate = 1)	
Solubility in Water			
Appearance and Odor			

Section IV — Fire and Explosion Hazard Data

Flash Point (Method Used)	Flammable Limits	LEL	UEL
Extinguishing Media			
Special Fire Fighting Procedures			
Unusual Fire and Explosion Hazards			

OSHA 174, Sept. 1985

5-8-86

Figure 4. Occupational Safety and Health Administration sample Material Safety Data Sheet (OSHA Form No. 174).

employees concerning the potential toxicity and hazards of chemicals worked with or purchased. Additionally, the MSDS serves as a safety tool for the proper handling of a given chemical or compound and contains the provisions necessary

Section V — Reactivity Data

Stability	Unstable		Conditions to Avoid
	Stable		

Incompatibility (*Materials to Avoid*)

Hazardous Decomposition or Byproducts

Hazardous Polymerization	May Occur		Conditions to Avoid
	Will Not Occur		

Section VI — Health Hazard Data

Route(s) of Entry: Inhalation? Skin? Ingestion?

Health Hazards (*Acute and Chronic*)

Carcinogenicity: NTP? IARC Monographs? OSHA Regulated?

Signs and Symptoms of Exposure

Medical Conditions Generally Aggravated by Exposure

Emergency and First Aid Procedures

Section VII — Precautions for Safe Handling and Use

Steps to Be Taken in Case Material Is Released or Spilled

Waste Disposal Method

Precautions to Be Taken in Handling and Storing

Other Precautions

Section VIII — Control Measures

Respiratory Protection (*Specify Type*)

Ventilation	Local Exhaust	Special
	Mechanical (*General*)	Other

Protective Gloves	Eye Protection

Other Protective Clothing or Equipment

Work/Hygienic Practices

Page 2

* USGPO 1986-491-529/45775

Figure 4 (continued).

to inform the employee about proper personal protection and engineering controls when working with the compound(s) of concern. Although the information contained within the MSDS is usually provided by the manufacturer, it is important to emphasize that the employer is responsible for the ultimate content of the data sheet. Accuracy and completeness of the MSDS lies within the scope of the employer's duty.

REGULATORY REQUIREMENTS

The original intention of the MSDS form was to enable employers in the shipbuilding industries to fulfill safety responsibilities to their employees.[10] Since the early 1980s, the regulatory framework in the occupational health and safety setting has increased greatly in the direction of new standards to protect the employee welfare. The OSHA Hazard Communication Standard,[11] or the Employee "Right to Know" Act, passed into law on November 25, 1983, was the next set of regulations that employed the MSDS as a hazard communication tool. The standard stipulated that copies of the MSDS for hazardous chemicals at a given work site must be readily accessible to employees engaged in work practices in that area. Since the worker relies on the MSDS as a source of quick and detailed information on a specific chemical or compound, the MSDS information area must be located in close proximity to the workers, they must have easy access to it at all times during the given workshift. This standard is explored in detail in Chapter 5. A third area of legislation that included the MSDS within the content of reporting requirements was the Superfund Amendments and Reauthorization Act (SARA) of 1986.[12] Under Section 311 of this bill, facilities covered under the Hazard Communication Standard must submit an MSDS or a list of each chemical or compound used at the facility to three parties (a state emergency response commission, a local emergency response committee, and the local fire department) for those hazardous chemicals that exceed an established threshold quantity. The purpose of these MSDS submittals is for the benefit of the hazardous materials incident responder. The responder can use the information from the MSDS to evaluate a given situation, and to help provide valuable facts to aid in personal protection during the actual response. In addition, the fire department can proactively utilize the MSDS to assist in developing a strategy for emergency response at the respective facility. One can conclude, based on regulatory action over the last decade, that not only has the MSDS become a communication tool for vital information, but it has also found its way into the reporting requirements area of the SARA laws.

APPROPRIATE APPLICATIONS AND INTERPRETATIONS

An understanding of how to interpret the data on the MSDS becomes essential if workplace exposures, accidents and injuries are to be avoided and/or treated effectively. Although the first aid and medical treatment section of the MSDS can be very helpful in the event of injury or overexposure, the form is actually provided to supply workers with the information necessary to prevent accidents and reduce chemical injuries in the workplace. Of particular importance within this chapter is the section of the MSDS that deals with health hazards. Particular substance and associated health hazards, both chronic and acute, along with symptoms of exposure, must be listed in this section. The route of exposure, be it dermal, inhalation, or ingestion, must also be outlined. Chemicals and compounds

Common Name of the Chemical

Name, Address, and Emergency Phone Number – of the chemical manufacturer, importer, or compounder

Signal Word — in order of descending seriousness:
DANGER, WARNING, CAUTION

Principal Hazards
Physical Hazards — Will it explode, catch fire, be reactive?
Health Hazards — Can it cause cancer, is it an eye irritant?

Precautionary Use Measures — Personal protective equipment or ventilation required; any special procedures

First-Aid Instructions

Storage and Handling Instructions

Figure 5. An example of label format. Exact order of terms is not important. Note that the label is oriented toward basic toxicological concepts: identifying adverse health effects, routes of exposure, etc.

considered health hazards include carcinogens, corrosives, toxic agents, irritants, sensitizers, teratogens, and mutagens. This section also must include any substances that are toxic to specific target organs, and the special organ (i.e., liver, kidney) or system (i.e., nervous, gastrointestinal) must be listed.

LABELS

It may be said that the MSDS is to the label as the foundation is to the house: it serves as a base upon which the label is built. The MSDS, which is intended to contain comprehensive information about the chemical, is the source of the information used in the labels appearing on chemical containers. Because it would be impractical to create a label containing all of the information on the MSDS, the label only contains a portion of the MSDSs information. The OSHA Hazard Communication Standard (or "Worker Right to Know," discussed in Chapter 5) requires only that the label contain the identity of the material; any applicable hazard warnings; and the name, address, and phone number of a responsible party from whom additional information can be obtained, if needed. The purpose of the label is to provide the end user with a quick source of the primary health hazards, etc., associated with the use of the product. Frequently, labels provide more; a common level of detail is shown in the example in Figure 5.

Labels come in many different formats and codes. Complying with OSHA's

Flammable Hazards
(Red Background)

0 Will not burn
1 Must be preheated to burn
2 Ignites when moderately heated
3 Ignites at normal temperatures
4 Extremely flammable

Health Hazards
(Blue Background)

0 Like ordinary material
1 Slightly hazardous
2 Hazardous - use breating apparatus
3 Extremely dangerous - use full protective clothing
4 Too dangerous to enter vapor or liquid

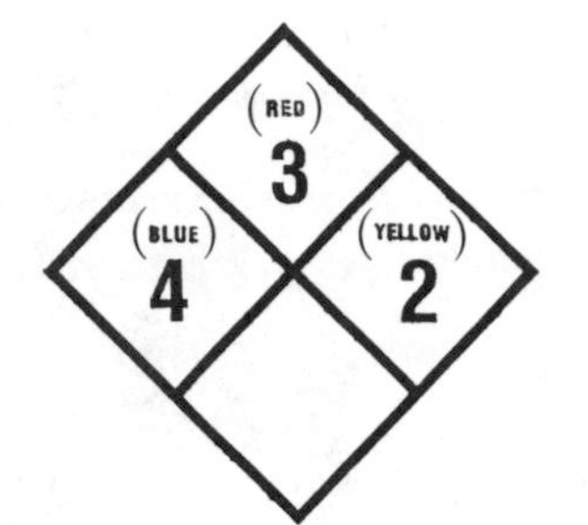

Reative Hazards
(Yellow Background)

0 Normally stable
1 Unstable if heated - use normal precautions
2 Violent chemical changes possible - use hose streams from distance
3 Strong shock or heat may detonate - vacate area if materials are exposed
4 May detonate - vacate area if materials are exposed to fire

Special Information
(White Background)

The letter "W" with a bar through it indicates that a material may have a hazardous reaction with water

The radioactive symbol indicates that a chemical is radioactive

Indicates that a chemical is an oxidizer

Figure 6. The NFPA 704 Diamond. This system is oriented toward the identification of fire hazards of materials. Numbers must be in a contrasting color (white or black). (Adapted from *The Fire Protection Handbook,* 16th ed., A. E. Cote, Ed., National Fire Protection Association, Quincy, MA, 1986.)

Hazard Communication Standard, any type of label is permitted as long as it contains the basic required information. Some labels are dependent upon words to describe the actual hazards, while other labels use a color and number coding system to help the user quickly identify the hazard. An example of a label using a specialized number coding system is shown in Figure 6.

Whether it be the word system of hazard warning or the number and color system, success in labeling is achieved only if the message is easily understood and contains accurate information. For this reason, a label can be understood well if the user possesses a basic understanding of the toxicological concepts presented earlier in this chapter.

SUMMARY

The MSDS and respective label for a hazardous agent can be an effective source of important information for the consumer. As has been stated within the context of this chapter, the job can only be done right if an individual has been provided with the right tools. The key elements are a basic understanding of the concepts, how to apply them to a real situation, and how to prepare oneself for such a situation. Provided that all of these goals have been achieved efficiently through training and education, the end user has successfully been given the means to properly do the job. Ultimately, the employer is responsible for hazard communication to the employee, and it is hoped that this short summary of the use and function of the MSDS has provided a better understanding of how that communication must work. If carried out properly, the Hazard Communication Standard and its appropriate requirements will indeed be the landmark in employee awareness of hazards for which the law was intended. The key, as in the case of daily life, is good communication.

REFERENCES

1. James, R. C. "General Principles of Toxicology," in *Industrial Toxicology*, P. L. Williams and J. L. Burson, Eds. (New York: Van Nostrand Reinhold, 1985), p. 7.
2. Loomis, T. A. *Essentials of Toxicology*, 3rd ed. (Philadelphia: Lea & Febiger, 1978), p. 4.
3. Casarett, L. J., and Bruce, M. C. "Origin and Scope of Toxicology," in *Casarett and Doull's Toxicology: The Basic Science of Poisons*, 2nd ed., J. Doull, C. Klaassen, and M. Amdur, Eds. (New York: Macmillan, 1980), p. 3.
4. Klaassen, C. D., and Doull, J. "Evaluation of Safety: Toxicologic Evaluation," in *Casarett and Doull's Toxicology: The Basic Science of Poisons,* 2nd ed., J. Doull, C. Klaassen, and M. Amdur, Eds. (New York: Macmillan, 1980), p. 22.
5. Klaassen, C. D., and Doull, J. "Evaluation of Safety: Toxicologic Evaluation," in *Casarett and Doull's Toxicology: The Basic Science of Poisons*, 2nd ed., J. Doull, C. Klaassen, and M. Amdur, Eds. (New York: Macmillan, 1980), p. 16.
6. Loomis, T. A. *Essentials of Toxicology,* 3rd ed. (Philadelphia: Lea & Febiger, 1978), p. 67.
7. Klaassen, C. D., and Doull, J. "Evaluation of Safety: Toxicologic Evaluation," in *Casarett and Doull's Toxicology: The Basic Science of Poisons,* 2nd ed., J. Doull, C. Klaassen, and M. Amdur, Eds. (New York: Macmillan, 1980), p. 14.
8. Ottoboni, M. A. *The Dose Makes the Poison* (Berkeley, CA: Vincente Books, 1984), p. 45.
9. "Occupational Safety and Health Administration Form Number 174: Material Safety Data Sheet," U.S. Department of Labor, U.S. Code of Federal Regulations, Vol. 29, Part 1910.1200 (1985).
10. "Safety and Health Standards for Shipyard Employment," U.S. Department of Labor, U.S. Code of Federal Regulations, Vol. 29, Part 1915.99 (1982).

11. "Occupational Safety and Health Standards Subpart Z — Toxic and Hazardous Substances, Hazard Communication Standard," U.S. Department of Labor, U.S. Code of Federal Regulations, Vol. 29, Part 1910.1200 (1983).
12. "Superfund Amendments and Reauthorization Act of 1986," H.R. 2005, U.S. House of Representatives Report 99-962 (1986).

CHAPTER 3

Occupational Epidemiology: Investigating Health Hazards in the Workplace

Braxton D. Mitchell

INTRODUCTION

Disease does not occur randomly in populations. Individuals vary in their susceptibility to disease, with some segments of the population clearly more "at risk" than others. Susceptibility to disease may be influenced by a wide variety of factors, ranging from the age, sex, and genetic constitution of the individual to more global factors relating to the individual's environment, such as lifestyle, diet, occupation, etc. Epidemiology involves the study of disease occurrence in populations with the aim of relating disease occurrence to characteristics of people and their environment. It is the use of populations, rather than individuals, as the unit of analysis that distinguishes epidemiology from clinical medicine and other biomedical disciplines.

Epidemiologic methods are a powerful tool for investigating the occurrence of potential health hazards in the workplace. The goals of occupational epidemiology are twofold: to monitor disease rates over time so that potential health problems may be recognized early; and to investigate the determinants of disease (i.e., to evaluate disease-exposure relationships). The term "disease" is used broadly in this chapter to include any adverse health outcome with possible relevance to occupational settings, such as injury and absenteeism.

This chapter provides an introduction to basic epidemiologic concepts and describes how these concepts may be applied to understanding disease occurrence in the workplace. First, issues related to the measurement of disease are discussed, and second, a brief description of the basic types of observational study designs is provided.

MEASUREMENT OF DISEASE: INCIDENCE AND PREVALENCE RATES

The major task of epidemiology is to quantify the occurrence of disease. For studies involving morbidity (i.e., illness or disability), two types of rates are commonly used: incidence and prevalence rates. *Incidence* measures the rate at which new cases of disease develop in the population during a specified time period. *Prevalence* measures the proportion of the population with the disease at a single point in time. The computational definitions for these two rates are given below:

$$\text{Incidence rate} = \frac{\text{Number of new cases}}{\text{Number of people at risk}} \text{ during a specified time interval} \quad (1)$$

$$\text{Prevalence rate} = \frac{\text{Number of existing cases}}{\text{Total population}} \text{ at a single point in time} \quad (2)$$

By definition, incidence rates are expressed relative to a time frame. The incidence of bladder cancer in U.S. males, for example, is estimated to be 15 cases per 100,000 population per year.[1] The calculation of incidence requires the measurement of disease status at baseline, to determine the number of disease-free individuals who are at risk for acquiring the disease, and at a follow-up time point, to determine the number of individuals who subsequently acquired the disease. Prevalence, on the other hand, is measured cross-sectionally, i.e., at a single point in time. Prevalence is thus the easier of the two measures to obtain and is often used when incidence rates are unavailable.

The distinction between incidence and prevalence is pivotal to epidemiology. Incidence, a measure of a population's risk of disease, is the desired measure for investigating relationships between disease and its causes (risk factors), and for making comparisons of the development of disease between different populations. In contrast, prevalence is a function, not only of the probability of developing disease, but also of the factors influencing the duration of the disease. An improved ability to detect the early stages of breast cancer, for example, would lead to an earlier age of diagnosis and hence a higher prevalence of the disease, since more women would be diagnosed. The incidence of breast cancer, however, would remain unchanged. Similarly, an improvement in the treatment for leukemia, e.g., one that extended the survival of patients with this disease, would also result in increased prevalence, but no change in incidence. In fact, under steady-state conditions (in which the incidence rate is constant over time and the population structure is stable), a mathematical relationship exists between prevalence and incidence:

$$\text{Prevalence} = \text{incidence} \times \text{disease duration} \tag{3}$$

RELATIVE RISK AND RISK DIFFERENCE

A common approach used in epidemiology is to compare the disease experience of a population having a particular characteristic to that of a population without that characteristic. If inferences are made concerning which population is more at risk for *developing* a disease (as opposed to which has a larger proportion of individuals who currently have the disease), the incidence rate, as discussed above, is the preferred measure. One might, for example, want to compare the incidence of lung cancer in workers exposed to arsenic with the lung cancer rate in unexposed workers. A measure of the association between the disease and the exposure of interest is the *relative risk.* Relative risk is defined as the incidence rate in the exposed population divided by the incidence rate in the unexposed population:

$$\text{Relative risk} = \frac{\text{Incidence rate in exposed}}{\text{Incidence rate in unexposed}} \tag{4}$$

The relative risk is used to quantify the disease risk in one group (the exposed) relative to the disease risk in another group (the unexposed). A relative risk of 1 occurs when the incidence rates in the two groups are identical; in this situation no association exists between disease and exposure. Relative risks >1 thus indicate a positive association between disease and exposure, while relative risks <1 indicate a negative association (i.e., less disease among exposed subjects). It should be noted that even in the absence of a disease-exposure association, relative risks exactly equal to 1 are rarely observed since the measurement of incidence rates is associated with some degree of sampling variability. Therefore, statistical procedures are used to determine whether the observed deviation of the relative risk from one is more than would be expected by chance.

Incidence rates may also be compared through the *risk difference,* defined as the incidence rate in the exposed population minus the incidence rate in the unexposed population:

$$\text{Risk difference} = \text{Incidence in exposed} - \text{Incidence in unexposed} \tag{5}$$

This measure corresponds to the rate of disease in the exposed population that is attributable to the exposure. By subtracting the mortality rates due to a variety of cancers in nonsmokers from the corresponding cancer mortality rates in smokers and multiplying these rate differences by the number of smokers in the U.S. population, this approach was used to estimate that there were approximately 122,000 tobacco-related cancer deaths occurring in the U.S. in 1978. Since there were approximately 400,000 total cancer deaths in the U.S. in 1978, it may be inferred that about 30% of all cancer deaths for that year were directly attributable to tobacco.[2]

When prevalence, rather than incidence, rates are available, the relative risk and risk difference cannot be calculated. Instead, the prevalence rate ratio (defined as the prevalence rate among the exposed divided by the prevalence rate among the unexposed) and the prevalence rate difference may be calculated. Since prevalence rates do not express risk, however, caution must be used in interpreting these measures.

DIRECT AGE ADJUSTMENT

When comparing disease rates between populations, it is frequently necessary to take into account differences in the age distributions of the comparison populations. One method is to compare *age-specific disease rates* between the different populations across several age strata. This approach is awkward because it requires aggregating comparisons across multiple age strata. An alternative method, known as "direct age adjustment," involves calculating a single "age-adjusted" summary disease rate for each comparison population. This method applies the age-specific disease rates of the comparison populations to the age distribution of some arbitrarily chosen "standard population." This procedure is demonstrated in Table 1.

In the example shown in Table 1, the "standard population" has been arbitrarily defined as the sum of the two comparison populations. The age-adjusted rate of respiratory disease for population 1 is obtained by computing the expected number of cases that would occur in each age stratum of the standard population if the standard population had the age-specific disease rates of population 1. The total number of expected cases in the standard population is then obtained by summing the expected number of cases across all age strata (assuming population 1 rates, for example, the expected number of cases in the standard population is: $(0.02 \times 1300) + (0.04 \times 1300) + (0.06 \times 1150) + (0.08 \times 1100) = 235$), and the age-adjusted rate of respiratory disease is obtained by dividing the expected number of cases in the standard population by the total number of individuals in the standard population. The age-adjusted rate is computed in similar fashion for population 2, i.e., by computing the expected number of cases in the standard population assuming the age-specific disease rates of population 2 (307.5). The summary age-adjusted rates of respiratory disease are thus 4.8 and 6.3%, respectively, for populations 1 and 2. Because these rates have been computed from the same standard population, they may be compared directly since adjustment has been made for the age differences between the two populations. Thus, it may be concluded that after

Table 1. Hypothetical Example Illustrating the Use of Direct Age Adjustment to Compare Respiratory Disease Rates in Two Populations

	Population 1		Population 2		Standard Population		
						Expected # of cases assuming rate in:	
Age	N	Rate	N	Rate	N	Pop. 1	Pop. 2
20—29	500	0.020	800	0.030	1300	26	39
30—39	600	0.040	700	0.060	1300	52	78
40—49	650	0.060	500	0.070	1150	69	80.5
50—59	700	0.080	400	0.100	1100	88	110
TOTAL	2450	0.053	2400	0.056	4850	235	307.5
Adjusted rate (population 1)			=	235/4850		=	0.048
Adjusted rate (population 2)			=	307.5/4850		=	0.063

adjusting for age differences between the two populations, population 2 has a higher rate of respiratory disease than population 1. Age-adjusted rates may be compared statistically to evaluate whether they differ by more than would be expected by chance alone.[3]

Age-adjusted rates are fictional in the sense that they are based on the age distribution of some arbitrary standard population. In the above example, we defined the standard population to be a composite of the two comparison populations. Another standard population that is commonly used for age adjustment is the 1980 U.S. population, as published in the U.S. Census.[4] The magnitude of the age-adjusted rates, however, will depend on the choice of the standard population used. Nevertheless, the comparisons afforded by this adjustment are valid since the age-specific rates of each comparison group are applied to the same standard population. Thus, in the above example, population 2 has consistently higher age-specific rates than population 1, and these differences result in a higher age-adjusted rate for this group. The similarity in crude (i.e., unadjusted) disease rates between populations 1 and 2 occurs only because population 2 is younger.

MORTALITY STUDIES: SMRs AND PMRs

A frequent endpoint used in many occupational studies is mortality (death). Often in mortality studies it is impractical to calculate age-adjusted mortality rates because of the small number of deaths in one or more of the comparison groups. One way to deal with this problem is to compare the observed number of deaths in the population of interest with the number of deaths that would be expected based on the mortality rates of some standard population, such as the general population. Consider the hypothetical example in Table 2 showing mortality among vinyl chloride workers. In this example, a total of 169 deaths are observed

Table 2. Calculation of the Standardized Mortality Ratio (SMR) in a Hypothetical Population of 3000 Vinyl Chloride Workers

Age	Observed Deaths	Expected Deaths
25—34	2	0.3
35—44	8	3.1
45—54	61	11.8
55—64	98	41.0
TOTAL	169	56.2

among 3000 workers over a 5-year period. The expected number of deaths in this population may be computed using age-specific mortality rates obtained from the general population (or for an arbitrarily chosen standard population). In Equation 6, hypothetical age-specific mortality rates from the general population have been used to generate the expected number of deaths in each age group by multiplying these "standard" death rates by the total number of vinyl chloride workers in each age stratum, and an expected number of 56.2 deaths is obtained for the entire worker population. This number represents the total number of deaths expected in the worker population if that population experienced the same death rates as the general population. The ratio of observed to expected number of deaths is therefore 169:56.2, or 3.01. The *standardized mortality ratio* (SMR) is defined as 100 times this ratio (SMR = 301 in this example), and its interpretation is analogous to that of the relative risk. Thus, SMRs >100 indicate an excess in the observed number of deaths, while SMRs <100 indicate a deficit.

$$\text{SMR} = \frac{\text{Total observed number of deaths}}{\text{Total expected number of deaths}} \times 100 \qquad (6)$$

$$= 169/56.2 \times 100 = 301$$

It is not always appropriate to use mortality rates from the general population in SMR calculations to obtain the expected number of deaths. In occupational settings, for example, employed populations are generally healthier than the general population. Thus, one might expect the employed population to have lower death rates for this reason. This potential source of bias is known as the *healthy worker effect*. This source of bias could be avoided if an internal rather than an external control (e.g., the U.S. general population) group were used, such as one consisting of workers in the same industry who were not subjected to the exposure of interest.

In addition to SMRs, another manner of evaluating cause-specific mortality differences between subgroups is to compare the proportion of mortality between these groups due to a specific cause. Such studies are called *proportionate mortality ratio (PMR) studies*. If 30% of deaths in one occupational group was due to cancer, compared to only 15% of deaths in another, for example, then this may suggest the presence of a cancer-related exposure in the first group. The advantage of this approach is that it is not necessary to obtain vital status information on all workers in each group, but only in the deceased to determine the cause of death. Since this information is readily obtainable from death certificates, proportional mortality studies are relatively quick and easy to perform. The drawback of this approach, however, is that it assumes that the rate of other causes of death between the two groups are comparable, which may not necessarily be true. For this reason, such studies must be interpreted with caution.

OBSERVATIONAL STUDIES

The ideal strategy for investigating the effects of an occupational exposure on a health outcome would be to assign workers randomly into two groups, exposed and unexposed, and to compare the subsequent incidence of disease between the two groups over time. Such an approach would be *experimental* since the conditions of the study are preset by the investigator. Insofar as workers choose their own occupations, however, this type of design is not feasible; the epidemiologist must therefore rely on *observational* study designs to evaluate disease-exposure relationships. In observational studies there is no intervention imposed by the investigator, and the potential observations are limited to the conditions that have already occurred. Despite this limitation, observational studies can provide valuable insight into disease etiology provided that much thought is given to which subjects will be studied and how the required information will be obtained.

There are a variety of means by which exposure may be assessed in observational studies. Actual dose to the workers, for example, may be estimated by the evaluation of toxic agents, or their metabolites, in blood, urine, or other body tissues. Alternatively, exposure levels in the environment may be determined through various industrial hygiene measurements or estimates. More easily obtainable but less direct measurements, such as employment in a particular job category or duration of employment, may also be used. For purposes of simplicity, exposure is treated for the remainder of this chapter as though it were measured as a categorical variable (e.g, subjects are simply considered as either exposed or unexposed). Also, for the following discussion, it is assumed that disease status is measured as a present/absent condition.

There are three general types of observational study designs: cross-sectional, cohort, and case-control. As summarized in Table 3, these may be distinguished by the time frame in which disease and exposure status are assessed and by the strategy of analysis. A common goal of all three study designs is to estimate the magnitude of the association between the disease and the exposure.

Cross-sectional studies differ in an important respect from cohort and case-control studies in that no attempt is made to determine the temporal relationship between exposure and disease. Cross-sectional studies involve a survey of the population at a single point in time. Current exposure and current disease status

Table 3. Characteristics of Epidemiologic Study Designs

Study Design	Time Span	Analysis
Cross-sectional	Disease and exposure assessed at a single point in time	Compare prevalence rates of disease between exposed and unexposed (prevalence rate ratio; odds ratio)
Cohort	Disease status assessed prospectively	Compare incidence rates of disease between exposed and unexposed (relative risk)
Case-control	Exposure status assessed retrospectively	Compare past exposure between diseased (cases) and nondiseased (controls) (odds ratio)

are assessed for all subjects and analyses are conducted to determine associations between current exposure and disease. *Cohort studies* involve follow-up of disease-free individuals over time so that subsequent disease status (i.e., incidence) may be compared between those who were exposed and unexposed at baseline. The temporal relationship between exposure and disease is also assessed in *case-control studies*. In the case-control approach, individuals with disease (cases) and individuals without disease (controls) are compared in terms of their prior exposure status. Exposure is thus assessed retrospectively.

CROSS-SECTIONAL STUDIES

Cross-sectional studies may be characterized according to the type of sampling design utilized. The simplest type of design is one in which subjects are selected for study without regard to either exposure or disease status (i.e., the sample is unstratified). For example, a sample is selected from all individuals who are members of a particular union (without knowledge of the individuals' exposure history or disease status). In unstratified samples exposure and disease status are measured only after the sample is defined. Alternatively, however, the sample may be stratified so that a predetermined number of diseased and nondiseased individuals, or exposed and unexposed individuals, is chosen for study. As discussed later, stratified samples are often efficient when either the exposure or the disease is very rare. Regardless of whether stratified or unstratified samples are used, however, the defining feature of a cross-sectional design is that both exposure and disease status are ascertained simultaneously on all subjects. Analyses are then conducted to determine associations between exposure and disease, and the magnitude of the association is evaluated by the prevalence rate ratio.

An exception to using prevalence rate ratios as a measure of association in cross-sectional studies occurs when the sample is stratified according to disease status. Such a sampling scheme is often used in the investigation of rare diseases. Consider a disease that has a prevalence rate of 1/1000. In an unstratified sample, 100,000 persons would have to be studied in order to obtain an expected number of 100 cases. A more efficient sampling design would be to identify a fixed number of diseased individuals and a fixed number of nondiseased individuals, and a reasonable analytic strategy would be to compare the frequency of exposure between the two groups. The prevalence rate, however, could not be determined from such a sample. If a total of 100 diseased and 100 nondiseased individuals were chosen for study, for example, the "prevalence" of disease in this sample would be 50% (100/200). Since this rate is determined completely by the sampling frame, it has no inherent meaning and thus it would not make sense to compute prevalence rates or the prevalence rate ratio. Instead, another measure of association, the *odds ratio*, is used for sampling designs that are stratified by disease status. This measure is discussed under "Case-Control Studies."

An important limitation of cross-sectional studies is that because they relate current exposure and current disease, it cannot be determined whether the exposure preceded the disease or the disease preceded the exposure. Cross-sectional studies can also generate misleading conclusions about exposure-disease relationships if individuals change their exposure category directly as a result of the disease process. Consider, for example, a hypothetical cross-sectional study of carpal tunnel syndrome (CTS) in automotive factory workers. CTS is a cumulative trauma disorder of the wrist. Suppose that a particular job category was associated with CTS but individuals in that job category who developed CTS subsequently switched job categories (perhaps to a job category that was less physically stressful). The prevalence rate of CTS in the suspected job category would be artificially low and, in fact, may be no higher than that of other job categories (whose rates would now be artificially too high). Thus, this type of study design may fail to detect or may underestimate a true association between job category and CTS.

COHORT STUDIES

Cohort, or incidence, studies provide the most direct way for investigating exposure-disease relationships because exposure status is assessed before the development of disease in all subjects in order that the temporal sequence of exposure and disease may be clearly discerned. The major limitations of cohort studies are most often practical, such as the time and expense required to perform them.

In most cohort studies data are gathered prospectively on a cohort of exposed and unexposed individuals. Occasionally it is possible to gather data retrospectively in a cohort study. Suppose, for example, that a cohort of individuals could be identified (perhaps from company employment records) who began working 5 or 10 years ago. If exposure information were available on these individuals at that time, the subsequent disease status of these subjects (classified retrospectively as exposed or unexposed) could be ascertained and their subsequent disease experience measured. Such a study is called a *retrospective cohort* or *historical cohort* study. It is important to realize that the study population in retrospective cohort studies consists of the entire retrospectively defined exposure cohort, not just the surviving members of the cohort. In such studies, outcome measurements of disease status must therefore be obtained on the entire cohort.

In cohort studies that entail a long period of observation, some subjects are invariably lost to follow-up for a variety of reasons: some move away, some die, and some may drop out of the study. When this occurs, calculation of incidence rates becomes problematic since subjects may be "at risk" of disease for varying lengths of time. The same problem occurs if subjects are enrolled in a study over an extended time period. One method of dealing with the problem of unequal periods of observation between study subjects is to express the denominator of the incidence rates in terms of person-time units, where one *person-year*, for example, equals one person followed for 1 year. Thus, if three cases of cancer were observed among four subjects who were followed for intervals of 3, 5, 6, and 9 years, respectively, the incidence rate would be 3/23 person-years. The use of person-years also allows for the fact that individuals may change exposure categories over a lifetime. A subject exposed for 5 years and unexposed for 5 years contributes 5 person-years to both the exposed and unexposed groups.

A useful measure that can be obtained from cohort studies is the *population attributable risk (PAR) percent*. This measure is defined as the proportion of disease in the total population attributable to the exposure.[5] The statement that 30% of all cancer deaths are attributable to tobacco is a statement of attributable risk. As seen in Equation 7, the PAR is a function both of the frequency of the exposure and the strength of the association between the exposure and the disease:

$$\text{PAR} = \frac{b \times (\text{RR} - 1)}{b \times (\text{RR}-1) + 1} \times 100\% \tag{7}$$

where b = frequency of exposure
RR = relative risk

In 1982 it was estimated that approximately one third of the adult population in the U.S. smoked cigarettes. Given a relative risk of smoking and lung cancer mortality of approximately 18.2, the PAR for smoking and lung cancer mortality is estimated at:

$$\frac{0.33 \times (18.2\text{-}1)}{0.33 \times (18.2\text{-}1) + 1} = 0.85, \text{ or } 85\%.$$

Table 4. 2 × 2 Table Showing Derivation of the Odds Ratio

	Disease	
	Yes (Cases)	No (Controls)
Exposed	a	b
Not Exposed	c	d

$$\text{Odds ratio} = \frac{\text{Odds of exposure among diseased}}{\text{Odds of exposure among nondiseased}} = \frac{\dfrac{a/(a+c)}{c/(a+c)}}{\dfrac{b/(b+d)}{d/(b+d)}} = \frac{a/c}{b/d} = \frac{ad}{bc}$$

Thus, 85% of the 95,000 lung cancer deaths occurring in the U.S. could have been prevented if people did not smoke.[6]

CASE-CONTROL STUDIES

The case-control approach is typically undertaken to investigate the relationship between exposures and rare diseases. Two groups of subjects are selected for study, one group with and the other group without the disease of interest. Information about past exposure is then compared between cases and controls. Because the sampling scheme for case-control studies depends on disease status, this is an efficient study design for studies involving rare diseases.

The measure of association used to evaluate disease-exposure relationships in case-control studies is the *odds ratio*. In case-control studies the odds ratio is defined as the ratio between the odds of exposure among cases and the odds of exposure among controls, where the odds of an event equals the probability of its occurrence divided by the probability of its nonoccurrence. In the cross-stratified data shown in Table 4, the odds of exposure among cases equal a/(a+c) ÷ c/(a+c), or a/c. Similarly, the odds of exposure among controls equal b/(b+d) ÷ d/(b+d), or b/d. The odds ratio is thus equal to a/c ÷ b/d, or the cross-products ratio, ad/bc. An important characteristic of the odds ratio derived from case-control studies is that under certain conditions it is an approximate estimate of the relative risk. Specifically, these conditions are that: the cases chosen for the case-control study are incident cases; the subjects are selected without bias; and the disease under study is rare.[7]

$$\text{Odds ratio} = \frac{\text{Odds of exposure among cases}}{\text{Odds of exposure among controls}} = \frac{ad}{bc} \qquad (8)$$

The use of the odds ratio is illustrated in the following hypothetical study of Alzheimer's disease. Suppose that 100 cases and 100 controls are chosen for study. From interviews with next of kin, it is established that 15 of the cases and 8 of the controls were employed in an industry that involved exposure to solvents. From these data, the following 2 × 2 table is constructed:

	Cases	**Controls**
Exposed	15	8
Unexposed	85	92
Total	100	100

The estimated odds ratio would be (15 × 92)/(85 × 8) = 2.03, thus indicating that the odds of exposure in cases are 2.03 times the odds of exposure in controls. Because the odds ratio is greater than 1 in these hypothetical data, this would provide evidence of an association between solvent industry employment and Alzheimer's disease.

The retrospective nature of case-control studies carries with it several limitations. The major difficulty usually lies in the choice of an appropriate control group since bias will be introduced into the study if the exposure of interest is associated with the criteria chosen for selecting the control group. For example, hospital or clinic-based controls are used, bias will arise if the exposure of interest (for instance, a health-related behavior) were associated with admission to the hospital or clinic admission. An additional problem affecting case-control studies is that it can sometimes be difficult to obtain accurate measures of exposure status in individuals when disease status is already known. A particular concern when past exposure is ascertained by interview is that cases may on average under- or overreport their prior exposure (compared to controls).

CONFOUNDING AND STRATIFICATION

Regardless of the study design used, a limitation of observational study designs is that the comparison groups may differ not only in terms of the exposure of interest but also in terms of a host of other variables. When comparing the prevalence or incidence of disease between two exposure groups, for example, it is impossible to attribute confidently a difference in disease experience between the two groups to the exposure of interest if the groups also differ with respect to other variables associated with the disease. This problem was discussed previously with regard to age differences in the context of age adjustment; other variables, however, may also be involved. Consider a hypothetical cohort study of arsenic exposure and lung cancer. If workers exposed to arsenic were more likely to smoke cigarettes than unexposed workers, and if the incidence rate for lung cancer were higher among the exposed workers, it would be unclear whether the excess incidence among the workers was attributable to arsenic exposure or to cigarette smoking. Cigarette smoking in this example is a possible confounding variable. In general, a *confounding* variable may be defined as any variable whose presence obscures the true association between the exposure and the disease. Thus, confounding variables have the properties of being associated both with the disease and with the exposure.

Provided the confounding variable has been accurately measured, there are ways to determine the independent effects of the exposure on the disease. The most straightforward method is to stratify, or subdivide, the analyses according to levels of the potential confounding variable. In the previous example, the effect of arsenic exposure on lung cancer could be examined independently of smoking status by comparing the incidence rates among workers exposed and unexposed to arsenic (separately for smokers and for nonsmokers). Smoking would be a confounding variable if the association between arsenic and lung cancer within each level of smoking differed from the overall (unstratified) measure of association. An example of confounding and stratification is illustrated by Table 5, using fictional data relating to arsenic exposure, lung cancer, and smoking status.

In this example, subjects are cross-stratified into eight cells (not including the "TOTAL" columns) according to exposure, disease, and smoking status. Smoking is a confounding variable because the smoking status-specific relative risks (0.99 for smokers and nonsmokers individually) differ substantially from the overall (crude) relative risk of 1.57. The strata-specific relative risks correctly indicate that there is no association between arsenic exposure and lung cancer in these data; the crude, or unadjusted, relative risk is misleading because it fails to take into account the fact that subjects exposed to arsenic are more likely to smoke than subjects not exposed to arsenic, and it is the smoking that is more directly associated with lung cancer. (The interested reader can verify that the relative risks for smoking and lung cancer in these hypothetical data are 5.68 for subjects exposed to arsenic and 5.67 for subjects not exposed to arsenic.)

Stratification may not be practical if there are several potential confounding variables whose effects must be controlled for simultaneously. For example, if one wanted to examine the association between exposure (two groups) and disease (two groups) while controlling for the joint confounding effects of smoking (two groups), sex (two groups), and age (four groups), a cross-stratified table with 64 cells would be needed ($2 \times 2 \times 2 \times 2 \times 4$). In these situations other statistical approaches must be used, such as statistical modeling.

Table 5. Hypothetical Example of How Smoking Might Confound the Effect of Arsenic Exposure in Causing Lung Cancer

	Smokers		Nonsmokers		Total	
	Lung Cancer		Lung Cancer		Lung Cancer	
	Yes	No	Yes	No	Yes	No
Exposed	193	35	7	40	200	75
Unexposed	657	115	143	810	800	925
Relative risk	0.99		0.99		1.57	

SUMMARY

This chapter introduced some of the basic epidemiologic measures of disease frequency (e.g., prevalence and incidence rates), measures of association (e.g., prevalence rate ratio, relative risk, and odds ratio), and some basic analytic tools commonly used in occupational epidemiology (e.g., age adjustment, SMR, and PMR). In addition, the three basic types of observational study designs were described (i.e., cross-sectional, cohort, and case-control). Finally, a brief description of confounding variables and methods for dealing with them was provided.

REFERENCES

1. Young, J.L., and E.S. Pollack. "The Incidence of Cancer in the United States," in *Cancer Epidemiology and Prevention,* Schottenfeld D., and J.F. Fraumeni, Jr., Eds. (Philadelphia: W.B. Saunders, 1982), p. 154.
2. Doll, R., and R. Peto. *The Causes of Cancer* (New York: Oxford University Press, 1981), pp. 1220-1223.
3. Kahn, H.A. *An Introduction to Epidemiologic Methods* (New York: Oxford University Press, 1983), p. 64-72.
4. 1980 Census of the Population, Vol. 1, Characteristics of the Population. Chapter B: General Population Characteristics. Part 1: United States Summary PC80-1-B1. Issued May 1983.
5. Lilienfeld A.M., and D.E. Lilienfeld. *Foundations of Epidemiology*, 2nd ed. (New York: Oxford University Press, 1980), p. 217.
6. "The Health Consequences of Smoking: Cancer: A Report of the Surgeon General," U.S. Dept. of Health and Human Services, U.S. Government Printing Office (1982), p. 23.

7. Schlesselman J.J. *Case-Control Studies: Design, Conduct, Analysis* (New York: Oxford University Press, 1982), pp. 33-39.

PART II

Occupational Health Standards

In the late 1700s, Sir Percival Pott, a London surgeon, discovered a high incidence of cancer of the scrotum in London chimney sweeps. In 1788, as a result of this, the Chimney Sweeps Act was passed. In 1905, Massachusetts became the first state to employ health inspectors to investigate occupational hazards. In 1908, the first Federal Compensation Act (covering civil service employees) was passed in the U.S. (this law gave certain employees the right to sue for compensation from occupational injuries). Compared to such humble legislative beginnings, businesses today are faced with a wide variety of laws. The four chapters in this section introduce occupational health legislation:

Occupational Health Laws, Standards, and Guidelines identifies and explains the major occupational health laws, regulations, and guidelines in the U.S.

Worker Right-to-Know and **Community Right-to-Know (SARA Title III)** explain two wide-reaching laws that require manufacturers and users of hazardous substances to inform their workers and the surrounding community about those chemical hazards.

Overview of Worker's Compensation presents a discussion of the Worker's Compensation system, including coverage, benefits, and state compliance.

CHAPTER 4

Occupational Health Laws, Standards, and Guidelines

Neil Zimmerman

INTRODUCTION

Why are occupational health laws, standards and guidelines needed? Who are the major players in their development, establishment, and enforcement? What are some of the major regulations and guidelines, what do they mean, and how are they used? These are some of the issues explored in this chapter.

All of life's activities, home life, sports, avocations, travel, work, military duty, whether by choice or by necessity, involve some risk of accident, injury, illness, or death. Although zero risk is a desirable goal, whether for prevention of falls, elimination of automobile accidents, or exposure to carcinogenic chemicals, it is an impossible goal. Therefore, socially and technically acceptable decisions must be made as to the amounts of allowable exposures to particular hazardous agents. The extent of these decisions form the basis for the regulations and guidelines under which society operates.

There is concern in the occupational area that a worker, regardless of occupation, should contribute well to some desired outcome and be compensated for his/her effort, but that this trade of effort for compensation should not include any physical or mental deterioration on the part of the worker.

This idea can be expressed in the "healthy worker equation":

Healthy Workers	+	Raw Materials	+	$	=	Still Healthy Workers	+	Finished Products	+	By-products	+	More $$

The worker's health should be the same on the right side of the equation as on the left, and at the conclusion of employment, health should not have been affected by the other variables.

The history of the occurrence of illness and injury as a result of occupational labor is as old as the history of the human race. Indeed, prehistoric man most likely suffered from occupational hazards involving the grinding of tools, the development of fire, and other endeavors of the times. What has changed the most over the centuries is the way in which individuals and society perceive such hazards, first with awareness, then concern and compensation, and eventually prevention and control.

The early Greeks and Romans wrote of their observations of occupational diseases, typically involving mining.[1] These observations were added to by Renaissance scientists.[1] With the onset of the Industrial Revolution of the 1700s, the plight of the worker became more obvious and complex as new hazards developed. This heightened awareness gradually led to the compensation of workers for their suffering (worker's compensation is covered elsewhere in this book).

Only in the last 50 years has this awareness and compensation been channeled into a more useful direction, prevention of injury and disease by controlling or eliminating hazards. A combination of voluntary and mandatory rules and regulations are necessary to ensure compliance with the recommended means of prevention. A body of voluntary or consensus standards developed and grew commensurately with industry. These standards serve as useful guidelines but are not legally binding. As voluntary standards are not always followed, mandatory or regulatory standards are also required. These standards have the authority of law.

In the U.S., safety and health legislation was enacted in the late 1800s and early 1900s, but lacked enforcability. For example, although the Bureau of Mines was established in 1910, it was not granted permission to inspect mines until 1941 and did not have the ability to enforce standards until 1952.[2] The Walsh-Healy Act of 1936 was the first attempt by the federal legislative branch to enforce occupational safety and health standards. It was somewhat limited in scope, however, as it applied only to those manufacturers with federal contracts in excess of $10,000.[2]

The increasing number of deaths, disabilities, and diseases per year — thousands of occupational deaths, hundreds of thousands of injuries — and their billion-dollar costs to society, culminated in the passage of the Williams-Steiger Occupational Safety and Health Act of 1970, to assure "... so far as possible, every working man and woman in the nation safe and healthful working conditions ...".[3]

So who are the major players in the occupational health regulations game? The major governmental agencies include the Occupational Safety and Health Administration (OSHA) and the National Institute for Occupational Safety and Health (NIOSH), with increasing activity and involvement by the Environmental Protection Agency (EPA). The major professional groups involved have been the professional industrial hygienists and safety professionals and their respective societies and associations: the American Industrial Hygiene Association (AIHA), the American Conference of Governmental Industrial Hygienists (ACGIH), and the American Society of Safety Engineers (ASSE). The voluntary groups include the American Society for Testing and Materials (ASTM) and the American National Standards Institute (ANSI). Workers' groups and unions such as the AFL-CIO and the Oil, Chemical and Atomic Workers International Union (OCAW), while not directly developing standards, have played a major role in pressuring their development and bringing forth critical information upon which decisions can be made.

THE GOVERNMENT'S ROLE — OSHA

Beginnings

A comprehensive federal occupational safety and health bill, first sponsored in 1968 by the administration of President Lyndon B. Johnson, was the direct result of the increasing outcry against the unnecessary deaths, injuries, and illnesses in the American workplace. Johnson's Secretary of Labor, Willard Wirtz, and others, spoke out in favor of this proposed legislation. Although Congress vetoed the bill in 1968, two important bills were passed in 1969: the Coal Mine Health and Safety Act and the Construction Safety Act.[2] A more comprehensive, general workplace bill was still needed, however. Disagreement in Congress over the extent of federal powers for regulating occupational health and safety finally culminated in a compromise between the Democratic-sponsored, generally prounion, Williams Senate bill and the Republican-sponsored, generally promanagement, Steiger House of Representatives bill. The final version of the Occupational Safety and Health Act, known as the OSHAct, which was closer to the more stringent Williams bill, was signed into law as Public Law 91-596 by President Nixon on December 29, 1970. The intent of Congress and the Act is stated in Section 2 of that law:

> "Sec. (2)(a) The Congress finds that personal injuries and illnesses arising out of work situations impose a substantial burden upon, and are a hindrance to, interstate commerce in terms of lost production, wage loss, medical expenses, and disability compensation payments."
>
> "(b) The Congress declares it to be its purpose and policy, through the exercise of its powers to regulate commerce among the several States and with foreign nations and to provide for the general welfare, to assure so far as possible every working man and woman in the Nation safe and healthful working conditions and to preserve our human resources."[3]

OSHA's Duties

The Secretary of Labor established the Occupational Safety and Health Administration (OSHA), which came into existence on April 28, 1971, the date the OSHAct became effective. The OSHAct authorized the newly created office of the Assistant Secretary of Labor to direct OSHA to[4]

- Promulgate standards for the protection of worker safety and health
- Enter and inspect workplaces
- Issue citations for violations of standards and recommend penalties
- Supervise and assist the States for occupational safety and health plans

All aspects of OSHA's duties are important in providing regulatory power to the OSHAct. Section 5 of the Act, although very brief, is a powerful section, setting out two duties for employers and one for employees. One of the two duties set out for employers (in paragraph 5 (a)(2)) simply states that each employer must comply with all standards promulgated under the OSHAct. The other duty, paragraph 5 (a)(1), states that each employer furnish a workplace

> "... free from recognized hazards that are causing or are likely to cause death or serious physical harm to his employees."

This statement, known as the General Duty Clause, is a comprehensive provision that can be used to cite employers for serious situations for which no standard exists. For example, if an employee dies or is seriously injured due to heat stress (for which there is no federal standard), the General Duty Clause can be invoked. Consensus standards or other reasonably accepted scientific information can be used as justification for applying the standard OSHA enforcement procedures of citations and penalties. Practically speaking, however, the enforcement of this provision is difficult and it is not frequently cited, being used mainly for unusual situations; however, its existence is an incentive to employers to make the effort to avoid such situations.

Paragraph (b) of Section 5 states that the employee must also comply with all OSHA rules and regulations applicable to his/her own working conditions or conduct. No provisions are made, however, for penalties to employees for noncompliance.

Standards Development

Three areas of OSHA standards development can be identified: standards initially incorporated into the OSHAct, emergency standards, and permanent standards. Initial standards were incorporated into the OSHAct because of the massive amount of issues that had to be immediately addressed. These included those laws already administered by the Department of Labor such as the Walsh-

Healy Federal Supply Contracts Act, the Longshoremen and Harbor Workers Compensation Act, and the National Foundation of Arts and Humanities Act.

In addition, many consensus and proprietary standards of various bodies were incorporated into the OSHA standards, such as ANSI, the National Fire Protection Association (NFPA), and the ACGIH.

Section 6 (c) of the OSHAct provides for a route of standards development in special cases or emergencies. Probably the best example of the need for and use of this emergency procedure was in the 1974 case of vinyl chloride exposure. Previously considered a chemical of rather low toxicity, a connection between vinyl chloride and a rare liver cancer known as angiosarcoma was discovered, and the federally regulated level was dropped from a concentration of 500 ppm to 1 ppm within a matter of months. An emergency standard, once enacted, must be replaced by a permanent standard, preferably after 6 months. Because of the major impact on business that an emergency standard can have, there have been very few additional successfully promulgated emergency standards.

Permanent standards are developed according to established procedures and all "interested parties" are given an opportunity for comment and input, as established by Section 6 (b) OSHAct rulemaking. A typical standards setting procedure begins with recommendations and background information on health hazards of a particular occupational hazard presented to OSHA by NIOSH. OSHA studies and selects issues for which standards are to be promulgated, either following NIOSH recommendations or through other channels. These decisions are then published as notices of intent in the *Federal Register.* After public hearings are held and comments have been received, OSHA makes any needed changes, and publishes the amended standard in the *Federal Register.* In 60 days it becomes the law.

OSHA standards may be classified in many ways. One type, the specification standard, gives the details of the manner in which the standard is to be complied with, such as equipment dimensions, materials, or ventilation design details. The opposite end of the spectrum is represented by the performance standard which indicates the end results or objectives but not the methodology by which it is to be accomplished, such as a chemical permissible exposure limit (PEL). The latter approach provides more flexibility for management. Two other categories of standards include one intended for a particular operation or industry, the vertical standard; the second category applies "across the board" to all workplaces, the horizontal standard.

The actual regulations that are the outcome of the OSHAct are listed in Title 29 in the Code of Federal Regulations (CFR). This title has a number of chapters that describe the intent of the OSHAct. For example, Chapter XX of the OSHAct deals with the review board for OSHA decisions, the Occupational Safety and Health Review Commission (OSHRC), while Chapter XVII deals with the specifics of OSHA. Each chapter is divided into parts, subparts, paragraphs, and subparagraphs; the most well known of these, at least in industrial hygiene circles, is Part 1910, the General Industry Standards, which contains the bulk of the safety and health standards for all general industries. Separate bodies of standards were promulgated for the construction industry, under Part 1926, and the maritime industry, under Parts 1915-1918.[5]

Standards Enforcement

Inspection is one of the most important powers granted to OSHA. An OSHA Compliance Officer has the right to enter and inspect any workplace at any reasonable hour during working periods.[4] These inspections are generally unannounced; in fact, there are stiff penalties for advanced warning of an inspection. As a result of the 1978 Supreme Court "Barlow's Decision," an employer may choose to bar OSHA from immediate entry pending the issuance of a warrant.[4] This, however, is not a recommended course of action, because it tends to present a more adversarial situation between management and the OSHA compliance officer. The Court decision, does, however, limit the scope of the inspection to the demands of the warrant.

Because of the extremely large number of potential inspections and the available number of inspection officers, inspection prioritization is necessary. The highest priority is a situation in which death or serious physical harm are imminent. The second priority is a site at which there has been a fatality or five or more serious injuries. Valid employee complaints rank next, but must be in writing, and signed (although informal complaints will still be followed up, at least with a letter to the employer), with the understanding that the employer, under penalty, cannot discriminate against a complainant. Specifically targeted high-risk industries are scheduled for inspection when possible. Although random inspections are also conducted, they have been given the lowest priority simply due to manpower limitations.[4]

An OSHA inspector conducts an Opening Conference with management upon arrival, during which the reasons for the inspection are explained. The walk-through survey, and when deemed necessary, workplace monitoring for chemicals, noise, etc., are then performed over one or more days. After completion of the inspection, a Closing Conference is held with management to explain what has been found. A separate meeting can be requested by the employees.

Based on the information gathered during the inspection, the compliance officer determines if the employer is in compliance with the various programs required by OSHA regulations such as monitoring, medical surveillance, protective device usage, work practice controls, education and training, and recordkeeping, as well as with the outcome of the evaluation of sampling data. If a violation is found, a citation is written and issued. Violations are categorized into: (1) imminent danger — the compliance officer feels there is reason to expect death or serious physical harm may occur immediately or before the hazard can be eliminated through normal procedures; (2) serious violation — there is substantial probability that death or serious physical harm could result and that the employer knew, or should have known, about the hazard; (3) nonserious or other violation — there is a direct relationship to job safety and health, but death or serious physical harm probably would not ensue; and (4) de minimis — the violation of a standard has no immediate or direct relationship to safety or health. Whether the employer willfully or repeatedly violates a standard or conversely shows good faith and cooperation will have a bearing upon the seriousness of the case and the level of fines. If it can be established that a standard was willfully violated, the violation is classified as serious. In addition, criminal charges and the possibility of imprisonment may develop, as in the case of a Chicago-area employer, Film Recovery Systems, whose owners were convicted on charges of involuntary manslaughter after the cyanide-related death of a worker.[6]

An employer may respond in a number of ways to a citation. Obviously, he can choose to abate the violation, following OSHA recommendations within a certain agreed upon abatement period, paying any fine. The employer also has the option of informally or formally contesting the citation, the amount of the fine, or the time for correction, or can request a variance from the standard.[3,4] Decisions can also be appealed, but this can be a lengthy and complicated process. The best avenue for maintaining the health and safety of the worker is for employers to comply with the standards.

It should be noted that while all private workplaces are covered by OSHA, many governmental agencies (such as the Department of Defense, the Coast Guard, NASA, etc.) establish their own safety and health standards or agree in principle to allow OSHA to enforce their safety and health regulations. Each service typically relies on OSHA's standards but adds its own interpretations and implementing guidelines. Many exposure standards are based on current ACGIH Threshold Limit Values (TLVs) rather than on OSHA's PELs.

Specific Standards

OSHA adopted two basic approaches in establishing standards for the protection of worker health. First, in addition to the large body of nationally accepted ANSI and NFPA consensus safety standards, OSHA obtained a large number of performance-based standards by establishing limits for air concentrations of approximately 400 chemical contaminants. OSHA's second approach was to develop a number of more complex and complete standards, covering a wide range of areas for which to provide worker protection. These standards are discussed chronologically, starting with the large number of air concentration limits, followed by the complete standards, and finally the recent updating of standards.

Shortly after OSHA was established, it adopted values for legal maximum allowable levels of contaminants in the air, PELs. To establish as many of these PELs as quickly as possible, values were adopted from the ACGIH 1968 TLV List and from the ANSI Z-37 Maximum Acceptable Concentrations. PELs for approximately 400 substances were listed in OSHA's General Industry Standard, Part 1910, Subpart Z, Section 1910.1000. TLVs with 8-hr time weighted average (TWA) concentrations or absolute ceiling concentrations were listed in Table Z-1 of Section 1910.1000. Mineral dust TLVs were listed separately in Table

Z-3. Data for these two tables were basically taken from the 1968 version of the TLVs, as these were the values referred to in the latest version of the Walsh-Healy Act. It is important to note that while TLVs are updated yearly by the ACGIH, the PELs, once established, remain law unless repromulgated. In addition to the TLVs, approximately 21 substances with ceiling values and maximum peak values as well as 8-hr TWAs were given in Table Z-2. These values were adopted from the ANSI Z-37 values.[7]

Following Section 1910.1000, standards were developed in much more detail for individual carcinogenic or highly toxic chemicals or substances such as asbestos, arsenic, lead, vinyl chloride, and benzene. These "complete" standards include not only the airborne concentration determined to provide a safe, healthful working environment, but also information and requirements for sampling and analysis of the chemical, appropriate engineering controls and personal protective devices, proper response for emergency spills or releases, medical surveillance programs, recordkeeping, and training. As these standards and their inherent issues became more complex, the length of the proceedings and documentation increased, and the number of standards issued decreased. Approximately 24 complete standards in all have been issued since the original 400 in the PEL lists.[3] One of OSHA's goals in the 1970s was to develop a complete standard for each chemical in the PEL list. This effort, known as the NIOSH/OSHA Standards Completion Program, remains unfulfilled. OSHA's current philosophy is to periodically update the PEL list and develop generic standards, such as sampling protocol and medical surveilance, that will apply to any chemical.

In addition to the increasing scientific complexity of standards development, the OSHA standards development process increasingly began to include court challenges to the need for and the technical and economic feasibility of the standards. As examples of issues developed and debated in the 1970s and resolved in the 1980s, the benzene and cotton dust standards are briefly discussed here due to their significance. The Supreme Court ruled in 1980[8] that OSHA's proposed reduction in allowable worker exposure to benzene (to 1 ppm) be overturned. This was based on a lack of sufficient scientific evidence of the need for a more stringent level (the issue of cost benefit was raised, but not debated). This decision was roundly thought to signal a major change in emphasis away from stricter workplace safety and health regulations. The reversal of this suspected trend came in 1981. The Supreme Court upheld a more stringent cotton dust standard previously challenged on the basis of scientific evidence and cost benefit.[9] This time the evidence was held to be sufficient and the issue of the cost was dealt with decisively in the written opinion of the High Court that it was not the intent of the Congress with the OSHAct to weigh human life or health against cost factors.

OSHA standards continued to generate controversy in the 1980s with the Hearing Conservation Amendment issuance (with its multireversals), Hazard Communication (with its many versions), asbestos (both an unsuccessful emergency standard and a subsequently successful permanent standard), field sanitation, ethylene oxide, benzene, and formaldehyde. A more detailed description of the complete history of OSHA and the development of all its major health standards can be found elsewhere.[10]

OSHA's anticipated standards agenda for the 1990s includes continued work on a generic standard that will provide for a more complete standard format for all chemicals with PELs. Other issues for future consideration include respiratory protection rules revisions, confined space and blood-borne disease final standards, a revised standard for benzene, and new standards for glycol ethers and methylene chloride.

New PELs

One of the last major changes of the 1980s was OSHA's PEL Project. This updating of PELs was long overdue since some were outdated and needed revision even as they were being adopted by OSHA in 1971. The TLVs of the ACGIH, which were the basis for many of the original PELs, are updated yearly, as mentioned earlier, and are based on detailed scientific review of the literature and current information. The TLVs are more fully explained later in this chapter. These TLVs were never intended to be legal limits separating safe and hazardous situations; rather, they are intended as guidelines to assist in an overall occupational health program. Because of the annual reevaluation and refinement of the TLVs, it was becoming clear by the early 1980s that many of the PELs no longer offered adequate protection and were in need of change. Many years of discussion and debate followed, with OSHA, practicing industrial hygienists, and corporations addressing this issue. During this time, many corporations and professional industrial hygienists adopted the philosophy and/or policy that lower, more stringent exposure limits (such as the most current TLV) should be adhered to rather than the PEL.

Finally, the PELs were changed on March 1, 1989, and include 164 new PELs for chemicals with no previous regulation, 212 more restrictive PELs and 160 unchanged PELs. Nine chemicals were still undergoing Section 6 (b) OSHAct standards rulemaking at issuance. Many of the changes were based on 1987 to 1988 TLVs and NIOSH Recommended Exposure Limits (RELs). A new Table Z-1-A, with both the new limits and the old or "transitional" limits, was issued to replace Tables Z-1, Z-2, and Z-3 of OSHA Regulations Section 1910.1000. The new limits became effective September 1, 1989, with compliance allowed by any means (engineering, administrative, or personal protective equipment) until the end of 1992, at which time they must be met by engineering controls unless interim adjustments are made to the compliance schedule.[11] The new PELs do not presently apply directly to the construction, maritime, and agriculture industries, however. How future changes will be dealt with is unclear; some of these new PELs already are or soon will be outdated.

NIOSH — THE GOVERNMENT'S RESEARCH AGENCY

In addition to the formation of OSHA, the second most important accomplishment of the OSHAct was the establishment of NIOSH. This federal agency is charged with conducting research to investigate and eliminate job hazards. To separate enforcement from research activities, NIOSH was assigned to the Department of Health, Education and Welfare (now known as Health and Human Services) and is administered through DHHS's Centers for Disease Control, a branch of the Public Health Service.

NIOSH was given the responsibility and authority to conduct research which can serve as the basis for standards that OSHA considers and promulgates. It also is charged with developing and supporting training programs for health professionals and testing/certification programs for various safety and health equipment.

Research Efforts

In addition to its own research labs in Cincinnati, OH and Morgantown, WV, where research studies on specific hazards, sampling techniques, control, and respiratory diseases are conducted, NIOSH sponsors a large amount of extramurally funded research in colleges and universities as well as in private facilities in a wide variety of subjects. NIOSH conducts applied research and investigatory activities in the field and in the laboratory. Usually at the request of management or labor, a Health Hazard Evaluation (HHE) can be conducted at a workplace to investigate a particular health problem.[3] This is not seen as an enforcement activity, although it is inherently assumed that recommendations derived from the study would be followed. Along with these short-term, single issue HHEs, NIOSH has been involved in long-term epidemiologically based field studies.

Standards Recommendations

The basic purpose of NIOSHs research efforts, in addition to contributing to the pool of scientific knowledge and literature, is to provide OSHA with the information needed to develop, promulgate, and enforce occupational health and safety standards. A large number of recommendations in the format of the complete OSHA standards are produced in "Criteria Documents," which include not only the NIOSH recommendation for the PEL or REL (Recommended Exposure Limit), but a detailed scientific literature review and other aspects of recommendations for standards. The majority of these Criteria Documents were published in the 1970s, and are somewhat outdated, but NIOSH continues to perform its role of presenting RELs and other scientific background information to OSHA in current in-depth publications as well as brief releases (Current Intelligence Bulletins, CIBs) on the most recent findings.

Training

Another important charge of NIOSH is to develop and support educational programs to ensure an adequate supply of trained occupational health professionals. These efforts are focused on established graduate school programs in Occupational Health through Educational Resource Centers (ERCs). NIOSH directly and through its ERCs has also been involved in numerous short course education programs as well as in promoting cross-training to develop occupational health awareness in the engineering, medical, and management professions.

Testing and Certification

NIOSH, under the authority of the OSHAct as well as other legislation such as the Coal Mine Health and Safety Act, develops and performs tests and certifies a variety of safety and health devices. NIOSH researchers determine if available devices such as respirators, hard hats, and safety goggles are appropriate and acceptable. In doing so, manufacturers are encouraged to develop and improve their own testing and quality control programs. For example, direct-reading gas and vapor detector tubes were at one time tested and certified by NIOSH, but the program developed into an industry-controlled and -monitored quality assurance program. NIOSH has also issued safety alerts relating to problems with respirator components, such as compressed air cylinders, that have resulted in equipment recalls and/or modifications, and ultimately, reduced numbers of accidents.

OTHER GOVERNMENTAL AGENCY ROLES IN OCCUPATIONAL HEALTH: EPA, MSHA

Chief among these agencies that impact on occupational health standards and activities is the Environmental Protection Agency (EPA) and its numerous environmental legislative activities.

The Toxic Substances Control Act

The TSCA, enacted by Congress in 1976, authorizes the EPA to require prenotification and testing of any new chemical prior to manufacturing. This authority includes the right to ban a chemical based on its toxicity testing results.

The Resource Conservation and Recovery Act

The RCRA, also enacted by Congress in 1976, established a universal hazardous waste management program. The concept of responsibility for hazardous material from "the cradle to the grave" automatically involved industrial hygienists and, indeed, all occupational health professionals (1) by sharing some of the responsibility while the hazardous material is interacting with or in the workplace (tracking, recordkeeping, etc.), and (2) protecting the workers who must treat, store, or transport the hazardous materials.

The Comprehensive Environmental Response, Compensation, and Liability Act of 1980

CERCLA, better known as Superfund, was enacted for emergencies and cleanup at uncontrolled waste sites. Although environmental in design, it has impact on occupational health since the workers at Superfund sites need protection. Often this protection requires the most conservative approach because the exposures are often unidentified.

The Superfund Amendments and Reauthorization Act of 1986

SARA, which reauthorized CERCLA, added major provisions that have specific impact on occupational health. New programs covered by this reauthorization include underground storage tanks, emergency planning, risk assessment, community right-to-know, radon and indoor air quality, and hazardous waste site worker training. With specific regard to hazardous waste workers, SARA mandated that OSHA promulgate a fully revised Hazardous Waste Operations and Emergency Response Standard (referred to as Hazwoper) under title 29 CFR 1910.120. It is designed to protect employees who may be exposed to hazardous substances during emergency incidents, cleanup operation and RCRA TSD (treatment, storage, or disposal) operations.

The Asbestos Hazard Emergency Response Act of 1986

AHERA regulates the management, and removal if necessary, of asbestos in school buildings. It defers to OSHA asbestos regulations for protection of asbestos workers, but it stipulates regulations for air sampling and final clearance of an asbestos abatement activity.

The Mining Safety and Health Administration (MSHA)

MSHA is a separate federal agency that warrants mention as it is OSHA's sister safety and health agency for all mining activities. Mining, which is not monitored by OSHA, includes all underground and surface areas from which a mineral is extracted as well as all mining-related activities. MSHA was established in 1978 by the Federal Mine Safety and Health Amendments Act, which transferred authority for all mine safety and health from the Mining Enforcement and Safety Administration in the Department of the Interior to the new MSHA, located in the Department of Labor. NIOSH is given similar research, field investigation and standards recommendation authority for MSHA as it has for OSHA. Many of its air concentration levels are also based on ACGIH TLVs.

PROFESSIONAL ASSOCIATIONS AND VOLUNTARY GUIDELINES

In addition to governmental agencies involved in setting mandatory occupational exposure limits, a number of private and professional groups are also involved in developing consensus standards. In fact, as has been previously discussed, the voluntary guidelines are often the basis for governmental regulations. The groups that are currently most active in the U.S. in the occupational health area are the ACGIH and the AIHA, and in the safety area, the ASSE. Members of these associations, as can be determined by the names of the groups, are industrial hygienists and safety professionals, professions that can be defined as follows. To paraphrase the AIHA's definition,[12] an industrial hygienist is a person with a college degree in one of the basic or applied sciences as well as special studies and training to enable him/her to recognize environmental factors and understand the effects on humans, to evaluate the magnitude of these stresses, and to control such stresses in order to alleviate their effects. A safety professional is a person whose job is to prevent accidents, harmful exposure to hazardous situations, and the personal injury or illness and property damage related to such accidents. Other groups, such as the Air and Waste Management Association (AWMA) and the American Society of Heating, Refrigeration and Air Conditioning Engineers (ASHRAE), are involved professionally, but to a lesser extent in establishing guidelines.

ACGIH

The American Conference of Governmental Industrial Hygienists, organized in 1938 by a group of industrial hygienists who worked in governmental agencies, is known and respected worldwide for its expertise in the area of occupational exposure limits and industrial ventilation as well as other topics. Its membership consists of over 3000 occupational health professionals in governmental agencies or educational institutions. The goals of the ACGIH are basically to encourage and promote sound industrial hygiene practices and to collect and disseminate useful information in the field. The ACGIH is most well known for its occupational exposure guidelines, TLVs. There are now over 600 TLVs for chemical substances, as well as TLVs for physical agents such as noise, heat stress, ionizing and nonionizing radiation, and vibration. Originally developed in 1946, the TLVs are updated yearly by a volunteer committee of professionals who review all current scientific literature. The name "threshold limit value" indicates the principle upon which they are based — that there is some threshold level of exposure to a hazardous agent below which the average, healthy adult worker may be repeatedly exposed for his working lifetime without adverse effect.[13] Although these guides are felt to apply to nearly all workers, the ACGIH adds disclaimers, including nonapplicability to hypersusceptible individuals who are more responsive to an agent, or individuals with a preexisting condition such as a medical problem, pregnancy, overweight, heavy smoker, etc., who may react differently. The TLVs are based on the best available scientific information, but are intended to be used only as guidelines in the control of health hazards, not as fine lines between safe and dangerous conditions. This original TLV concept is important to consider when dealing with the OSHA PELs as regulatory levels. There are three categories of TLVs:[13]

1. The Threshold Limit Value-Time Weighted Average (TLV-TWA) is a time-weighted average concentration for a normal 8-hr/day, 40 hr/workweek to which nearly all workers may be repeatedly exposed without adverse effects; the TWA aspect of this guideline means that the actual concentration can fluctuate above the stated TLV during the day, as long as it also balances the average with sufficient time below the TLV; this concept applies best to agents that are more likely to build up slowly in the body, causing chronic poisoning.

2. The Threshold Limit Value-Short-Term Exposure Limit (TLV-STEL) is the concentration, averaged over 15 min, that will not cause irritation, chronic or irreversible tissue damage, or narcosis to the degree of increasing the likelihood of injury or reducing work efficiency; this concept applies best when there are acute effects from a substance whose toxic effects are primarily chronic.

3. The Threshold Limit Value-Ceiling (TLV-C) is the concentration that should never be exceeded during any part of the workday; here the averaging concept does not apply — it is most applicable for substances that produce immediate irritation or acute poisoning.

A "skin" notation is also listed for some chemicals, which indicates that the cutaneous route of entry has the potential for a major contribution to overall body burden and exposure, along with other routes of entry.

The Biological Exposure Indices (BEIs), established by the ACGIH in 1984, serve as reference values for the evaluation of potential health hazards in the workplace by actual determination of body burden due to environmental exposure, not by the environmental exposure alone. BEIs represent the level of a contaminant (or its metabolite) most likely to be observed in specimens collected from a healthy worker who has been exposed to inhalation exposure at the TLV. BEIs are not meant to indicate a sharp distinction between safe and dangerous exposures, nor are they intended for diagnosis of occupational illness. Typical determinants used for BEIs are urine, blood, and exhaled breath, although other tissues are sometimes used.[13]

Both TLVs (chemical and physical agents) and BEIs are presented in more detail, including their scientific rationale, in the ACGIH publication, "Documentation of the TLVs," which should be consulted prior to interpreting the results of air monitoring compared to a TLV.[14] Another area in which the ACGIH claims considerable expertise is industrial ventilation. Its industrial ventilation manual[15] is updated biannually and is the basis for many standards or voluntary guidelines throughout the world.

AIHA

The American Industrial Hygiene Association is a nonprofit, professional society that was organized in 1939 with the purpose of promoting the recognition, evaluation, and control of environmental stresses arising in or from the workplace or its products, and to encourage increased knowledge of industrial hygiene. Its more than 8000 members practice industrial hygiene in industry, government, labor, academic, and research institutions. Like the ACGIH, it is widely known for its development, by volunteer committee, of occupational guidelines, including Hygienic Guides, Workplace Environmental Exposure Levels, Emergency Response Planning Guidelines and Occupational Exposure Work Practice Guidelines.

The AIHA has been publishing its Hygienic Guides for a number of years. There are approximately 150 Guides, some of which are in need of updating. They take the form of a detailed Material Safety Data Sheet (MSDS) but actually predate the widespread popularity of the MSDS. The exposure limits listed in these Guides are the current PELs or TLVs rather than a new value.[16] Because not all chemicals have established exposure limits, AIHA filled the void by developing its Workplace Environmental Exposure Levels (WEELs) which focus on environmental agents and stresses for which there are no legal or authoritative limits. Some 35 WEELs have been published, including as examples urea, triethylene, glycol diacrylate, piperadine, and benzoyl chloride. These publications are toxicologically based and intended to assist the occupational health professional.[17]

Another set of AIHA guides, the Emergency Response Planning Guidelines (ERPGs), are intended for the health professional who is planning for or involved in emergency response actions. Twenty-five ERPGs have been published through 1990; those already released include chlorine, sulfuric acid, ammonia, formaldehyde, hydrogen fluoride, hydrogen chloride, and phosgene. ERPGs provide estimates of concentrations at which one might anticipate observing adverse effects in the *general population* resulting from a short-term exposure, as would be the case in a chemical spill or other chemical emergency. This is in contrast to occupational exposure guidelines which are normally based on long-term exposures and chronic health effects. The levels are termed ERPG-1, ERPG-2, and ERPG-3:[18]

- ERPG-1 is the maximum airborne concentration below which it is believed that nearly all individuals could be exposed for up to 1 hr without experiencing other than mild transient adverse health effects.

- ERPG-2 is the maximum concentration below which it is believed that nearly all individuals could be exposed for up to 1 hr without experiencing or developing irreversible or other serious health effects or symptoms that could impair an individual's ability to take protective action.

- ERPG-3 is the maximum concentration below which it is believed that nearly all individuals could be exposed for up to one hour without experiencing or developing life-threatening health effects.

As more ERPGs are developed and continue to become more well known, they could prove to be one of the most useful of all the exposure guidelines.

A new guideline series is also being developed by the Occupational Health Standards Committee of AIHA, called Occupational Exposure and Work Practice Guidelines. They are similar in depth and coverage to the NIOSH Criteria Documents, and will deal with current controversial substances for which practicing hygienists need more knowledge in order to make management decisions on handling them. The first two guides are state of the art reviews of the health issues surrounding formaldehyde and fibrous glass, with others being planned for the future.[19]

ASSE

The American Society of Safety Engineers, established in 1911, is the oldest U.S. professional society dedicated to the improvement of worker safety and health, and has more than 24,000 members who are deeply involved in the safety standards setting process, with volunteer participation on a number of ANSI, ASTM, and NFPA standards committees. ASSE serves as the secretariat or standards generating and coordinating committee for four ANSI committees: Confined Space, Railing and Toeboard, Eye Protection and Fall Protection. Members also serve as reviewers for consensus and regulatory standards developed outside of ASSE. Through these activities they have contributed a significant positive influence to the level of U.S. safety standards.

Standards Setting Organizations

The most active groups in the occupational health standards setting arena have been ASTM and ANSI. ASTM develops standards for a wide range of disciplines, but with regard to occupational health has been involved in the development of analytical methods for air sampling. ANSI, through its Z-37 Committee, has developed "acceptable concentrations" for a number of chemicals, many of which were incorporated into Table Z-2 of the OSHA PELs. ANSI's acceptable concentrations consist of an 8-hr TWA as well as ceiling and peak values; however, the number of chemicals covered by ANSI is somewhat limited and has not been updated regularly. ANSI is very active in areas such as ventilation, respiratory protection, confined space standards, personal protective equipment, and other safety standards. The NFPA, although mainly involved with fire safety guidelines, has developed standards for determining the effectiveness of protective clothing materials and for ventilation systems.

Other Professional Societies

Other professional societies, such as the Air Pollution Control Association (APCA) and ASHRAE, have also shown an interest in areas impacting on occupational health issues and guidelines. APCA, traditionally interested in airborne environmental pollution issues, has (as of Fall 1989) changed its name to the Air and Waste Management Association (AWMA) to reflect the widening interests of its membership. Although AWMA does not propose guidelines, it serves as a forum for discussion. For example, at the AWMA Total Exposure Assessment Meeting held in 1989, discussion centered around the need to consider exposure to environmental contaminants from a total population dose standpoint including contributions of air, water, food, indoor (residential and commercial), industrial, and personal (tobacco smoke).

ASHRAE, with its activity in fan ratings and duct design, has long been involved with industrial ventilation guidelines. ASHRAE has also been involved in recommendations for ventilation rates and human comfort in buildings. With the increased interest and emphasis on indoor air quality (IAQ), ASHRAE's new Ventilation Standard 62-1989 will influence occupational health evaluations in many indoor environments.

THE INDIVIDUAL PROFESSIONALS

The beginning of this chapter posed the question: who are the major players in occupational health regulations? Although the roles of a number of public and private organizations have been discussed, it is actually the safety and health professionals within the organizations who are the major players. While an explanation of professionalism, certification, and accreditation in industrial hygiene and safety may not seem directly related to occupational health regulations, it is certainly indirectly related, as it influences the caliber of the individuals who set and enforce these regulations. However, the existence of standards, laws, regulations, and guidelines does not guarantee the safety of the workforce. The regulations require enforcement and compliance, but more importantly, they must be questioned. Are the standards adequate? Are they based on the most current information? Are the effects on special groups considered? For carcinogens with possibly no threshold level, are the risk estimates for numbers of predicted cases of cancer acceptable?

The capabilities of the professionals to answer and to provide our working population with the safest and healthiest work environment possible depend on the level of training and professionalism attained. The American Board of Industrial

Hygiene (ABIH) and the Board of Certified Safety Professionals (BCSP) both attempt to promote the highest levels of professionalism through their certification and other activities.

ABIH

The American Board of Industrial Hygiene was established in 1960 to improve the practice and educational standards of the industrial hygiene profession. In carrying out its objectives, ABIH certifies individuals as to education, experience, and professional ability. The ABIH issues three categories of certification as an assurance that the individuals thus certified possess a high level of professional competence in industrial hygiene. The Certified Industrial Hygienist (CIH) designation is attained after proving ability in either the Comprehensive Practice of Industrial Hygiene or in an Aspect area such as Acoustics, Air Pollution, Chemical, Engineering, Radiological, or Toxicological. This requires a minimum of 5 years of experience and the passage of a two-part comprehensive examination. Prior to attaining CIH status, an individual with at least 1 year of experience can sit for the first part or "core" examination to achieve the status of Industrial Hygienist in Training (IHIT). There are currently approximately 3500 CIHs and 650 IHITs. For an individual who lacks the training and education necessary to attain these designations, a third category, the Occupational Health and Safety Technologist (OHST) (jointly sponsored by ABIH and BCSP since 1985), is available, which also requires 5 years of experience and passage of an examination. This individual is then recognized as having proficiency in a particular phase of industrial hygiene or safety, performing these tasks under the supervision of a CIH or Certified Safety Professional (CSP). There are currently 800 OHSTs.

Designation as a CIH also qualifies the individual as a Diplomate of the American Academy of Industrial Hygiene (AAIH), a separate organization whose purpose is to raise the level of competence of industrial hygienists and secure wide recognition of the need for high quality industrial hygiene practice. To further these goals, AAIH sponsors a Professional Conference in Industrial Hygiene and is involved with accreditation of industrial hygiene graduate programs.

BCSP

The Board of Certified Safety Professionals was established in 1968 to ensure high standards and the professional status of individuals in the safety profession. A CSP is an individual, who by virtue of academic training and education, professional experience (minimum of 4 years), and passage of a two-part examination, is designated by the BCSP to have attained high competence in the field of safety. There are currently 6600 CSPs. Prior to obtaining such status, an individual with an accredited academic degree in safety who passes the core examination can be designated as an Associate Safety Professional (ASP).

SUMMARY

To summarize, the successful prevention of occupational disease, illness, and injury requires the participation by competent safety and health professionals who can assess and understand the various characteristics and conditions of workplaces, who can evaluate the level of exposure to various hazardous agents, who can understand and interpret the comparisons between the workplace evaluations and established occupational health guidelines and regulations, and who can then take the necessary action to provide a safe and healthy workplace. This requires the most current and carefully developed regulations and guidelines possible, since they are ultimately the benchmarks with which decisions are made influencing safety and health in the workplace.

REFERENCES

1. Clayton, G. D. "Introduction," in *The Industrial Environment — Its Evaluation and Control*, National Institute for Occupational Safety and Health Publication No. 74-117, U.S. Government Printing Office (1973).
2. Trasko, V. M. "The Design and Operation of Occupational Health Programs in Governmental Agencies," in *The Industrial Environment — Its Evaluation and Control,* National Institute for Occupational Safety and Health Publication No. 74-117, U.S. Government Printing Office (1973).
3. 91st U.S. Congress, The Occupational Safety and Health Act of 1970, S.2193, Public Law 91-596, U.S. Government Printing Office (1977).
4. U.S. Department of Labor, All About OSHA, No. 2056, U.S. Government Printing Office (1985).
5. U.S. Department of Labor, 29 CFR 1910, 1915-18, 1926, General Industrial, Maritime and Construction Standards, U.S. Government Printing Office (1989).
6. Burnett, J. "Corporate Murder Verdict May Not Become Trend, Say Legal Experts," *Occup. Health Safety*, 54(10):22-26, 58-59 (1985).
7. Cook, W. A. Occupational Exposure Limits — Worldwide (Akron, OH: American Industrial Hygiene Association, 1987).
8. "Supreme Court Invalidates Standard for Benzene, Says Risk Must be Shown," *Occup. Safety Health Rep.,* 10(6):147—148 (1980).
9. "Supreme Court Rejects Cost Benefit, Upholds OSHA's Cotton Dust Standard," *Occup. Safety Health Rep.,* 11(3):53—54 (1981).
10. Mintz, B. W. "Occupational Safety and Health: The Federal Regulatory Program — A History," *in Fundamentals of Industrial Hygiene,* 3rd ed., B. A. Plog, Ed. (Chicago: National Safety Council, 1988).
11. U.S. Department of Labor, 29 CFR 1910.1000, Air Contaminants — Permissible Exposure Limits, OSHA 3112, U.S. Government Printing Office (1989).
12. American Industrial Hygiene Association, Membership Directory, AIHA, Akron, OH (1989).

13. American Conference of Governmental Industrial Hygienists, Threshold Limit Values and Biological Exposure Indices for 1989-90, ACGIH, Cincinnati, OH (1989).
14. American Conference of Governmental Industrial Hygienists, Documentation of the Threshold Limit Values, 5th ed., ACGIH, Cincinnati, OH (1986).
15. American Conferences of Governmental Industrial Hygienists, Industrial Ventilation — A Manual of Recommended Practice, 20th ed., ACGIH, Cincinnati, OH (1988).
16. American Industrial Hygiene Association, Hygienic Guide Series, AIHA, Akron, OH (1989).
17. American Industrial Hygiene Association, Workplace Environmental Exposure Levels, AIHA, Akron, OH (1989).
18. American Industrial Hygiene Association, Emergency Response Planning Guidelines, AIHA, Akron, OH (1989).
19. American Industrial Hygiene Association, Occupational Exposure Work Practice Guidelines, AIHA, Akron, OH (1989).

CHAPTER 5

Worker Right-To-Know

James E. Brower

INTRODUCTION TO RIGHT-TO-KNOW REGULATIONS

There are two federal laws, commonly referred to as right-to-know laws, which relate to information on hazardous substances. The Occupational Safety and Health Administration (OSHA) Hazard Communication Standard[1] refers to workers right-to-know. The Superfund Amendments Reauthorization Act (SARA) Title III[2] is the Emergency Planning and Community Right-to-Know Act (EPCRA). These laws require that manufacturers and users of hazardous substances provide information about the hazardous materials at their facility to workers and to community planning and emergency response organizations.

Prior to the passage of the Occupational Safety and Health Act of 1970 (OSHAct) that formed OSHA, informing workers about the hazards of materials they were using was largely a voluntary response by industry management. In 1969, amendments to the Longshoremen's Act required the use of material safety data sheets (MSDSs) to convey hazard information to specific maritime industries such as shipbuilders. In most industries workers were generally unaware of the content of the specific chemicals they were using and their hazards. The flow of chemical hazard information was largely influenced by several factors[3] (Table 1).

The demand for chemical information was often countered by pressure to protect manufacturing trade secrets. In addition, the existing toxicity and health hazard data were often limited. Communication of hazard information was often reactive, i.e., once an accident occurred, chemical hazard information would follow. Hazard information was heavily concerned with prevention of accidents

Table 1. Factors Influencing the Flow of Chemical Hazard Information

Market forces (i.e., economic benefits to the manufacturer)
Trade secret protection
Limited availability of toxicity data
Emergency situations (i.e., after an accident occurred)
Potential for high hazards
Warnings from safety and health professionals
Worker demands for information
Liabilities stemming from court litigation

such as fires, explosions, acute poisonings, or personal injury. Therefore, safety training concentrated on these risks while chronic or long-term health effects were not as strongly emphasized as they are today.

COVERAGE

Although the OSHA Hazard Communication Standard originally covered only manufacturing industries when it was promulgated in November 1983, it was expanded on May 23, 1988 to include all private sector industries that manufacture, import, distribute, or otherwise use hazardous chemicals. In addition, federal agencies have adopted and applied the standard. Most states have adopted laws that cover state and local facilities not preempted under the OSHA standard. Thus, in practice, few public and private workplaces that use chemicals are exempt from worker right-to-know laws.

The purpose of the OSHA Hazard Communication Standard is to inform the worker of the materials hazards as well as to inform them of usage and handling procedures which will maximize their protection. The objective is to dispel the "mystery" surrounding potential chemical hazards. This objective is achieved via three forms of information which employers must supply to their employees:

- Fully labeled chemical containers
- Detailed Material Safety Data Sheets (MSDS)
- Training in the requirements of the standard and safe use of chemicals

The Hazard Communication Standard is intended, however, to form one part of an integrated program for the protection of worker health and safety. Other parts of the health and safety program include management's regulating of allowed exposures to levels below the OSHA Permissible Exposure Limit (PEL); providing necessary safety equipment and clothing, utilizing appropriate engineering controls, and implementing emergency plans in case of spills, leaks, fires, or other accidents.

Table 2. Objectives of the Hazard Communication Standard

1. To ensure the evaluation of chemicals and determine their hazards
2. To inform workers of the hazards to which they may be exposed
3. To preempt state laws covering worker right-to-know in order to ensure uniformity of information and requirements

Table 3. Hazard Communication Issues

Who is to inform?
Who is to be informed?
What materials are subject to the regulation?
What information is to be transmitted?
How is the information to be transmitted?
How is the information standardized?

OBJECTIVES AND ISSUES

OSHA's Hazard Communication Standard has promulgated three objectives (see above Table 2) and six issues implied by these objectives (Table 3). These issues determine the method of implementation by the employer.

The sixth issue pertaining to the standardization of information should be clarified first. Unlike other OSHA standards, the Hazard Communication Standard is "performance" standards oriented rather than "specification" oriented. In a specification standard detailed procedures and requirements are given such as the type of respirator for a given chemical or ventilation rate for a chemical hood. A generic performance standard provides criteria for satisfaction of the regulation rather than specification of detailed requirements. In this case, compliance with the standard is measured by comparing the employer's performance against these criteria rather than identifying specific measurement standards.

For example, OSHA PELs[4] are specification standards. The employer is in violation of the law if workers are exposed to 8-hr average concentrations that exceed the specified level for that chemical. In the Hazard Communication Standard no detailed list of regulated hazardous chemicals is provided other than the floor of OSHA's PELs, the American Conference of Governmental Industrial Hygienists (ACGIH) Threshold Limit Values (TLVs),[5] and known carcinogens;[6,7] however, the standard provides specific criteria for determining if any chemical or chemical mixture that is not included in the floor list should be defined as hazardous. Thus, in effect, tens of thousands of chemicals and chemical mixtures are covered under the OSHA standard rather than just the approximately 800 or so chemicals explicitly given in various lists (some state regulations may have additional requirements). Similarly, the OSHA standard does not provide specifications for labeling chemicals as does the U.S. Department of Transportation (DOT) for shipping hazardous materials. OSHA provides guidelines for required labeling information, but does not require a standard label format, specific warning words, or symbols, as does the DOT.

Employers not covered by OSHA may be more stringently regulated by state laws. For example, New York State specifies that any chemical listed in the National Institutes of Occupational Safety and Health (NIOSH) Registry of Toxic Effects of Chemical Substances (RTECS),[8] be covered under its right-to-know law. However, RTECS provides data for over 70,000 chemicals, which includes nontoxic chemicals such as glucose, as well as the most toxic, such as dioxins.

Who Is To Inform?

Who transmits hazard information is a key issue and is clearly identified by OSHA. Manufacturers and importers are primarily responsible for evaluating and identifying the hazards and safety measures related to each chemical they produce. They are also required to transmit this information to *all* users of their products. The secondary responsibility for disseminating this information (supplied by the manufacturer or importer) falls on the employer who uses chemicals.

Who Is To Be Informed?

This appears obvious, but several nonusers of chemicals also require information. The target persons are workers who use or are potentially exposed to the chemical. Although the original standard was restricted to workers employed in the industrial manufacturing sector, the standard has been expanded to include essentially all private industries that employ ten or more persons where chemicals are used. Employees of the federal government are covered through FEOSH (Federal Employee Occupational Safety and Health program) which has adopted the private industry OSHA standard. In addition, most states have adopted their own worker right-to-know laws that cover those institutions not preempted under the federal standard. In many states, for example, public schools, state colleges and universities, municipal sewage treatment plants, and other state and local agencies are covered under state programs. Most of these laws are similar in scope to the federal standard but may vary in specific requirements such as the chemicals covered, frequency of training, and reporting requirements. Although laboratories are exempt from certain requirements, OSHA has a separate standard that addresses the safety and health issues for laboratory workers; therefore, there are few workers who use chemicals that are not covered under federal or state worker right-to-know laws. State laws may be implemented under different agencies, such as labor or health departments.

In addition to the worker (the person ultimately targeted for health and safety information), others need this information to help protect the worker, including occupational physicians, nurses, industrial hygienists, safety professionals, train-

Table 4. OSHA's Criteria for Hazardous Materials Covered by the Hazard Communication Standard

1. OSHA-regulated substances in Subpart Z of CFR 1910 (i.e., all chemicals having PELs)
2. All chemical substances and physical agents having TLVs published annually by the ACGIH
3. Chemicals listed in the NTP *Annual Report on Carcinogens*
4. Chemicals listed in IARC Monographs
5. Any chemical that has been determined to be a health hazard
6. Any chemical that has been determined to be a physical hazard
7. Mixtures of chemicals containing 1% or more of a hazardous chemical or 0.1% of a carcinogen

Table 5. Items Exempted from the Requirements of the OSHA Hazard Communication Standard

1. Radioactive materials regulated by the Nuclear Regulatory Commission
2. Consumer products such as tobacco and cosmetics
3. Consumer pharmaceuticals and all tableted drugs
4. Finished articles that may contain hazardous chemicals such as light bulbs, batteries, circuit boards, etc.
5. Pesticide labeling covered by U.S. EPA regulations

ers, and employee supervisors. These persons play the key role in transmitting, interpretating, and implementing of health and safety information for the worker.

What Materials Are Covered?

The materials covered under the standard are not specifically listed, but include a broad range of chemicals and chemical mixtures. The OSHA standard is intended to cover any chemical that is a physical or a health hazard; however, it does not cover all hazardous materials. OSHA has not compiled a comprehensive list of chemicals covered under the standard. Published lists of chemicals explicitly and implicitly covered by the standard are available,[9] however. Although extensive, such lists may aid in screening hazardous chemicals if they are included, but they do not cover all possible chemicals that may be hazardous by OSHA's criteria. The standard includes all chemicals specifically regulated by OSHA as well as those that fall under its criteria as hazardous. Table 4 is a list of criteria for hazardous materials covered by the OSHA standard and Table 5 lists those items exempted under the OSHA standard.

What Information Is Transmitted?

The OSHA standard requires specific types of information related to hazardous chemicals to be transmitted to employers and users, of which six basic types are included in Table 6. The information is disseminated as discussed in the following

Table 6. Types of Information Required for Transmission

Material identification
Manufacturer's identification
Material properties
Hazard information
Safety and protection information
Emergency information

Table 7. Requirements for Transmitting Hazard Information

Specific hazard determination procedures
Develop a written hazard communication plan
Labeling requirements for containers
MSDS procedures
Employee training requirements
Release of trade secret information

Table 8. Methods for Transmitting Hazard Information

Labeling of containers by the manufacturer
MSDS sent by the manufacturer or vendor
Training of the worker by the employer

section. The criteria for each information category is specified in the OSHA Hazard Communication Standard in various levels of detail.

How Is Information Transmitted?

OSHA has six requirements for communicating hazard information (Table 7). The determination of hazard properties and safety procedures is the responsibility of the manufacturer or importer, as discussed earlier. This information is then disseminated to the employer and ultimately to the worker. To assure that the employer communicates this information as required, a detailed hazard communication plan must be written which describes all the elements and procedures of the implemented hazard communication program. The amount of detail included depends on whether the employer is a manufacturer of the chemical or an end user and on the extent of potential chemical exposure.

Information is transmitted by three means as specified by OSHA (Table 8). Chemical containers must be fully labeled by the manufacturer or importer. MSDSs prepared by manufacturers must be cross-referenced to the label and are intended to contain more detail on hazards and properties of the material than can be found on the label. Training contributes verbal information to workers on generic chemical health and safety hazards and practices, instructions on reading and interpreting labels and MSDSs, and workplace practices for safe use of specific materials. Requirements for hazard communication are detailed below (MSDSs are fully described in Chapter 2).

Table 9. Hazard Determination Process

1. Identify and characterize the chemical
2. Identify all health and physical hazards associated with use of that chemical
3. Determine measures required to use the chemical safely and protect the worker

Table 10. Physical Hazards

1. Combustible liquids
2. Flammable materials
3. Explosives
4. Organic peroxides
5. Oxidizers
6. Pyrophoric materials
7. Unstable or reactive substances
8. Water-reactive substances

Table 11. Chemical Health Hazards

1. Carcinogens
2. Corrosive substances
3. Acutely toxic chemicals
4. Irritants
5. Sensitizers
6. Chemicals toxic to specific target organs

PROGRAM COMPONENTS

Hazard Determination

The first major process of hazard communication is hazard determination. The function of this process is to identify and characterize hazards (Table 9), which is the responsibility of manufacturers and importers under the OSHA standard. A chemical is defined as hazardous if it is a physical or a health hazard. Physical hazards as defined by OSHA are given in Table 10.

The criteria for a health hazard are much broader than those of a physical hazard and are listed in Table 11. OSHA defines a health hazard as a chemical for which there is statistically significant evidence based on at least one study conducted in accordance with established scientific principles that acute or chronic health effects may occur in exposed persons. In most cases, studies use laboratory animals. Criteria for characterizing each of these health hazards are detailed in two appendices to the standard.

The carcinogenicity of a chemical also must be considered in the hazard assessment. If the substance is listed in either the International Agency for Research on Cancer (IARC) *Monographs*[6] or the National Toxicology Program (NTP) *Annual Report on Carcinogens*[7] it must then be considered a carcinogen.

In addition, if the chemical has been shown to be carcinogenic in one species of animal in a valid scientific study it is also considered a carcinogen. In the case of chemical mixtures, any mixture that contains a concentration of 0.1% or more of a carcinogen is considered carcinogenic.

Mixtures are treated differently than pure chemicals. If a mixture has been tested as a whole regarding its health and physical hazards, these results may be utilized in the determination of the health and physical hazards of the mixture. For mixtures that have not been tested, the hazards of the mixture shall be assumed to present the same health hazards as each of its components. For example, if one chemical component is known to be a neurotoxin and a second component is a renal toxin, the mixture would be determined to be both neuro- and nephrotoxic.

Components with concentrations less than 1% of the total weight are not generally considered hazardous unless there is evidence that lower concentrations are hazardous. For a large number of chemicals, this 1% rule does not apply since these chemicals are known to be toxic in parts per million (ppm) concentrations (e.g., dioxins, PCBs). This makes it difficult to apply the 1% rule in practice; therefore, many hazard assessments evaluate all components of the mixture even if the components <1%.

Physical hazards for chemical mixtures may either be tested as a whole or its hazards estimated using valid scientific data and calculations. Reference sources for such methods may be found in References 10 and 11. Procedures for evaluating the hazards of the chemicals must be described in writing and available upon request.

Written Hazard Communication Program

All employers who fall under the jurisdiction of the Hazard Communication Standard must develop and implement a written hazard communication program. Elements of the program are listed in Table 12. This written program must be made available upon request to workers, their representative(s), and OSHA representatives.

Labeling and hazard warning procedures must be clearly specified (see next section). A complete list of all chemicals in the workplace must be compiled, and the names must correspond or be cross-indexed to the name found on the container label or MSDS. For some industries this list may be extensive so as to form an appendix to the written program. In addition, there may be separate lists for different workplaces that must be referenced in the employers' written program.

Labeling and Warning Requirements

The chemical manufacturer, importer, or distributor is required to fully label all hazardous material containers before that container is distributed to workers or other employers using that material. Each container label must contain the information in Table 13.

Table 12. Elements of an Employer's Hazard Communication Program

1. A description of how containers will be labeled and hazard warnings implemented
2. A list of chemicals known to be present in the workplace
3. The procedures for making MSDSs available
4. A description of the methods used to inform and train employees
5. A description of how workers will be informed of hazards for nonroutine tasks
6. An explanation for informing contractors about workplace hazards

Table 13. Minimum Label Requirements

1. The identity of the chemical
2. Appropriate hazard warnings
3. The name and address of the manufacturer, importer or responsible party

Containers are broadly defined to include any receptacle: bottles, jars, sacks, boxes, drums, tanks, vats, or vessels. Pipes whose contents may vary need not be labeled but the hazards of these materials must be addressed as part of the written and training program. Although not covered under the Hazard Communication Standard, vehicle containers such as truck or railcar tanks must be placarded according to DOT requirements.

All employers using hazardous materials must ensure that each container in the workplace is labeled, tagged, or marked with the identity of the chemical and its appropriate hazard warnings. The substance name on the container may be the formal chemical name, a common name, or the tradename; however, the name on the container must correspond to the name given on the MSDS in order to ensure that the label, MSDS, and workplace chemical list can be cross-referenced. Special alternative warnings may be used on containers such as process containers and pipes. In such cases, process sheets, placards, batch tickets, or tags may be used to convey the identity information and hazard warnings. Portable and temporary containers intended for immediate use need not be fully labeled, if a single employee is involved in the movement or transfer of a chemical from a labeled container; however, such containers not for immediate use, or to be used by other workers, must be labeled or tagged with the appropriate material name and hazard warnings.

Since the standard is performance oriented, OSHA does not require that labels conform to a specific format, standard warning words, warning symbols, or codes as required by the DOT for shipping labels. OSHA does, however, have specific labeling requirements for 14 regulated materials specified in the Code of Federal Regulations (29 CFR 1910.1001—1910.1048). In addition, certain substances covered under other federal labeling laws take precedence over the generic performance requirements of OSHA. Specific labeling requirements for those materials must be consulted when preparing labels (see Table 14).

Several voluntary labeling systems and standards exist, such as the American National Standards Institute (ANSI Z129.1),[12] the National Fire Protection Asso-

Table 14. Materials Having Specific Labeling Requirements Under Other Regulations

Pesticides
Foods, drugs, and cosmetics
Alcohol and tobacco products
Consumer products and hazardous materials regulated by the Consumer Product Safety Commission

```
                    ACETONE

DANGER
               FLAMMABLE LIQUID
          TOXIC IF INHALED OR INGESTED
        AVOID PROLONGED CONTACT WITH SKIN
                HARMFUL TO EYES
            TOXIC TO NERVOUS SYSTEM

Work in well ventilated area.
Avoid fire and other sources of ignition.
Avoid prolonged contact with skin

               Ace Solvents Inc.
               347 N. 128 Avenue
               New York, NY 10083
```

Figure 1. An example of a label meeting minimum OSHA requirements.

ciation (NFPA),[10] and the National Paint and Coatings Association Hazardous Materials Information System (HMIS). Details on various labeling systems are summarized by O'Connor and Lirtzman[13] and Lowry and Lowry.[14] Figure 1 illustrates an acceptable OSHA-required label.

Some manufacturers use the numerical rating systems of the NFPA and the HMIS to summarize hazards (Table 15). The NFPA rating system is presented in the form of a diamond and the HMIS in the form of colored bars. Although the rating systems are similar, the worker and health and safety professionals must be aware that NFPA health hazard ratings are related to the dangers of certain materials in a fire situation. NFPA has assigned nonfire health hazard ratings for many (but not all) of its listed chemicals; however, the list is limited and does not include most chemicals that are not fire, explosion, or reactive hazards. In contrast, the HMIS system primarily applies criteria for hazard ratings, hazard symbols, and protective equipment graphics to labels. Since this system is adopted by many manufacturers and distributors, there is no single list or summary of chemicals with HMIS ratings. Individual chemical labels and MSDSs must be examined to obtain these ratings. Since there may be some latitude in professional judgment, particularly for health ratings, the HMIS ratings on labels and MSDSs may not be consistent between brand names for the *same chemical.*

Table 15. NFPA and HMIS Hazards Rating System

Rating	Health Hazard	Fire Hazard	Reactivity
4	Deadly	Flash point below 73°F	May detonate
3	Extreme danger	Flash point below 100°F	Shock and heat may detonate
2	Hazardous	Flash point 100—200°F	Violent chemical change
1	Slightly hazardous	Flash point above 200°F	Unstable if heated
0	Nonhazardous	Will not burn	Stable

Some labels have become so complex and detailed that they are essentially equivalent to printing an MSDS on the container. Since the average consumer does not tend to read labels, the real problem is placing too much information on a label. A balance between sufficient information that will be heeded by the worker and too much detail that will be ignored is difficult to achieve. In practice, detailed information should be placed on the MSDS, and the label should serve only as a functional warning system, as is the intent of the OSHA Hazard Communication Standard.

Hazard warnings may be in the form of words, pictures, symbols, or a combination. Explanation of these words, pictures, and symbols should be included in the workers' training. The label in Figure 1 contains the signal word DANGER and the warning hazard TOXIC. In addition to these warnings, specific hazards from inhalation, ingestion, and skin or eye contact are given. The primary physical hazard, FLAMMABLE, is also listed.

Material Safety Data Sheets

These documents are the primary means for transmitting detailed hazard information to employees. MSDSs must be accessible to all workers who are actually or potentially exposed to hazardous materials, without impediments from supervisors or management. The law does not require that copies of MSDSs be distributed to each employee, only that he or she have easy access to them upon request. The MSDS is the focus of the employer hazard communication program. The MSDS provides the information needed for preparing labels and training workers. Besides providing information to workers as required by OSHA, the MSDS is a document critical to health and safety professionals for the prevention and treatment of exposures to hazards.

As with labeling, MSDSs are required to be prepared in English by the manu-

Table 16. OSHA Required Contents of an MSDS

The identity of the material as listed on the container label
The chemical and common names of all ingredients determined to be health hazards
Physical and chemical characteristics of the material
All known physical hazards
All known health hazards
The OSHA PEL, ACGIH TLV, or other applicable exposure limits
Carcinogenic status as determined by IARC or NTP
Precautions for safe handling
Appropriate control measures
Emergency first aid
The date the MSDS was prepared or updated
The name, address, and telephone number of the chemical manufacturer or importer

facturer, importer, or distributor of the hazardous material. It is also the responsibility of the manufacturer to distribute the MSDS to all purchasers upon initial order and then whenever the MSDS is updated. However, it is the employer's responsibility to assure that all hazardous materials in the workplace have a corresponding MSDS. MSDSs must be prepared for all hazardous materials including those having special labeling requirements as listed above. Table 16 lists the OSHA required contents of a MSDS and an example is given in Figure 2.

Because workers should have easy intellectual and physical access to MSDSs, manufacturers must make the MSDS as readable as possible. As the result of legal and liability interpretations by industry, however, many MSDSs are prepared that defeat that purpose. To protect against liability, manufacturers may prepare MSDSs on substances generally considered harmless (such as glucose or grape sugar) or MSDSs that are seemingly contradictory. For example, an MSDS for isopropyl alcohol stated, "avoid skin contact," although the chemical is used as a skin disinfectant. Other MSDSs are so detailed that they may be of an excessive length (some are eight to ten pages long) and/or filled with highly technical toxicology data only interpretable by a professional toxicologist. In some cases, the container label may be of more practical use to the worker than the MSDS. Because of the wide variations in MSDS quality, length, and detail, specific effort is required to train workers in reading and interpreting MSDSs.

Standard voluntary MSDS forms are available from OSHA and several private sources. To replace the old OSHA Form 20, OSHA has published Form 174, which contains the basic elements required in the standard. Several items in the form, though implied in the standard's criteria, are not specified as requirements. In addition, there may be relevant hazard information for a given material that is not itemized on the OSHA form.

If there is no information available on any of these items, the MSDS must state that no information is available. No blank spaces are permitted. Any abbreviations used such as NA or ND should be defined on the MSDS since these have alternate meanings such "not applicable" or "not determined." Since OSHA requires that all hazards be determined and that missing items on the MSDS may not be applicable or not determined, these terms should be used where appropriate.

MATERIAL SAFETY DATA SHEET

EASTMAN KODAK COMPANY
343 State Street
Rochester, New York 14650

For Emergency Health, Safety, and Environmental Information, call 716-722-5151
For all other purposes, call 800-225-5352, in New York State call 716-458-4014

Date of Revision: 04/15/88 Kodak Accession Number: 900763

SECTION I. IDENTIFICATION

- Product Name: Acetic Acid
- Synonym(s): Glacial Acetic Acid; Ethanoic Acid
- Formula: C2 H4 O2
- CAT No(s): 102 1538; 102 1872; 107 6728; 108 4615; 108 4631; 110 4017; 114 6117; 114 6141; 114 9889; 129 4438; 136 1807; 136 1815; 136 1823; 138 1953; 165 3914; 186 1889; 186 1921; 180 7825
- Chem. No(s): 00763; 13001
- Kodak's Internal Hazard Rating Codes: R: 2 S: 3 F: 2 C: 0

SECTION II. PRODUCT AND COMPONENT HAZARD DATA

COMPONENT(S):	Percent	ACGIH TLV(R)	CAS Reg. No.
Acetic Acid	ca. 100	10 ppm	64-19-7

SECTION III. PHYSICAL DATA

- Appearance and Odor: Colorless liquid; strong pungent, vinegar-like odor
- Boiling Point: 118 C (245 F)
- Vapor Pressure: 11.4 mmHg at 20 C (68 F)
- Evaporation Rate (n-butyl acetate = 1): 0.97
- Volatile Fraction by Weight: ca. 100 %
- Specific Gravity (Water = 1): 1.05
- Solubility in Water: Complete

SECTION IV. FIRE AND EXPLOSION HAZARD DATA

- Flash Point: 39 C (103 F) Tag closed cup
- Flash Point: 43 C (109 F) Tag open cup
- Autoignition Temperature: 516 C (960 F)
- Flammable Limits in Air (mg/L): Lower: 143 at 60 C (140 F)
  Upper: 378 at 93 C (199 F)
- Extinguishing Media: Water spray; Dry chemical; Carbon dioxide; Alcohol foam; Foam
- Special Fire Fighting Procedures: Wear self-contained breathing apparatus and protective clothing.
- Unusual Fire and Explosion Hazards: None

R-0002.400D 88-4545

Figure 2A. Example of an MSDS.

-2-

==

SECTION V. REACTIVITY DATA

- Stability: Stable
- Incompatibility: Alkali, amines, alcohols, strong oxidizers
- Hazardous Decomposition Products: Combustion will produce carbon dioxide and probably carbon monoxide.
- Hazardous Polymerization: Will not occur.

==

SECTION VI. TOXICITY AND HEALTH HAZARD DATA

A. EXPOSURE LIMITS: 10 ppm, TWA 8 hour, STEL 15 ppm, ACGIH 1988-89.
OSHA PEL 10 ppm.

B. EXPOSURE EFFECTS:

Inhalation: Acetic acid vapor is irritating to the upper respiratory tract. Unacclimatized humans experience extreme eye and nasal irritation at concentrations in excess of 25 ppm. Fifty ppm is intolerable; however, acclimatized workers may tolerate concentrations up to 30 ppm. Exposures to such vapor concentrations have produced neither severe systemic injury nor death. This is most probably due to the fact that acetic acid is readily metabolized within the body. Repeated exposures to high vapor concentrations may produce respiratory tract irritation with pharyngeal edema, chronic bronchitis, discoloration of the teeth, and thickening of the skin.

Eyes: Severe eye burns can result from direct contact with the concentrated liquid. Vapors are very irritating to the eyes.

Skin: Direct skin contact with the concentrated acid is the most frequent occupational accident with this compound. The concentrated acid produces severe skin burns. These are deep burns and usually slough in a day or two. Concentrations below approximately 50 % acid are moderately irritating to the skin and usually cause minimal injury if promptly removed from the skin. Sensitivity dermatitis has been reported.

Ingestion: The ingestion of concentrated acetic acid produces burns of the upper digestive tract. This is characterized by severe pain in the mouth, pharynx, esophagus, and stomach. There may be immediate vomiting with diarrhea and possible bloody stools. The ingestion of as little as 1.0 mL of 100 % (glacial) acetic acid has resulted in perforation of the esophagus. Severe intestinal irritation with gross bleeding, collapse, and death has been reported.

==

R-0002.400D 88-4545

Figure 2A (continued).

==

SECTION VI. TOXICITY AND HEALTH HAZARD DATA (Con't)

C. FIRST AID:
Inhalation: Remove to fresh air. If not breathing, give artificial respiration, preferably mouth-to-mouth. If breathing is difficult, give oxygen. CALL A PHYSICIAN
Eyes: Immediately flush eyes with plenty of water for at least 15 minutes and get prompt medical attention.
Skin: Immediately flush with plenty of water for at least 15 minutes while removing contaminated clothing and shoes. Get medical attention if symptoms persist. Wash contaminated clothing before reuse.
Ingestion: Do not induce vomiting. If conscious give one glass of milk or water. Never give anything by mouth to an unconscious person. CALL A PHYSICIAN AT ONCE.

==

SECTION VII. VENTILATION AND PERSONAL PROTECTION

A. VENTILATION:
Good general ventilation* should be used. Local exhaust ventilation or an enclosed handling system may be needed to control air contamination below the TLV.

*Typically 10 room volumes per hour is considered good general ventilation: Ventilation rates should be matched to conditions of use.

B. RESPIRATORY PROTECTION:
A NIOSH approved organic vapor respirator should be worn if needed. If respirators are used, a program should be instituted to assure compliance with OSHA standard 29CFR 1910.134.

C. SKIN AND EYE PROTECTION:
Wear goggles or face shield, rubber gloves and protective clothing.

==

SECTION VIII. SPECIAL STORAGE AND HANDLING PRECAUTIONS

Material is classified as a combustible liquid. Keep away from heat and flame.
Keep from contact with oxidizing materials, alkali, amines, and alcohols.

==

SECTION IX. SPILL, LEAK, AND DISPOSAL PROCEDURES

Remove all sources of ignition.
Neutralize with baking soda (sodium bicarbonate).
Small Amount - flush material to sewer with large amounts of water.
Large Amount - absorb material in vermiculite or other suitable absorbent and place in impervious container.
Dispose in an approved incinerator or contract with licensed chemical waste disposal service.
Discharge, treatment, or disposal may be subject to federal, state, or local laws.

==

The information contained herein is furnished without warranty of any kind. Users should consider these data only as a supplement to other information gathered by them and must make independent determinations of the suitability and completeness of information from all sources to assure proper use and disposal of these materials and the safety and health of employees and customers.

==

R-0002.400D 88-4545 @900763*

Figure 2A (continued).

MATERIAL SAFETY DATA SHEET

EASTMAN KODAK COMPANY
343 State Street
Rochester, New York 14650

For Emergency Health, Safety, and Environmental Information, call 716-722-5151
For all other purposes, call 800-225-5352, in New York State call 716-458-4014

Date of Revision: 10/03/89 Kodak Accession Number: 901778

SECTION I. IDENTIFICATION

- Product Name: Butylbenzene
- Synonym(s): 1-Phenylbutane
- Formula: C10 H14
- CAT No(s): 113 3628; 113 3636; 113 3644
- Chem. No(s): 01778
- Kodak's Internal Hazard Rating Codes: R: 1 S: 2 F: 2 C: 0

SECTION II. PRODUCT AND COMPONENT HAZARD DATA

COMPONENT(S):	Percent	ACGIH TLV(R)	CAS Reg. No.
Butylbenzene	ca. 100	---	104-51-8

SECTION III. PHYSICAL DATA

- Appearance: Colorless liquid
- Boiling Point: 182 C (360 F)
- Vapor Pressure: 0.9 mmHg at 20 C
- Evaporation Rate (n-butyl acetate = 1): Not Available
- Volatile Fraction by Weight: ca. 100 %
- Specific Gravity (Water = 1): 0.86
- Solubility in Water (by Weight): Negligible

SECTION IV. FIRE AND EXPLOSION HAZARD DATA

- Flash Point: 59 C (139 F) Setaflash closed cup
- Extinguishing Media: Water spray; Dry chemical; Carbon dioxide; Foam
- Special Fire Fighting Procedures: Wear self-contained breathing apparatus and protective clothing. Use water spray to keep fire-exposed containers cool.
- Unusual Fire and Explosion Hazards: None

SECTION V. REACTIVITY DATA

- Stability: Stable
- Incompatibility: Strong oxidizers
- Hazardous Decomposition Products: Combustion will produce carbon dioxide and probably carbon monoxide.
- Hazardous Polymerization: Will not occur.

R-0063.500B 86-6889

Figure 2B (continued).

Employers may want to transfer information from the supplier's MSDS to a standard system of their own such as a computer database which would automate transfer of MSDSs to workers and other employees. Besides automation, the MSDS would be distributed in a standardized format that may be less confusing to the worker. The original MSDS must be kept on file as documentation, and the employer making the transformation would become responsible for the changed content of the new MSDS.

An accurate identity of the material must be included on the MSDS. This includes the material name, which must correspond to the product's label. The specific hazardous chemical or chemicals, in the case of mixtures, must be given. The specific chemical name is one that will allow a chemist to identify the structure of the substance(s) in the product. Although not required by OSHA, the compound's unique Chemical Abstract Service (CAS) number is commonly given along with the chemical name.

The chemical name of all hazardous ingredients of mixtures must be listed; however, the actual composition or formula of the mixture need not be specified, although many manufacturers voluntarily include this information. For each chemical listed the OSHA PEL, ACGIH TLV, or other exposure limit must be given where one exists. If the components of the mixture are a trade secret, the ingredients need not be disclosed on the MSDS; however, the potential or real hazards and appropriate precautionary measures must be given. When this information is required by a health and safety professional such as a physician, industrial hygienist, or occupational nurse, trade secret information can be obtained for specified purposes.

OSHA requires that the relevant physical and chemical properties of the material be listed on the MSDS; however, no list of required properties is specified. Table 17 is a list of the properties generally given on the MSDS. Additional properties may be listed if the manufacturer determines that a given property relates to the substances hazards, safe use, protection, or emergency response.

Physical hazards information includes fire and explosion hazard data and reactivity data (seeTable 18).

All pertinent health hazard data must be included on the MSDS. It must be stated if the material is a carcinogen. Other required health hazard information is given in Table 19.

Precautions for safe handling and use may include some items not explicitly listed by OSHA. Items required by OSHA are listed in Table 20.

Although waste disposal and shipping methods are not specifically required, they should be included as a procedure for safe handling. The Environmental Protection Agency's (EPA) hazardous waste regulations should be consulted for guidance. Shipping requirements are regulated by the DOT, and shipping and packaging requirements are commonly included on many MSDSs under handling and use precautions. The MSDS is legally intended for worker protection, however, it is becoming a tool for all disciplines in environmental and hazardous materials management. For example, it is a key source of information for emergency planning and community right-to-know requirements.

Table 17. Physical Properties Given on an MSDS

Boiling point
Melting point
Vapor pressure
Vapor density
Evaporation rate
Solubility in water
Specific gravity
Appearance
Odor

Table 18. Fire and Reactivity Hazard Information Given on an MSDS

Fire Hazards
Flash point
Flammable limits (as lower and upper explosive limits, LEL/UEL)
Extinguishing media
Special fire fighting procedures
Fire and explosion hazards
Reactivity Hazards
Stability information
Conditions to avoid
Incompatabilities
Hazardous decomposition by-products
Hazardous polymerization

Table 19. Health Hazard Data Given on an MSDS

Routes of entry (inhalation, ingestion, skin)
Acute health hazards
Chronic health hazards
Signs and symptoms of exposure
Medical condition aggravated by exposure
Emergency and first aid procedures

Table 20. Hazards Precautions

Applicable precautions for safe handling and use
Appropriate hygienic practices
Protective measures during repair and maintenance of contaminated equipment
Procedures for clean up of spills and leaks

Table 21. Applicable Control Measures

Appropriate engineering controls (e.g., ventilation)
Respiratory protection
Protective clothing and equipment
Specific work practices

Table 22. Training Requirements

Discussion of the requirements of the standard
The rights of the workers under the standard
The location of the written hazard communication program
Details of the hazard communication program
The location of the workplace hazardous chemical list
Access to MSDSs
Operations in their work area where hazardous chemicals are present
Methods and observations used to detect the presence of hazardous chemicals
Physical and health hazards of chemicals in the workplace
Protective measures

The MSDS also must contain information on any generally applicable control measures such as listed in Table 21.

Training Requirements

Employers must provide training to all employees who are potentially exposed to hazardous materials. Training and information must be given upon initial assignment and whenever a new hazard is introduced into the workplace. Information required for the training program includes those listed in Table 22.

REFERENCES

1. *Hazard Communication*, OSHA 29 CFR 1910.1200.
2. *Superfund Amendments Reauthorization Act. Title III — Emergency Planning and Community Right-to-Know*, 40 CFR Parts 300 and 355.
3. Brower, J.E. Ed. *Hazard Communication: Issues and Implementation,* ASTM Special Technical Publication STP 932 (Philadelphia: American Society for Testing and Materials, 1986).
4. *Limits for Air Contaminants* OSHA 29 CFR 1910.1000 Subpart Z.
5. *TLVs Threshold Limit Values and Biological Exposure Indices* (Cincinnati, OH: American Conference of Governmental Industrial Hygienists, 1991).
6. *IARC Monographs on the Evaluation of Carcinogenic Risks to Humans*, Vol. 1-49 (Lyon, France: World Health Organization, International Agency for Research on Cancer).
7. *Annual Report on Carcinogens* (Washington, D.C.: National Toxicology Programs, U.S. Department of Health and Human Services, 1989).

8. *Registry of Toxic Effects of Chemical Substances. RTECS* (Washington, D.C.: National Institutes of Occupational Safety and Health, 1991).
9. Clansky, K.B., Ed. *Chemical Guide to the OSHA Hazard Communication Standard* (Burlingame, CA: Roytech Publications, 1989).
10. *Fire Protection Guide on Hazardous Materials,* 10th ed. (Boston: National Fire Protection Association, 1990).
11. Bretherick, L. *Handbook of Reactive Chemical Hazards,* 3rd ed. (London: Oxford University Press, 1984).
12. *Standard for Hazardous Materials - Precautionary Labeling*, ANSIZ129.1-1982 (New York: American National Standards Institute, 1982).
13. O'Connor, C.J., and S.I. Lirtzman. *Handbook of Chemical Industry Labeling* (Park Ridge, NJ: Noyes Publications, 1984).
14. Lowry, G.G., and R.C. Lowry. *Lowrys' Handbook of Right-to-Know and Emergency Planning* (Chelsea, MI: Lewis Publishers, 1988).

CHAPTER 6

Community Right-To-Know (SARA Title III)

James E. Brower

INTRODUCTION TO COMMUNITY RIGHT-TO-KNOW REGULATIONS

Chemical accidents are frequently reported in the news media. It was not until the industrial accident at Bhopal, India, in 1984, however, that successful efforts to prepare for such accidents were initiated by the U.S. Congress. In the Bhopal incident, over 2000 deaths and several thousand injuries resulted from an accidental release of the toxic chemical, methyl isocyanate, manufactured by Union Carbide.[1,2] The focus of world attention and the American press on that accident highlighted the importance of emergency planning and informing local communities of potential hazards from facilities in the immediate area. It is important to know what toxic chemicals exist and their potential for release, when people should be evacuated and the actions needed to protect the public.

There had long been a need for the federal government to enact emergency planning and right-to-know legislation because states and local governments lacked resources, adequate laws or ordinances, and uniform standards for industries in case of accidental toxic release. The Superfund Amendments Reauthorization Act of 1986 (SARA), Public Law 99-499 as Title III of that Act,[3] satisfied that need. This law is included as a separate title with the Superfund Law which pertains to hazardous waste sites. Title III is officially called the "Emergency Planning and Community Right-to-Know Act," or EPCRA, and is not to be confused with those portions of SARA that pertain to clean up of hazardous wastes. SARA Title III or EPCRA consists of three subtitles listed in Table 1.

Table 1. Three Subtitles of the Emergency Planning and Community Right-to-Know Act

Subtitle A establishes the functions of state and local emergency response and planning organizations
Subtitle B requires information on the presence and release of hazardous substances at facilities
Subtitle C contains provisions and requirements related to trade secret claims, enforcement, and citizen suits

EPCRA requires two fundamental actions:

1. Manufacturers and users of hazardous substances must provide information about hazardous materials contained within their facility to community planning and emergency response organizations.
2. State and local governments must undertake planning measures to respond to emergencies involving those hazards.

SARA Title III is administered under the U.S. Environmental Protection Agency (EPA) with implementation enforced by state and local committees. The Occupational Safety and Health Administration's (OSHA) Hazard Communication Standard[4] and EPCRA regulations are quite separate in their intent, scope, and regulatory authorities (see Chapter 5), yet, they are linked because the primary base of chemicals includes those covered in OSHA's standard, and the key source of hazards information in both laws is the Material Safety Data Sheet (MSDS). The primary concern of EPCRA is the release of and potential public exposure to workplace chemicals. Since this concern over workplace chemicals traditionally has been the province of the industrial hygienist, community right-to-know issues are linked with worker right-to-know issues.[5]

COVERAGE

Facilities that use extremely hazardous substances that exceed a specified quantity or use less hazardous chemicals that exceed 10,000 lb must comply with the reporting requirements of this law. The EPA has published a list of extremely hazardous substances and has established a "threshold planning quantity" for each substance on the list.[6] This quantity takes into account the amount of the chemical release in an accident that could result in a hazard to the exposed public. This list and its planning quantities is used to trigger emergency planning in states and local communities. In addition, facilities that release chemicals to the environment have reporting requirements that are triggered by "reportable quantities" given for all chemicals of concern.

All facilities using hazardous materials must notify their State Emergency Response Commission (SERC) of such use in excess of the threshold planning quantity or risk a fine. In addition, facilities must notify the Local Emergency Planning Committee (LEPC) of the name of the person who will participate in the emergency planning process as facility emergency coordinator, and of any relevant changes that occur at the facility and provide them with information necessary for developing and implementing the local emergency plan. The LEPC is composed of local government officials, emergency response personnel, health officials, and interested citizens and is responsible for local implementation of this law.

EMERGENCY PLANNING REQUIREMENTS

Planning Committees

Although federal law is administered and enforced by the EPA, direct implementation of the community right-to-know provisions are the responsibility of state and local governments. The SERC, which is appointed by the governor, is responsible for appointing LEPCs and for supervising and coordinating their activities. The SERC establishes procedures for receiving and processing information requests from the public, including chemical inventory information. The SERC also designates emergency planning districts to facilitate preparation and implementation of local emergency plans.

The LEPC is responsible for implementation of EPCRA requirements for its emergency planning district. The LEPC includes owners and operators of facilities, elected state and local officials, law enforcement and emergency response groups, health and environmental personnel, news media, and community groups. The LEPC has an information coordinator for receiving and processing information requests. For further details on the functions of these organizations, consult the law[3] or References 7 through 11.

Local Emergency Response Plan

Each LEPC must prepare an emergency plan that is reviewed and updated annually. The LEPC must identify the resources to agencies and organizations necessary to implement and exercise the emergency response plan (see Table 2). This plan is intended to protect the public in the event of a toxic chemical release. A key to this plan is information from facilities that may potentially release toxic materials. The local emergency plan not only identifies potential hazards but specifies notification procedures, evacuation process, response resources, training requirements, and drills.

Site Emergency Plan

Although the law does not provide specific requirements for a facility specific emergency response plan, most of the SERC and LEPC request that facilities

Table 2. Local Emergency Response Plan Requirements

- Identify all applicable facilities in planning district
- Identify routes likely to be used for transport of extremely hazardous materials
- Identify facilities that may contribute or be subject to additional risk due to proximity of facility to institutions such as hospitals
- Describe methods and procedures that must be followed by facility operators and local emergency personnel in response to release of a hazardous substance
- Designate a community emergency coordinator
- Describe procedures that provide reliable, effective, and timely notification of a hazardous release to appropriate officials and to the public
- Describe methods for determining occurrence of a release and area or population likely to be affected by release
- Describe emergency equipment and facilities in community and identify persons responsible for equipment and facilities
- Prepare evacuation plans with provisions for precautionary evacuation and alternative routes
- Prepare training programs, including schedules for training emergency response and medical personnel
- Describe procedures and schedules for exercising and implementating plan

storing a significant amount of highly hazardous materials submit a facility plan. These facility plans generally follow the provisions for the local emergency response plan given in Table 2. These site plans should contain identification of the specific chemical hazards for each building or process at the facility and should specify contacts and functions for each task, including the owner/operator responsibilities, as well as local agencies and response personnel that would be involved in an emergency release at the facility.

CHEMICAL REPORTING

EPCRA requires three types of reports to be submitted to the LEPC and SERC:

- The MSDS or lists of chemicals used or stored at the facility
- An inventory of hazardous materials
- A toxic chemical release report

These reports must be submitted to the LEPC, SERC, and/or the local fire department as specified below. Other individuals or organizations desiring this information can obtain these reports by written requests to the LEPC or SERC, as this information is in the public domain.

Threshold Planning and Reportable Quantities

There is a specific list of chemicals designated as extremely hazardous.[6] The criteria for identifying extremely hazardous substances are based on acute toxicity data from laboratory animals and are given in Table 3. These criteria are intended only as screening tools for identifying highly acute toxic chemicals that may

Table 3. Criteria for Acutely Toxic Chemicals[a]

Route of Exposure	Acute Toxicity Measure	Criteria Value
Inhalation	Median lethal concentration in air (LC_{50})[b]	≤0.5 mg/L
Dermal	Medial lethal dose (LD_{50})[c]	≤50 mg/kg
Oral	Medial lethal dose (LD_{50})	≤25 mg/kg

[a] Chemicals that may present severe health hazards to humans during a chemical accident or other emergency.

[b] LC_{50} is the concentration of a chemical in air which results in the death of 50% of a population of tested animals.

[c] LD_{50} is the dose of a chemical which results in the death of 50% of a population of tested animals.

present severe health hazards to humans following short term exposure to a chemical during an accident or other emergency. The primary route of exposure of concern is through inhalation of the toxic chemical. Oral and dermal toxicity are used to communicate concern about inhalation hazards and for identifying compounds with the potential for acute toxicity. In cases where lethal dose or lethal air concentration data are lacking, the lowest lethal dose reported is used to identify highly toxic chemicals.

If continuous or frequently recurring releases of a substance are known or are expected to cause adverse health effects, the EPA may include this chemical in the list, even though the substance may not have a high acute toxicity. Additional criteria for inclusion on the highly hazardous list are given in Tables 4 and 5. These criteria apply to concentrations of a chemical likely to cause adverse health effects beyond the facility boundaries. Also listed are chemicals (e.g., ammonia) that have less acute toxicity than those in Table 3, and are produced in large quantities.

Extremely hazardous chemicals have a specific quantity that triggers the inventory reporting requirements, i.e., the threshold planning quantity (TPQ). Any facility that has one or more of the chemicals on this list in quantities equal to or greater than the TPQ must notify the LEPC and SERC. Highly hazardous chemicals, which number nearly 400 substances, may have TPQs ranging from 2 to 10,000 lb. Toxic chemicals not on the list and having no assigned TPQ are given a default TPQ value of 10,000 lb.

All chemicals on the list of extremely hazardous substances also have a reportable quantity (RQ). This is the quantity that must be reported if there is a release or spill greater than or equal to that amount. Separate lists are referenced in SARA Title III that include RQs for other hazardous chemicals. These are the Section 313 chemicals for release reporting and Section 304 chemicals subject to spill reporting. These chemicals, while not listed as extremely hazardous, have hazardous properties as listed in Tables 4 and 5.

Table 4. Characterization of Physical and Health Hazards for SARA Title III Reporting

OSHA Health Hazard	EPA SARA Health Hazard
Highly toxic	Acute hazard
Toxic	Acute hazard
Corrosive	Acute hazard
Irritant	Acute hazard
Sensitizer	Acute hazard
Blood toxin	Acute hazard, chronic hazard
Eye hazard	Acute hazard, chronic hazard
Mucous membranes	Acute hazard, chronic hazard
Kidney toxin	Acute hazard, chronic hazard
Liver toxin	Acute hazard, chronic hazard
Lung toxin	Acute hazard, chronic hazard
Nervous system toxin	Acute hazard, chronic hazard
Reproductive toxin	Acute hazard, chronic hazard
Carcinogen	Chronic hazard
OSHA Physical Hazard	**EPA SARA Physical Hazard**
Combustible liquid	Fire hazard
Compressed gas	Sudden release of pressure
Corrosive	Acute hazard
Explosive	Sudden release of pressure
Flammable	Fire hazard
Organic peroxide	Reactive hazard
Oxidizer	Fire hazard
Pyrophoric	Fire hazard
Unstable, reactive	Reactive hazard
Water reactive	Reactive hazard

Table 5. Criteria for Highly Hazardous Chemicals

Causes adverse acute health effects in humans
Causes cancer or teratogenic effects
Causes serious or irreversible reproductive disorders
Causes serious or irreversible neurological disorders
Causes serious heritable genetic mutations
Causes serious or irreversible chronic health effects
Production of a toxic material having a high production volume that may be of concern
A toxic chemical that is persistent in the environment or one which can bioaccumulate and have a significant adverse impact.

MSDS List

For each chemical covered under the OSHA Hazard Communication Standard (see Chapter 5), employers must have an MSDSs available. Employers are required to submit a copy of each MSDS (or a list of all MSDSs) to the LEPC, SERC, and the fire department serving each workplace. Within 3 months of receipt of any new hazardous material or information regarding a new hazard, the new MSDS (or a revised list) must be submitted to the above groups. Since a facility may have 1000 or more chemicals, the LEPC, SERC, and fire departments often prefer receiving a list of chemicals over the actual MSDS in order to reduce their paper burden. The primary intent of this submission is to notify the agencies and fire departments of those firms using toxic chemicals. The lists are used to identify sites requiring emergency planning.

Hazardous Materials Inventory

Facilities required to prepare or maintain MSDSs under OSHA requirements must annually prepare and submit an emergency and hazardous chemical inventory form, a two-tiered system of reporting. The annual report, which is submitted to the LEPC, SERC, and local fire department, is commonly referred to as a Tier I report (see Figure 1). This report includes summary information on all chemicals that exceed the TPQ or 10,000 lb and provides an overview of potential chemical hazards by health and physical hazard categories. The EPA SARA hazard categories are listed in Table 5. The Tier I report does not itemize specific chemicals, however, pure chemicals as well as chemicals found in mixtures must be included. This report must contain the information shown in Table 6.

The maximum and average daily amounts are reported in ranges of quantities listed in Table 7. The maximum amount of chemicals present represents the maximum amount the facility has on site at any given time. For example, assume formaldehyde is used as formalin solution and the total amount of formalin used is 62,000 lb/year. The TPQ for formaldehyde is 500 lb (Table 8). Note that all quantities are reported in pounds even though the chemical may be a liquid, a solid, or a gas. Therefore, conversion from gallons, tons, cubic feet, etc., to pounds is needed. In the case of a mixture (e.g., formalin, which is 40% formaldehyde), the amount of the specific chemical in the mixture must be computed (i.e., 62,000 lb of formalin = 24,800 lb of formaldehyde). If a tank holds 10,000 lb of formalin, but is filled to only 8000 lb at each monthly delivery, then 3200 lb of formaldehyde (8000 × 40%) is the maximum amount present.

The average daily amount is the average amount on hand each day for 1 year. One may total the amounts on hand each day and divide by the number of days the chemical is present on site. If daily amounts are not monitored, other estimations may be used. In the previous example, approximately 2000 lb are consumed per month, leaving an average excess of 1200 lb in the tank at the end of the month. Thus, an average daily amount of 2200 lb of formaldehyde (i.e., [3200 + 1200]/2 = 2200) is estimated for an average month over the year.

Certain facilities may be required to provide more detailed information on specific chemical hazards by the LEPC, SERC, or local fire department. This report is known as the Tier II report (Figure 2). At the prerogative of a facility, a

Page ____ of ____ pages
Form Approved OMB No. 2050-0072

Tier One EMERGENCY AND HAZARDOUS CHEMICAL INVENTORY
Aggregate Information by Hazard Type

FOR OFFICIAL USE ONLY: ID # ____ Date Received ____

Important: Read instructions before completing form

Reporting Period From January 1 to December 31, 19____

Facility Identification

Name ____
Street Address ____
City ____ State ____ Zip ____
A. ☐ entire facility B. ☐ establishment within a facility
SIC Code ☐☐☐☐ Dun & Brad Number ☐☐-☐☐☐-☐☐☐☐

Emergency Contacts

Name ____
Title ____
Phone ( )
24 Hour Phone ( )

Name ____
Title ____
Phone ( )
24 Hour Phone ( )

Owner/Operator

Name ____
Mail Address ____
Phone ( )

☐ Check if form is identical to form submitted last year.

☐ Check if site plan is attached

	Hazard Type	*Max Amount**	*Average Daily Amount**	*Number of Days On-Site*	*General Location*
Physical Hazards	Fire	☐☐	☐☐	☐☐☐	____
	Sudden Release of Pressure	☐☐	☐☐	☐☐☐	____
	Reactivity	☐☐	☐☐	☐☐☐	____
Health Hazards	Immediate (acute)	☐☐	☐☐	☐☐☐	____
	Delayed (Chronic)	☐☐	☐☐	☐☐☐	____

Certification *(Read and sign after completing all sections)*

I certify under penalty of law that I have personally examined and am familiar with the information submitted in this and all attached documents, and that based on my inquiry of those individuals responsible for obtaining the information, I believe that the submitted information is true, accurate and complete.

Name and official title of owner/operator OR owner/operator's authorized representative

Signature ____ Date signed ____

*** Reporting Ranges**

Range Code	Weight Range in Pounds From...	To...
01	0	99
02	100	999
03	1000	9,999
04	10,000	99,999
05	100,000	999,999
06	1,000,000	9,999,999
07	10,000,000	49,999,999
08	50,000,000	99,999,999
09	100,000,000	499,999,999
10	500,000,000	999,999,999
11	1 billion	higher than 1 billion

Figure 1. Tier I emergency and hazardous chemical inventory form.

Table 6. Requirements of Tier I Reports

Estimated maximum amount of hazardous chemicals in each hazard category present at any time during the past year
Estimated average daily amount of hazardous chemicals for each hazard category
The general location of the hazardous chemicals in each category

Table 7. Range Values of Quantities for Inventories of Hazardous Chemicals

Range Value	Weight Range (lb) From	Weight Range (lb) To
00	0	99
01	100	999
02	1,000	9,999
03	10,000	99,999
04	100,000	999,999
05	1,000,000	9,999,999
06	10,000,000	49,999,999
07	50,000,000	99,999,999
08	100,000,000	499,999,999
09	500,000,000	999,999,999
10	1 billion	>1 billion

Table 8. Example List of Extremely Hazardous Substances, Threshold Planning Quantities, and Reportable Quantities

Chemical Name	CAS No.	Ambient State[a]	TPQ	RQ
Ammonia	7664-41-7	Gas	500	100
Arsine	7784-42-1	Gas	100	1
Benzotrichloride	99-07-7	Liquid	100	1
Bromine	7726-95-8	Liquid	500	1
Carbon disulfide	75-15-0	Liquid	10,000	100
Chlorine	7782-50-5	Gas	100	10
Dichlorvos	62-73-7	Liquid	1000	1
Digitoxin	71-63-6	Solid	100/10,000[b]	1
Dinitrocreosol	534-52-1	Solid	10/10,000[b]	10
Formaldehyde	50-00-0	Gas	500	1000
Isobenzan	297-78-9	Solid	100	1
Lindane	58-89-9	Solid	1000	1
Methyl isocyanate	624-83-9	Liquid	500	1
Methyl mercaptan	74-93-1	Gas	500	1
Phenol	108-95-2	Solid	500/10,000[b]	1000
Sulfur dioxide	7446-09-5	Gas	500	1
Tetraethyl lead	78-00-2	Liquid	100	10
Trichlorophenylsilane	96-13-5	Liquid	2	1

[a] Physical state under normal temperature and pressure.
[b] The larger TPQ is for solids with a particulate size >100 μM.

Page ____ of ____ pages
Form Approved OMB No. 2050-0072

Tier Two
EMERGENCY AND HAZARDOUS CHEMICAL INVENTORY
Specific Information by Chemical

Facility Identification
Name ____
Street Address ____
City ____ State ____ Zip ____
A ☐ entire facility B. ☐ establishment within a facility
SIC Code ☐☐☐☐ Dun & Brad Number ☐☐-☐☐☐-☐☐☐☐

FOR OFFICIAL USE ONLY
ID # ____
Date Received ____

Owner/Operator Name
Name ____ Phone ( )
Mail Address ____

Emergency Contact
Name ____ Title ____
Phone ( ) 24 Hr. Phone ( )
Name ____ Title ____
Phone ( ) 24 Hr. Phone ( )

Important: Read all instructions before completing form
Reporting Period From January 1 to December 31, 19____ ☐ Check if form is identical to form submitted last year

Chemical Description	Physical and Health Hazards (check all that apply)	Inventory: Max. Daily Amount (code)	Avg. Daily Amount (code)	No. of Days On-site (days)	Storage Codes and Locations (Non-Confidential): *Storage Code*	*Storage Locations*
CAS ☐☐☐☐☐☐ ☐☐ ☐ Trade Secret ☐ Chem. Name ____ *Check all that apply:* ☐ Pure ☐ Mix ☐ Solid ☐ Liquid ☐ Gas ☐ EHS	☐ Fire ☐ Sudden Release of Pressure ☐ Reactivity ☐ Immediate (acute) ☐ Delayed (chronic)	☐☐	☐☐	☐☐☐		
CAS ☐☐☐☐☐☐ ☐☐ ☐ Trade Secret ☐ Chem. Name ____ *Check all that apply:* ☐ Pure ☐ Mix ☐ Solid ☐ Liquid ☐ Gas ☐ EHS	☐ Fire ☐ Sudden Release of Pressure ☐ Reactivity ☐ Immediate (acute) ☐ Delayed (chronic)	☐☐	☐☐	☐☐☐		
CAS ☐☐☐☐☐☐ ☐☐ ☐ Trade Secret ☐ Chem. Name ____ *Check all that apply:* ☐ Pure ☐ Mix ☐ Solid ☐ Liquid ☐ Gas ☐ EHS	☐ Fire ☐ Sudden Release of Pressure ☐ Reactivity ☐ Immediate (acute) ☐ Delayed (chronic)	☐☐	☐☐	☐☐☐		

Certification *(Read and sign after completing all sections)*
I certify under penalty of law that I have personally examined and am familiar with the information submitted in this and all attached documents, and that based on my inquiry of those individuals responsible for obtaining the information, I believe that the submitted information is true, accurate, and complete.

Name and official title of owner/operator OR owner/operator's authorized representative ____ Signature ____ Date signed ____

Optional Attachments *(Check one)*
☐ I have attached a site plan
☐ I have attached a list of site coordinate abbreviations

Figure 2. Tier II emergency and hazardous chemical inventory form.

Table 9. Information Required in a Tier II Inventory Report

The identity of the substance including chemical or common name of the substance and its CAS number
An estimate, in ranges, of the maximum amount of the substance present at the facility during the preceding calendar year
An estimate, in ranges, of the average daily amount of the substance present at a facility during the previous calendar year
A brief description of the primary physical and chemical hazards of the substance
A brief description of the manner in which the substance is stored
Where the substance is located at the facility
Whether the owner or operator elects to withhold information pertaining to allowable trade secret exemptions

Tier II report may be prepared in place of the required annual Tier I report, as long as the Tier II is acceptable to the SERC. In fact, the information in a Tier I report is a summary of the more detailed Tier II information. If a facility is likely to be required to prepare a Tier II report it may be more efficient to prepare the more detailed report first. The information required in the Tier II report is given in Table 9. Note that in addition to the types of information required for the Tier I report, specific ranges of the maximum and average quantities on hand (Table 7) must be given for each chemical and storage situation (Tables 10 and 11). Tier I and II report forms are available from the EPA and contain detailed instructions for the calculation and compilation of information for these forms.

RELEASE REPORTING

The reporting of the release of toxic chemicals covers two categories: emergency or accidental and normal or planned. A release is broadly defined as any spilling, pumping, pouring, emitting, emptying, injecting, escaping, leaching, dumping, or disposing of any hazardous chemical into the environment. Accidental releases must be reported immediately to federal, state, and local officials. Normal or planned releases must be reported annually. For example, a facility may have a normal release of ammonia vented to the atmosphere from a chemical process. If this release is legally permitted and results from normal operation of the facility, it is considered a planned release; however, if an operator turns the wrong valve, or the process malfunctions causing an excess release of ammonia that exceeds the reportable quantity, that incident must be reported as an accidental release.

The chemicals covered by this requirement are those listed under "Toxic Chemicals Subject to Section 313 of the Emergency Planning and Community Right-to-Know Act of 1986" and any subsequent revisions.[3] Facilities must report if 25,000 lb of toxic chemicals are manufactured or processed per year and if

Table 10. Storage Codes for Inventories of Hazardous Chemicals

Codes	Types of Storage
A	Above-ground tank
B	Below-ground tank
C	Tank inside building
D	Steel drum
E	Plastic or nonmetallic drum
F	Can
G	Carboy
H	Silo
I	Fiber drum
J	Bag
K	Box
L	Cylinder
M	Glass bottles or jugs
N	Plastic bottles or jugs
O	Tote bin
P	Tank wagon
Q	Rail car
R	Other

Table 11. Storage Condition Codes for Inventories of Hazardous Chemicals

Storage Codes	Storage Conditions
	Pressure Conditions
1	Ambient pressure
2	Greater than ambient pressure
3	Less than ambient pressure
	Temperature Conditions
4	Ambient temperature
5	Above ambient temperature
6	Below ambient temperature
7	Cryogenic conditions

10,000 lb of chemicals are otherwise used. More detail concerning the requirements and reporting instructions are available in References 10 through 17.

Accidental Releases and Emergency Notification

If any hazardous material is spilled or released to the environment in excess of the RQ, the incident must be reported immediately to the LEPC community emergency coordinator, the SERC, and the National Response Center. The information required to be reported is given in Table 12. Accidental release reporting is required under CERCLA (i.e., Superfund Law) and EPCRA (i.e., SARA Title

Table 12. Immediate Emergency Notification Requirements

Name of the chemical or identity of any substance involved
Identify the substance as extremely hazardous
An estimate of the quantity released to the environment
Time and duration of the release
Medium or media into which the release occurred
Any known or expected acute or chronic health risks and appropriate response advice
Proper precautions to take against the release
Evacuation recommendations
Name and telephone number of the person(s) to be contacted for further information.

Table 13. Emergency Release Notification Exemptions[a]

Federally permitted releases
Releases that result in exposure only within the boundaries of the facility
Releases in amounts less than the reportable quantity
Releases from a facility that produces, uses, or stores no hazardous materials
Continuous releases requiring annual reporting that are stable in quantity and rate
Application and releases of registered pesticides

[a] Exemptions under SARA Title III.

III). The RQs under these two laws are similar and in the future the individual lists will be consolidated. In general, SARA Title III accidental and emergency release reporting involves a release of all RQs except those given in Table 13. Releases that result in only on-site potential exposures or environmental contamination may require reporting under CERCLA.

When an extremely hazardous substance is released, specific procedures must be followed. Immediately after the release, the facility operator must notify the community emergency coordinator for the LEPC and SERC by telephone, radio, or in person. In case of transportation (or storage accidents related to transportation) notice must be given by dialing 911 or the local operator. The National Response Center is notified by dialing 800-424-8802 or 202-267-2675. The information provided may depend on what is known at the time of the incident; however, limited data should not interfere with timely notification and should not cause delays in responding to the emergency.

As soon after the release as possible a written follow-up emergency notice regarding the release must be provided containing the information in Table 12 and any information concerning actions taken to respond to the release and to contain it. There are several exemptions to emergency notification, listed in Table 13. Although these releases may not require notification under EPCRA, they may need to be reported under other Superfund regulations. For these requirements, consult Section 103 of CERCLA.

Toxic Chemical Release Inventory Report

A manufacturing facility with ten or more employees, in SIC Codes 20 through

Table 14. Standard Industrial Codes (SIC) for Release Reporting Requirements Under SARA Title III

SIC Code	Industry
20	Food and kindred products
21	Tobacco manufacturers
22	Textile mill products
23	Apparel and other finished fabric products
24	Lumber and wood products
25	Furniture and fixtures
26	Paper and allied products
27	Printing and publishing
28	Chemicals and allied products
29	Petroleum refining industries
30	Rubber and miscellaneous plastics products
31	Leather and leather products
32	Stone, clay, glass, and concrete products
33	Primary metal industries
34	Fabricated metal products
35	Machinery
36	Electrical power, machinery, equipment, and supplies
37	Transportation equipment
38	Measuring, analyzing, and controlling instruments
39	Miscellaneous manufacturing industries

Table 15. Required Information for the Toxic Chemical Release Form

Name, location, and principal business activity of the facility
Appropriate certification signed by a senior management official certifying the accuracy and completeness of the report
Note if the toxic chemical released is manufactured, processed, or otherwise used and the general categories of use
Estimated maximum amounts of the toxic chemical present at the facility at any time during the year
The waste treatment or disposal methods employed for each waste stream and estimated efficiency of the treatment
The annual quantity of the toxic chemical entering the environment for each medium

29 (Table 14), and covered by the inventory requirements stated above, must complete a toxic release form for each chemical release listed on the inventory of hazardous chemicals. This form, commonly known as Form R, must be submitted to the EPA and SERC annually by July 1. This form is a complex five page form requiring a facility description, off-site toxic waste dumps, chemical specific information on each release for each release media, and supplemental information related to the release (see Table 15).

This reporting requirement applies to each listed toxic chemical manufactured or processed in quantities >25,000 lb or for chemicals otherwise used at the facility in quantities that exceed the TPQ, or 10,000 lb. Facilities not in these manufacturing SIC codes may be required to complete Form R if either the EPA or SERC determines a need. The need is based on the toxic chemical inventory reports (i.e.,

Tier I or II), proximity to other facilities that release toxic chemicals, proximity to population centers or institutions, and the history of chemical releases at that facility.

Information on Form R must include those items listed in Table 15. Release reporting is required whether the releases are due to accidental spills, leaks, or equipment malfunctions, or normal processes or discharges. Releases must be reported for all environmental media including air, water, and ground releases. The EPA has provided a detailed description of the procedures for estimating toxic chemical releases.[17]

These toxic chemical release forms are meant to provide information to the LEPC, SERC, EPA, and to the general public living or working near the facility. Data on these forms are intended to inform the public about releases of toxic chemicals to the environment in order to assist agencies and scientists in research and data gathering, and to aid in developing appropriate regulations, standards, and guidelines. Data collected on these forms are available on a public database through the National Library of Medicine.

Trade Secrets

Extensive provisions are made in the law regarding protection of trade secrets. These detailed provisions should be consulted for companies to which this applies. In order to qualify for these exemptions, the substance in question must be a trade secret that is not readily discovered through reverse engineering.

Specific provisions are made in the law regarding the acquisition of information by health professionals in medical emergencies. If the complete description of the chemical is needed for diagnosis or treatment of an individual exposed to that chemical, that trade secret information must be provided to the health professional requesting the information. The health professional is required in writing not to use the obtained information for any purpose other than diagnosis or treatment of the patient. This release of information applies to medical emergencies as well as preventive measures for assessing exposure, sampling, medical surveillance, and the health effects of a potentially exposed population.

REFERENCES

1. Bhopal report, *Chem. Eng. News* 63(6):14-15, (1985).
2. Shrivastava, P. *Bhopal: The Anatomy of a Crisis* (Cambridge, MA; Ballinger Publishing, 1987).
3. Superfund Amendments Reauthorization Act. Title III — Emergency Planning and Community Right-to-Know, 40 CFR Parts 300 and 355.
4. Hazard Communication, OSHA 29 CFR 1910.1200.
5. Brower, J. E., Ed. *Hazard Communication: Issues and Implementation,* ASTM STP 932 (Philadelphia: American Society for Testing and Materials, 1986).

6. The List of Extremely Hazardous Substances and Their Threshold Planning Quantities, 40 CFR Part 355. Appendix A.
7. Arbuckle, J. G., T. A. Vanderver, and P. A. J. Wilson. *SARA Title III Law and Regulations* (Rockville, MD: Government Institutes, Inc.).
8. *Community Right-to-Know Manual* (Washington, D.C.: Thompson Publishing Group).
9. *Community Awareness and Emergency Response Handbook* (Washington, D.C.: Chemical Manufacturers Association, 1985).
10. Lowry, G. G., and R. C. Lowry. *Lowrys' Handbook of Right-to-Know and Emergency Planning* (Chelsea, MI: Lewis Publishers, 1988).
11. *SARA Title III Compliance Guidebook* (Rockville, MD: Government Institutes, 1988).
12. *Hazardous Materials Emergency Planning Guide* (Washington D.C.: National Response Team).
13. O'Reilly, J. T. *Emergency Response to Chemical Accidents: Planning and Coordinating Solutions* (New York: McGraw-Hill, 1987).
14. *Emergency Response Guidebook* (Washington, D.C.: U.S. Department of Transportation, 1990).
15. *Toxic Chemical Release Inventory: Questions and Answers,* EPA 560/4-89-002 (Washington, D.C.: Environmental Protection Agency, April 1988).
16. *Technical Guidance for Hazards Analysis: Emergency Planning for Extremely Hazardous Substances* (Washington, D.C.: Environmental Protection Agency, Federal Emergency Management Agency, and Department of Transportation, December 1987).
17. *Estimating Releases and Waste Treatment Efficiency for the Toxic Chemical Release Inventory Form* (Washington, D.C.: Environmental Protection Agency, 1987).

CHAPTER 7

Workers' Compensation

Glenn Shor

INTRODUCTION

Under state Workers' Compensation Acts, persons injured in the course of work are entitled to partial income replacement and full medical care. Workers' compensation is intended to be a no-fault system that provides "quick and sure benefits" that adequately compensate for disability. Employees are relieved from the burden of proving that their injury was due to the negligence of their employers, but surrender the right to sue their employers for damages in civil or criminal court. To employers, workers' compensation substitutes a regular, fixed, and predictable compensation premium for uncertain, potentially ruinous liability judgments. By incorporating some portion of the social costs of employment injuries to the firms in which the injuries take place, there are also theoretical incentives to prevent injuries.

THE ORIGINS OF INJURY COMPENSATION

At the turn of the 20th century, employees injured in an industrial accident in the U.S. had no automatic entitlement to compensation or medical care. The basis of liability was negligence, under a system modeled on English common law. Employers had a basic duty to act with due care for employee safety, to furnish a sufficient number of safe tools and equipment, and to have a sufficient number of qualified employees to do the job. Employers were responsible for issuing and enforcing safety rules, and had a duty to warn workers of unusual hazards.[1]

In theory, when the employer did not meet these standards of conduct, an injured employee could sue for damages. In order to hold the employer liable, however, the injured worker (or his dependents in case of death) had to prove that the injury or death was caused by some particular negligent act of the employer; if no negligence could be established, no damages could be recovered. One obstacle to proving negligence was convincing other workers who witnessed the events leading up to the injury to testify on behalf of the injured worker in court. Having little protection under governmental or union programs, workers who testified risked losing their own jobs. When lawsuits for negligence were brought under this system, common law gave employers three strong defenses: that workers had assumed the risks of the jobs, that a fellow worker was at fault rather than the employer, or that a worker's own contributory negligence was responsible. While these defenses pleased employers and insurers, keeping them insulated from workers' claims, the social costs of welfare, medical care and rehabilitation for the disabled eventually prompted some states to ease claimants' burden of proof.[2] The combination of this weakening of employers' defenses, the humanitarian concerns of Progressive era reformers (such as the Wobblies), and a fear that the continuing burden of occupational injury might provoke radical political action led to the passage of workers' (then workmen's) compensation laws in 40 states between 1911 and 1920. In 1948, Mississippi became the last state to establish a state workers' compensation program for occupational injury.[3]

THE STRUCTURE OF WORKERS' COMPENSATION

There are currently 53 different workers' compensation systems in the U.S. Each state and the District of Columbia has its own, and there are separate systems to handle the injury claims of federal workers, and longshore and harbor workers. (The territories of Guam, Puerto Rico, and the Virgin Islands also have unique systems.) While there are many similarities in the scope and intent of these laws, there are also significant differences.

Nearly all states have compulsory workers' compensation systems, i.e., all employers are required to cover eligible employees for the compensation of job-related injury and laws are constantly changing. In three states (New Jersey, South Carolina, and Texas) coverage is elective, permitting an employer to reject the law's provisions. However, in rejecting the law, employers also lose three important common law defenses, and in practice few invoke this option.

Most workers' compensation statutes require employers to purchase insurance or obtain permission from the relevant state authority to self-insure their potential liability. In all but six states, insurance is offered by private insurance carriers, and

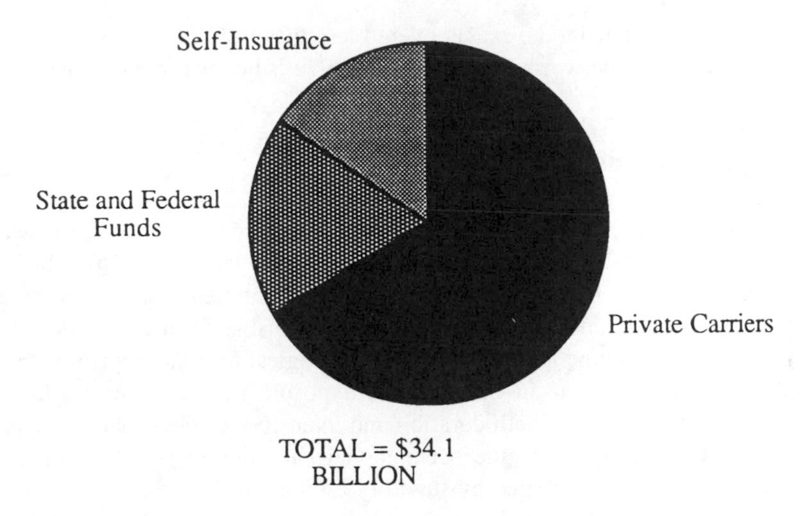

Figure 1. Employer costs by type of insurer, U.S., 1986. Total = $34.1 billion. (Adapted from Nelson, W. J., Jr., *Social Security Bull.* 52(3) (1989).

in 14 states, state insurance funds compete for workers' compensation business. In Nevada, North Dakota, Ohio, Washington, West Virginia, and Wyoming, there are "exclusive" state insurance funds through which employers must insure.[6] Individual employers may receive permission to self-insure in all states except North Dakota, Texas, and Wyoming; the employers choosing self-insurance are usually large companies that can put aside sufficient reserves for future claims, and that can establish or contract for their own claims adjustment and safety engineering services.

Employers paid an estimated $34.1 billion in 1986 for workers' compensation coverage. Approximately two thirds of the costs went to private insurance carriers, 15% was paid as benefits and administrative costs by self-insured employers, and 18% was paid to state and federal insurance funds (see Figure 1).

COVERAGE

Approximately 86 million U.S. workers, or 87% of the workforce, were covered by workers' compensation laws in 1986. The gaps in coverage are found primarily among domestic workers, agricultural workers, casual laborers, and among workers in small firms, nonprofit institutions, and state and local governments.[4] Numerical exemptions under the law apply in 14 states; coverage is restricted to employers having greater than four employees (three states), to employers having greater than three employees (two states), and to employers having greater than two employees (nine states). The extent of coverage of the workforce varies by state. Agricultural workers are covered on the same basis as other employees in 12 states and the District of Columbia. As of 1984, two jurisdictions (Louisiana and Texas) covered less than 70% of the workforce, 21 jurisdictions covered between 70 and 85% and 29 jurisdictions covered over 85%.[6]

Coverage for Occupational Disease

All state workers' compensation statutes now include coverage for diseases arising out of and in the course of employment. However, several "roadblocks" constrain individuals from filing and proving occupational disease claims.[7] In many cases, occupational diseases are indistinguishable from nonwork-related conditions. There may be a paucity of epidemiological or toxicological research to support a link between an occupational exposure and an illness, a lack of recorded exposure data (on both duration and intensity), or physicians may have insufficient training to determine occupational causation. These limitations of medical science are exacerbated by statutory limitations that restrict case filings by requiring claims to be filed within a specific time period after exposure to hazardous chemical or substances, rather than within a time period after manifestation of an illness, or by statutes that only recognize limited lists of diseases. While there were efforts during the late 1970s and early 1980s within the states to reduce these statutory restrictions, many of these reform campaigns stalled in the absence of continued federal attention to the problem.[8]

BENEFITS

Under workers' compensation statutes, workers are restricted from suing their employers for the effects of job-related injury. The benefits provided under the various acts, then, are intended to compensate for economic losses of the workers. Benefits are of three major types: (1) medical benefits, to cure or relieve the injured worker from the effects of the injury; (2) cash benefits, which replace lost present and future eamings due to impairment or disability, or provide survivor benefits to families of fatally injured workers; and (3) rehabilitation benefits, which are intended to assist seriously disabled workers with both medical and vocational services in order to return them to gainful employment.

In 1986, workers injured in the course of employment (and dependents of fatally injured workers) received over $25 billion in benefits; two thirds of the total was paid as partial compensation for lost wages, while the remainder was paid for medical expenses. Figures 2A and 2B show the distributions of benefits by type in the U.S. for the period 1970—1986 in current dollars and by percentage of distribution.

States generally provide that medical benefits will be paid without limits on duration or amount.[9] Each state has fixed its own schedule of disability and death benefits. States differ not only on the minimum and maximum levels of weekly benefits, but on the waiting periods between injury and first payment, and the

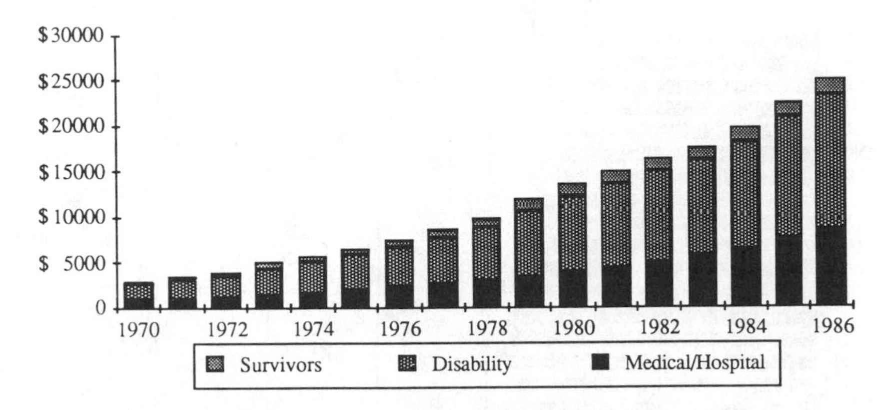

Figure 2A. Benefit payment by type, U.S., 1970—1986. (Adapted from Nelson, W. J., Jr., *Social Security Bull.* 52(3) (1989).

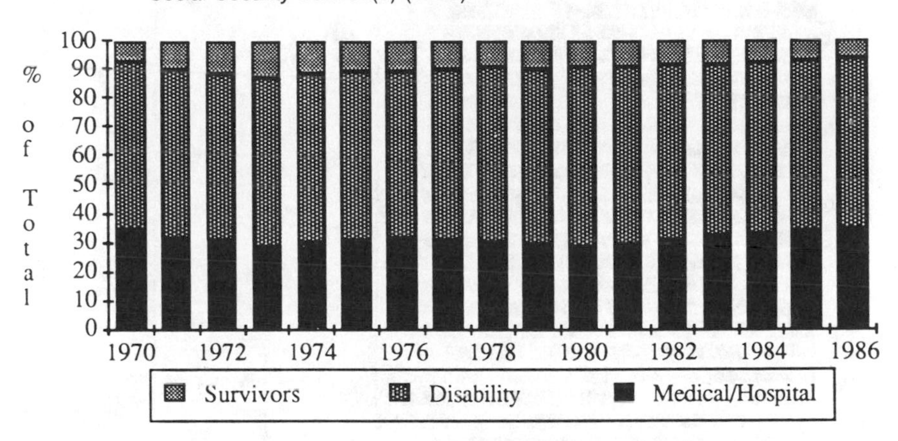

Figure 2B. Benefit payment by type, percentage distribution, U.S., 1970—1986. (Adapted from Nelson, W. J., Jr., *Social Security Bull.* 52(3) (1989).

maximum (dollar or time) limits on payments. Currently, all but eight states (Alaska, Arizona, California, Georgia, Indiana, Nebraska, New York, and Tennessee) adjust the benefit levels for temporary total disability by changes in the state average weekly wage (SAWW). Approximately 30 states maintain these maximum benefits at or above the SAWW. In eight states, including California and Texas, maximum benefits replace 60% or less of the average disabled worker's wages. Figure 3 shows maximum benefit levels by state during 1989.

Although the maximum levels for temporary total disability are the most easily used comparison of state workers' compensation benefits, they are somewhat misleading as a measure of employer costs for the system. Part of the reason is that while temporary total disability cases account for most of the cases involving cash benefits to workers, the per case payout for permanent disability is much higher. Figure 4A shows the distribution of the number of cases that include benefit payments to workers. Figure 4B shows the distribution of the costs of these cases.

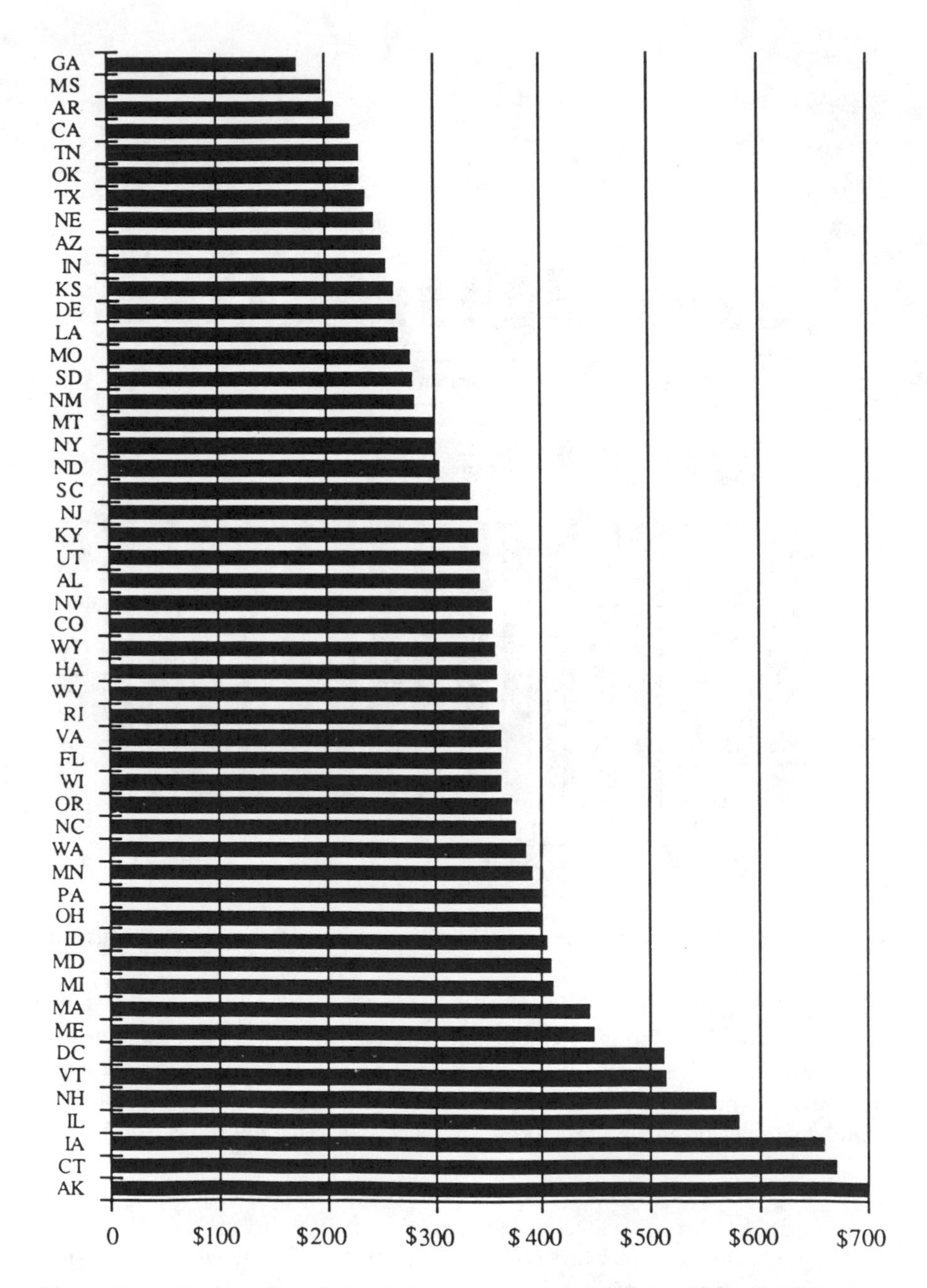

Figure 3. Maximum benefit levels for temporary total disability, U.S., 1988. (Adapted from Nelson, W. J., Jr., *Social Security Bull.* 52(3) (1989).

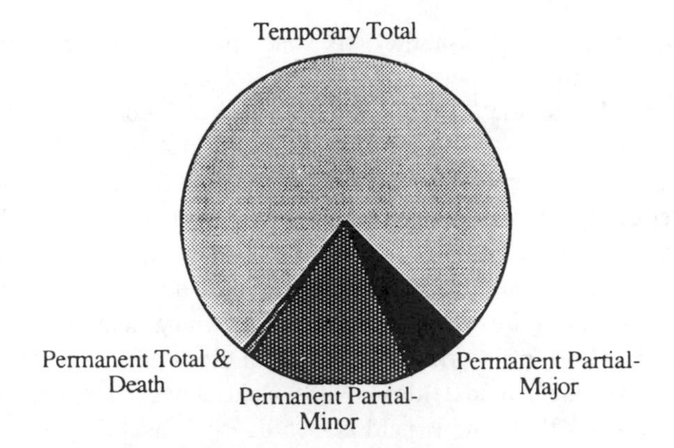

Figure 4A. Distribution of compensable cases, U.S., 1982. (Adapted from Nelson, W. J., Jr., *Social Security Bull.* 52(3) (1989).

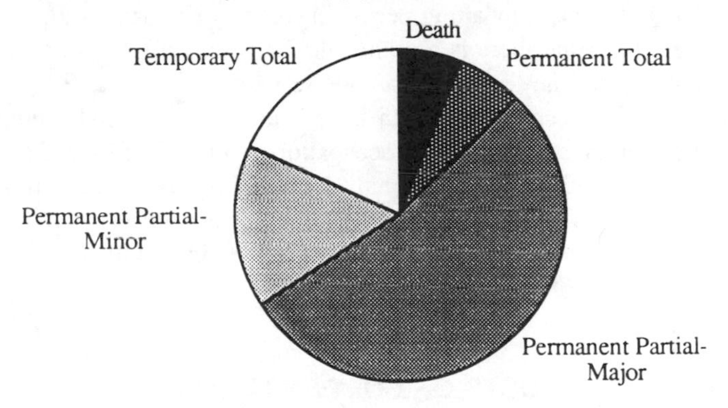

Figure 4B. Distribution of benefits, U.S., 1982. (Adapted from Nelson, W. J., Jr., *Social Security Bull.* 52(3) (1989).

While 76% of compensable cases are for temporary total disability, they account for less than 18% of cost. In contrast, permanent partial disability cases are less than 23% of the total, but they account for nearly 70% of the costs of disabling injury.

Benefit Adequacy

In theory, workers' compensation is intended to adequately compensate those injured on the job. In current practice, workers' compensation payments generally replace two thirds of the gross wage loss during disability, but because of various restrictions and obstacles to payment, this amount is often reduced. The main restrictions involve maximum benefit levels (discussed above), maximum periods of compensation, and waiting periods. Thus, the adequacy of compensation payments varies by jurisdiction.

Fourteen states put limits on the total number of weeks that a person can receive benefits for temporary total disability; six states placed limits on the number of weeks of payments to those permanently and totally disabled.[10]

Waiting periods are the "deductibles" of workers' compensation; they are intended to provide return-to-work incentives by not paying disability benefits to workers for the first several days they do not report for work. (They are also intended to reduce the administrative burden of high volumes of small claims.) In the U.S., current waiting periods range from 3 to 7 days, with compensation retroactive if disability continues longer than an indicated period after injury. For instance, California requires the payment of temporary disability benefits only after a worker has been off work for 3 days, with those days paid retroactively if the disability lasts longer than 21 days. North Dakota has a 4-day waiting period, but compensation for that time is paid after 5 days of disability. Mississippi has a 5-day waiting period, payable retroactively after 14 days of disability. Georgia has a 7-day waiting period, with compensation retroactive after 28 days of disability. In 17 states, there is no waiting period in cases of permanent total disability; such laws recognize that there is no rationale for penalizing workers when there is no expectation that they will be retuming to work.[11]

The disparity between the states in benefit adequacy is readily apparent by looking at the average cost per case in occupational fatalities (Figure 5). This 1988 information, tabulated by the National Council on Compensation Insurance (NCCI) for 44 states (and the District of Columbia) that allow private insurance coverage, ranges from a low of just under $26,000 in Tennessee to a high of $256,900 in Delaware.

ALLOCATING THE COSTS OF COMPENSATION

Unlike programs of pure social insurance, where there are no price differentials by the type of work, workers' compensation systems in the U.S. have retained a relationship between the hazardousness of particular industries and the costs of coverage. Under the American system, the insurance premium varies with the degree of hazard of a particular industry, and for some firms, varies with the individual firm's claims experience as well.[12]

The rationale for relating compensation premium to hazard is to instill incentives to reduce injury and illness, and thereby reduce one's compensation costs. Theoretically, this should apply to whole industries as well as individual firms. Where employers in a given industry can take positive action to reduce injuries and claims across the board, this may eventually translate into lower premium rates; however, there is a time lag between changes in the injury profile and the adjustment of rates. In California, for example, the rates for an industrial classification are based on the average of injury costs during the previous 4-year period; even a dramatic 1-year downturn in injuries will have only minor effect on rates for several years to come.

Nevertheless, for individual firms, there are two types of "merit rating" used to

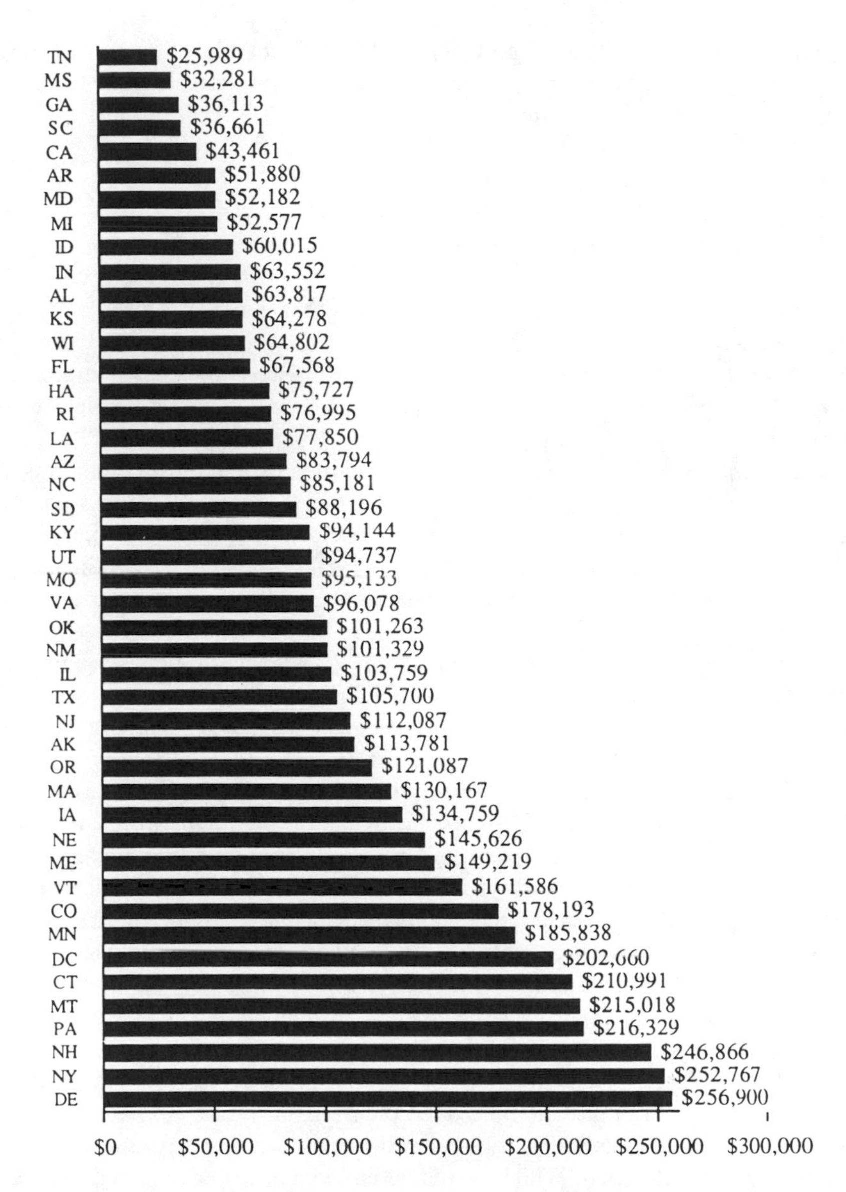

Figure 5. Average cost of death cases, by state. (From National Council on Compensation Insurance.)

adjust premium rates. "Experience rating" adjusts premiums based on a firm's claims costs relative to others in its classification during a recent time period. "Schedule rating" looks prospectively, and attempts to reward firms for showing evidence of physical or administrative attention to known hazards. In some states, schedules are used to give premium reductions to firms setting up joint labor-management health and safety committees, passing inspection as meeting certain regulatory standards, or showing evidence of significant education and training programs. Often these merit rating plans are restricted to relatively large firms, or to firms with a minimum premium level.

THE ROLE OF THE INDUSTRIAL HYGIENIST IN WORKERS' COMPENSATION

In most states, there are well-documented increases in the number of workers' compensation cases related to occupational disease and industrial health hazards.[13] To help deal with this rapidly emerging issue, insurance companies that provide workers' compensation coverage are increasingly hiring industrial hygienists as part of their safety and health, or loss control, departments. There are several organizational models of these loss control departments; the most common is to primarily staff regional or district offices with safety engineers or loss control generalists, and to have a central office that includes industrial hygienists and other specialized professionals. Some insurers prefer to train their own industrial hygienists from their ranks of generalists, but others choose to hire experienced people.

Industrial hygienists in workers' compensation insurance companies generally divide their time between two activities: underwriting inspections and "service." During the underwriting phase, an insurer decides whether to become the carrier for the specific company and, when that decision is reached, the insurer attempts to classify the employer for the purpose of determining a proper premium rate. Generally, the more hazardous the work, the higher the premium. Most states have means by which to modify premium within an industrial grouping. Increasingly, when employers in prespecified types of work apply for workers' compensation insurance, they will be given an "occupational health questionnaire" to identify the types and degree of chemical, physical, or other occupational disease exposures to which employees are potentially subject at work. In most cases, an insurance company safety engineer will compile an initial evaluation of a prospective client. The industrial hygienist is sometimes called upon to interpret the answers on the questionnaire, and possibly to conduct a walk-through inspection or records review to verify the information provided.

The service role of an industrial hygienist in a workers' compensation insurer is complicated and varied. While the industrial hygienist or safety engineer may be trained to recognize and control hazards, they may confront practical limits on their advice in the field. They often find that recommending changes to an employer involves salesmanship and negotiation. The limited number of industrial hygienists means that they often cannot follow up on their recommendations. On a walk-through inspection, an industrial hygienist may find 20 problems, but know that the problems will be addressed in phases, or not at all.

Industrial hygienists may also be called as expert witnesses in various types of workers' compensation cases, particularly those related to occupational diseases or cumulative trauma disorders. Industrial hygienists may be expected to provide data on existing exposures through air monitoring or sampling, or may be requested to estimate past exposures through their understanding of prior work practices or materials usage.

Industrial hygienists may be expected to train insured employees in mandated educational programs, such as how to read and understand material safety data sheets (MSDS), or in the proper use of personal protective equipment while handling hazardous substances.

In 1983, the two major trade associations representing workers' compensation insurers estimated that their members employed about 8600 loss control specialists, while independent firms added another 1000, for a total of 9600. These specialists conducted over 1.5 million visits to policyholders, analyzed 177,000 samples of suspected toxic substances, and held training programs for over 40,000 policyholders.[14] Surveys of California workers' compensation insurers in the late 1970s indicated that about 6.5% of loss control personnel hired by compensation carriers were industrial hygienists. Applied to the national figures, there were approximately 625 industrial hygiene specialists in these companies. A review of the 1983 membership list of the American Industrial Hygiene Association (AIHA) reveals that 5% of their members (248 of 5981) were employed by insurance carriers.

Examples of Industrial Hygienists in Workers' Compensation Insurance

Company A is a large national multiline insurer that in 1984 was among the five largest workers' compensation insurance groups in the U.S. It had 391 professional staffers in its loss control department spread around the country in a "pyramid" structure. As a baseline, all employees would be expected to have a Bachelor of Science or equivalent degree. Upon hiring, the person would receive generalized training in health and safety, through an "exposure awareness training program" consisting of written and audiovisual materials, and some classroom work. Approximately 10% of these employees would go on to an in-depth study consisting of several modules: Introduction to Industrial Hygiene Concepts, Introduction to Industrial Hygiene Instruments, How to Control Industrial Hazard Exposures, and Exposure Measurement and Sampling. A third tier of ten experts were known as "advanced lead representatives" who had an additional week or more of specialized training at the insurers' accredited industrial hygiene laboratory. Finally, the top tier was the home office staff, consisting of five certified industrial hygienists (CIH), two CIH in training, and two industrial hygienists technologists.

Company B, in contrast, kept its safety engineers and health professionals separate, and very few people rose through the ranks to become industrial hygiene experts. Work styles were radically different, with safety personnel expected to carry 250 different accounts concurrently, while industrial hygienists might be expected to do only 30 in-depth surveys per year. Safety personnel would spend 80% of their time in the field visiting insured companies, checking records, and conducting training of company management and personnel. Industrial hygienists would spend more time in the home office, responding to incoming questions and setting company policies for the underwriting of potential occupational disease exposure risks.

STATE COMPLIANCE WITH THE NATIONAL COMMISSION RECOMMENDATIONS

In 1972, the National Commission on State Workmen's Compensation Laws, mandated by the OSHAct, recommended that if states could not voluntarily meet certain minimum standards in their workers' compensation systems, Congress should impose such standards.

In the OSHAct, Congress found that in light of the growth and changing nature of the labor force, increases in medical knowledge, changes in the hazards associated with various types of employment, new technology creating new risks to health and safety, and increases in the general level of wages and the cost of living, "serious questions" concerning the fairness and adequacy of existing compensation programs were being raised. In recognizing the links between the preventive and rehabilitative aspects of workers' overall health and safety, Congress saw that any federal intervention into standards of job health and safety would also need to address the aftermath of injuries and illnesses.[15]

Congress declared that the "full protection of American workers from job-related injury or death requires an adequate, prompt and equitable system of workmen's compensation as well as an effective program of occupational health and safety regulation."[16] In order to determine whether the state laws were providing an "adequate, prompt, and equitable system of compensation" for work-related injury or death, Congress established a 15-member national commission appointed by the president, with broad interest group representation. The secretaries of Labor, Health, Education and Welfare (HEW) and Commerce all served as ex officio members. The group was given the task of comprehensively studying and evaluating 16 areas of the law, shown in Table 1.

The final report, submitted on schedule in July 1972, concluded that the protection furnished by workers' compensation to American workers was "inadequate and inequitable," but that with widespread improvements there could be a "significant role for a modern workmen's compensation program" run under state auspices. The Commission detailed 19 "essential recommendations" — eight concerning coverage, nine concerning income benefits, and two concerning medical benefits — that would constitute a minimum state response toward achieving a nationally adequate program standard. Sixty-five other recommendations were also made.

In 1976, the 50 states and the District of Columbia collectively complied with 557.5 of 969 recommendations (57.5%). In the shadow of Congressional hearings during the Carter Administration, the number rose to 604 (62.3%) in 1978 and 621

Table 1. Areas of Inquiry of National Commission on State Workers' Compensation Law, Established by OSHAct of 1970

The amount and duration of permanent and temporary disability benefits and how maximum limitations were determined
The amount and duration of medical benefits and provisions insuring adequate medical care and free choice of physicians
The extent of coverage of workers, including exemptions based on numbers or type of employment
Standards for determining which injuries or diseases should be deemed compensable
Rehabilitation
Coverage under second or subsequent injury funds
Time limits on filing claims
Waiting periods
Compulsory or elective coverage
Administration
Legal expenses
The feasibility and desirability of a uniform system of reporting information concerning job-related injuries and diseases and the operation of workers' compensation laws
The resolution of conflict of laws, extraterritoriality, and similar problems arising from claims with multistate aspects
The extent to which private insurance carriers were excluded from supplying workers' compensation coverage and the desirability of such exclusionary practices

(64.0%) by the beginning of 1981. Since the start of the Reagan Administration in 1981, however, the compliance ratio has not risen. Today, 16 years after the Commission report, no state meets even the 19 essential recommendations. The average state meets just over 12. Only New Hampshire, the District of Columbia, Ohio, and Wisconsin meet 15 of the standards. At the other extreme, Arkansas, Georgia, Idaho, Kansas, Michigan, Mississippi, New Mexico, Oklahoma, Tennessee, Texas, and Wyoming each meet less than 10 of the essentials. While these recommendations may not be the best measure of adequacy, they still constitute a benchmark for the progress of reform.

CONCLUSIONS

Workers' compensation programs were begun with the intent of providing workers with quick and sure benefits in the event of an occupational injury or illness, and of giving employers a predictable cost of insuring against that liability. In the U.S., workers' compensation systems were established on a state by state basis, with diverse structures of public and private insurance, and with varying requirements for coverage. Benefit types and benefit levels differ across states; while some states may furnish higher maximum benefit levels, others may be more liberal in the scope of injuries and illnesses that are typically allowed. The costs of workers' compensation are typically allocated through classifying employers into nature of work categories, and basing premium rates on the relative degree of hazard within the industrial grouping. Mechanisms of retrospective and prospective review of claims and hazard reduction are used to adjust rates for an individual firm, although these direct incentives are often unavailable to small employers. With the increasing scope of workers' compensation laws allowing for more occupational diseases, there is an increasing role for industrial hygienists and other occupational health professionals in insurance companies covering employment injuries.

During passage of the federal Occupational Safety and Health Act, a national effort was begun to identify the problems inherent in existing programs, and to bring pressure to improve them. That effort has stalled. Many states still fail to cover significant groups of workers, fail to provide adequate levels of compensation, or both. Some of the most significant problem areas are in the larger states. A renewed effort of research, education, and action is needed to assist in making workers' compensation become a positive force for occupational safety and health.

REFERENCES

1. U.S. Congress, Office of Technology Assessment, Preventing Illness and Injury in the Workplace, U.S. Government Printing Office, (April 1985).
2. See, for example, California Statutes 1907, pp. 119-120.
3. See, for example, State of California, Laws of 1911, Chapter 399.
4. Nelson, W. J., Jr. "Workers' Compensation: Coverage, Benefits, and Costs, 1986," *Social Security Bull.* 52(3) (1989).
5. U.S. Department of Labor, Employment Standards Administration, "State Workers' Compensation Laws," Table 2, Numerical Exemptions (January 1989).
6. Nelson, W. J., Jr. "Workers' Compensation: 1980-84 Benchmark Revisions," *Social Security Bull.* 51(7):6 (1988).
7. Boden, L. "Workers' Compensation," in *Occupational Health: Recognizing and Preventing Work-Related Disease,* Levy, B., and D. Wegman, Eds. (Boston: Little, Brown, 1983), p. 446.
8. For a current listing of state law regarding occupational disease compensation, see U.S. Chamber of Commerce, *Analysis of Workers' Compensation Laws,* Washington, D.C. Annual.
9. In Arkansas, employer liability for medical care ceases 6 months after injury where no time is lost from work, or 6 months after a claimant returns to work, or a maximum of $10,000 has been paid, unless the employer waives rights or the Commission extends time and dollar amounts. In New Jersey, employer liability ceases after $100 has been paid for medical care; employee must petition for further treatment. All other states provide full medical benefits without time or monetary limitations. U.S. Department of Labor, Employment Standards Administration, "State Workers' Compensation Laws," January 1989, Table 5a.
10. U.S. Department of Labor, Employment Standards Administration, State Workers' Compensation Laws, Tables 5-6 (January 1989).
11. U.S. Department of Labor, Employment Standards Administration, State Workers' Compensation Laws, Table 14 (January 1989).

12. Larson, A. *Workmen's Compensation for Occupational Injuries and Death, Desk Edition* (1989), Sect. 3.20.
13. See, for example, Shor, G. Occupational Disease Compensation in California, California Policy Seminar, University of California (1987).
14. U.S. Congress, Office of Technology Assessment, *Preventing Injury and Illness in the Workplace,* OTA-H-256 (April 1985), p. 308.
15. Occupational Safety and Health Act, 84 Stat. 1590, Section 27 (a)(1)(B).

PART III

Controlling Hazards in the Work Environment

Industrial hygienists work to recognize, evaluate, and control hazards in the workplace. Once workplace hazards, the applicable law, legal standards, or recommendations have been identified, the hazard often must be controlled to bring workplace exposures to a safe level.

This section of the book deals with the control of many different types of workplace hazards. Obviously, the utility of each type of control will vary from workplace to workplace.

Ventilation for Contaminant Control deals with the use of ventilation to control airborne hazards in the workplace. Ventilation is often referred to as an engineering control.

Heat Stress explains the body's limitations when working in hot environments, and discusses control and measurement methods.

Effective Industrial Plant Noise Abatement and **Hearing Conservation and Employee Protection** both discuss occupational hazards related to noise exposure. The first chapter introduces basic noise concepts while the second deals with the specifics of hearing conservation.

Trip, Slip, and Fall Prevention examines the accidents that account for a high percentage of workplace injuries. This chapter examines their causes and presents methods for their control.

Fire Protection discusses important features of a fire protection program.

CHAPTER 8

Ventilation for Contaminant Control

Richard P. Garrison

INTRODUCTION

Ventilation Control

Airborne contaminants can pose significant health and safety hazards in a workplace. In many situations, it is not possible to eliminate potentially hazardous substances, to prevent their release, or to isolate personnel in order to prevent exposure. It may also be undesirable or inappropriate, for many reasons, to rely upon personal protective equipment or other administrative procedures to help assure worker health and safety. It *is* possible, however, in many situations to control the movement of air in the workplace, and thereby to control the movement and concentration of airborne contaminants.

There are other benefits to be derived by providing effective ventilation in the workplace. For some situations, these benefits may justify investment in ventilation control. These additional purposes include replenishing oxygen, controlling odors, regulating ambient temperature and humidity, transporting materials, preventing fires, improving product quality, recovering valuable materials, enhancing employee and public relations, and complying with governmental regulations and insurance requirements.

This chapter focuses upon what is called "industrial ventilation," which in the context of occupational health means the controlled movement of air (removal and replacement) to help achieve a safe and healthful workplace. There are several references on industrial ventilation that provide excellent discussions on subjects covered briefly (or not at all) in this chapter.[1-6]

All of these books provide useful information on a variety of subjects important to ventilation for contaminant control. The books by Alden and Kane, Burton, the British Occupational Hygiene Society, and the American Conference of Governmental Industrial Hygienists (ACGIH) also emphasize design methodology and provide design data.[1,3-5] Burton's text is structured as a self-study guide for learning the fundamentals of industrial ventilation.[3]

There are two basic approaches to ventilation control, dilution ventilation and local exhaust ventilation (LEV). They are fundamentally different in concept but are not mutually exclusive. Both are likely to be found in workplace environments and can operate simultaneously in the same environment. Figure 1 illustrates these types of systems. Table 1 provides a comparison of dilution and local exhaust ventilation.

Natural Ventilation

Mechanical ventilation requires the use of air moving equipment, e.g., fans or blowers. Natural ventilation involves air movement that does not require mechanical devices.

The two primary sources of natural ventilation are wind and convection. Wind creates pressure differences around buildings and can cause air movement inside buildings. Convection is caused by the tendency of warm air to rise because it is of lower density than cooler air. The rise/fall of warm/cool air causes air masses to move and mix within a workplace. Natural ventilation can involve significant amounts of air movement, particularly in older buildings which may be tall, loosely constructed, and poorly insulated. Single-story buildings of newer design are not as greatly affected by natural ventilation.

Natural ventilation typically is difficult to predict, measure, and even understand. It is changeable without warning and generally should not be relied upon for protection against hazardous substances; however, it should be utilized and taken advantage of whenever possible. All further discussion here of ventilation for contaminant control pertains to mechanical ventilation.

PRESSURES AND VELOCITIES

The concepts of pressure and velocity are fundamental to understanding the operation of LEV systems. Molecules of gases normally move about in random directions. When they impact a surface, the force of the impact (per unit of surface area) is called pressure. Air is composed primarily of two gases, oxygen and nitrogen. The weight of these gases, forming a "blanket" around the earth, causes atmospheric pressure. Atmospheric pressure is relatively stable, but it does change with altitude and the movements of large air masses (e.g., weather).

When air is still, it exerts a pressure called "static" pressure (SP). This pressure is exerted uniformly in all directions. If SP is less than atmospheric pressure, it is described as being "negative." If it is above atmospheric, it is called "positive" SP.

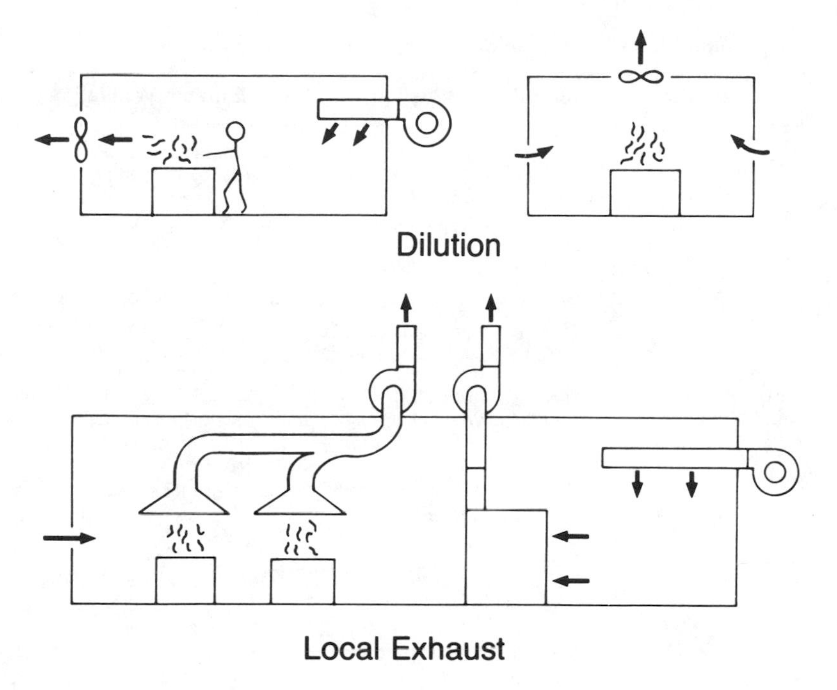

Figure 1. Dilution and local exhaust ventilation systems with replacement air.

Table 1. Comparison of Dilution and Local Exhaust Ventilation

Characteristic	Dilution Ventilation	Local Exhaust Ventilation (LEV)
Basic principle	Contaminants migrate from source and are diluted to acceptable levels	Contaminants captured/ contained at source and removed within duct
Conditions conducive to use	Low contaminant toxicity/ fire hazard Gaseous contaminants Employee distant from source Low, steady emission rates Large, numerous, movable emission sources Moderate climates	High contaminant toxicity/ fire hazard Any contaminants Employee close to source High, variable emission rate Need for contaminant air cleaning Extreme climates
System components	Wall/ceiling exhaust fans Baffles or partitions to direct air movement Replacement air system	Exhaust hood(s): capture, enclosure, and receiving Exhaust duct Exhaust fan and stack Replacement air system
Volume flow rate (Q) and ventilation time	Steady-state $Q = \frac{GC}{K}$	Capture hood: Q = function of hood shape, capture distance and capture velocity

Table 1 (continued). Comparison of Dilution and Local Exhaust Ventilation

Characteristic	Dilution Ventilation	Local Exhaust Ventilation (LEV)
	Purging $T = -\frac{KP}{Q} \ln\left(\frac{C_2}{C_1}\right)$	Enclosure hood: Q = face(s) velocity × opening(s) area
	Build-up $T = -\frac{KP}{Q} \ln\left(\frac{G - \frac{QC}{K}}{G}\right)$	Receiving hood: Q = function of hood shape, size, and orientation to source
System pressure losses	Usually do not apply for ductless wall/ceiling exhaust fan systems Replacement air system same as for LEV	Component (hood, duct, air cleaner) losses estimated as functions of airflow velocity pressure; system loss determines fan selection and motor horse-power
Air moving devices	Usually axial fans for wall/ceiling exhaust Replacement air natural or mechanical (centrifugal fan if long duct)	Centrifugal fans for high SP losses (e.g., slot hoods, long /small duct, air cleaners) Axial fans may be acceptable (e.g., canopy hoods, short/ large duct)
Air cleaning devices	Typically not utilized for dilution exhaust Replacement air may be filtered	Wide variety of equipment for particulates (dust, mist, fume) and gases/vapors Replacement air may be filtered
Performance testing	Airflow patterns near source, in room at doorways (large scale-smoke bombs; small scale-smoke tubes) Fan Q, HP Room SP Replacement air Q, distribution	Hood face/slot velocities Hood airflow patterns (smoke tubes) Duct velocities Fan Q, RPM, SP, HP Hood SP, duct SP Room SP, airflow patterns Stack discharge characteristics Crossdraft velocities Replacement air Q, distribution

Air movement is caused by a static pressure differential, with air moving from locations of higher SP to those of lower SP.

Air exerts another type of pressure when it is moving, "velocity" pressure (VP). This pressure acts only in the direction of movement and the amount of pressure is a function of the speed of movement. The combination (sum) of SP and VP is defined as "total" pressure (TP).

Pressures are sometimes measured in units of force per unit area (e.g., pounds per square inch, psi). Atmospheric pressure is roughly 14.7 psi. Pressure is often measured by causing displacement of a fluid column. When two different pressures (i.e., a pressure differential) are applied to the ends of a U-shaped tube containing a fluid, the fluid will displace to the point at which the weight of the displaced fluid is equal to the force of the pressure differential. This principle has led to pressure units such as inches of mercury (e.g., atmospheric pressure = 29.5 in. Hg = 760 mmHg).

Pressure variations in ventilation systems are quite small compared to atmospheric pressure. This, and convenience of measurement, have led to common use of the units "inches of water" to measure SP, VP, and TP in ventilation systems (13.6 in. H_2O = 1.0 in. Hg). Manometers are devices incorporating a fluid-filled column to measure pressure. The fluid may be water or red-colored oil, but in either case, reading out in inches of water. Sometimes mechanical gauges are used to measure pressure, and they too will typically read out in inches of water.

Velocity is the speed at which air moves. The most common units are feet per minute (fpm) or meters per second (m/s). A velocity of 4000 fpm will correspond to a VP of about 1.0 in. H_2O. Velocities may be measured in several ways (discussed later). Several velocities are important in characterizing and evaluating the performance of LEV systems.

Face velocity is the average velocity across the opening(s) of an exhaust hood. It is measured in the plane of the opening. A well-designed hood usually will have a relatively uniform face velocity, but velocity may be distributed quite unevenly across a hood opening. Face velocity is an important design parameter for enclosure-type hoods (discussed later).

Capture velocity is defined as the velocity required to overcome the natural movement of contaminants upon their release, and interfering crossdrafts, so that contaminants will be drawn into an exhaust hood. Capture velocity is important in the design of capture hoods (discussed later).

Transport velocity is the airflow velocity inside duct necessary to maintain the movement of particulates, primarily dusts, through the duct without settling out. This is an important parameter in duct design.

DILUTION VENTILATION

Basic Concepts

Most airborne contaminants are considered to have an acceptable workplace concentration, i.e., a level at which exposure is not believed to cause a health, fire, or explosion hazard. Dilution ventilation strives to achieve contaminant control by causing the mixing of fresh air with contaminants to be at or below acceptable concentration limits.

Dilution ventilation does not prevent fugitive emissions from a contaminant source. Rather, it attempts to control the migration of contaminants into work areas, to dilute contaminant concentrations as quickly as possible, to remove contaminated air, and to assure the introduction of adequate amounts of fresh replacement air.

Conditions conducive to dilution ventilation include the presence of contaminants having relatively low toxicity or fire hazard, employees not working close to contaminant sources, relatively low and steady rates of contaminant emission, emission sources of very large size or many small sources or movable sources, codes or standards that require dilution ventilation, and moderate climates. The primary disadvantages of dilution ventilation are the likelihood of high concentrations near emission sources and the possibility of very high air flow rate requirements (and associated cost), especially when acceptable contaminant levels are very low and/or emission rates are high.

Design Considerations

It can be relatively difficult to design effective dilution ventilation. This results from a typically wide range of alternative approaches and from difficulty in predicting how a given design will perform in an actual workplace. Dilution ventilation is usually accomplished by the placement of exhaust fans in walls and/or ceilings to remove contaminated air and by the introduction of equal amounts of replacement air through walls or directly into work areas. Replacement air may need to be conditioned, and this becomes the primary operating cost of dilution ventilation.

Volume flow rate (Q) is one of the most important design considerations necessary for both dilution and local exhaust ventilation. Table 1 highlights several relationships that are used to calculate Q for different situations: steady-state, purging, and build-up. In steady-state, contaminants are released at a constant rate (G) and a constant contaminant concentration (C) is achieved. Purging is a situation in which a high initial concentration drops (with G negligible) over a period of time (T). Build-up involves contaminant concentrations increasing with time (G is not negligible). The time-dependent cases (purging and build-up) also depend upon the volume (P) of the workplace.

It can be quite difficult to determine appropriate values for some of the parameters needed for design calculations. T, C, and P can be measured or specified relatively easily; however, the contaminant release rate, G, is not usually known precisely. Estimates can be made from known consumption (evaporation) rates of specific substances (e.g., pints per hour of cleaning solvents). It can be even more difficult to specify the parameter K.

K is a combination of a safety (contingency design) factor and an air mixing (ventilation effectiveness) factor. It represents the fact that the design method provides only a rough approximation, at best, of the ventilation flow rate needed for control, and that uncertainties can be dealt with by increasing the total dilution airflow. It also represents the fact that the effectiveness of air mixing (dilution) will vary with the ventilation design.

Dilution ventilation design involves careful location of exhaust fans and replacement air outlets. If exhaust fans are close to the contaminant source(s) and replacement air is supplied throughout the workplace, it is possible to achieve dilution with less airflow than if the exhaust is far from source with poor replacement air distribution. The K factor is usually described to range from 3 to 10, with lower values for ventilation systems assumed to have effective mixing and low toxicity hazard, and higher values for situations with ineffective mixing and higher toxicity. Unfortunately, very little specific data and few guidelines are available to select a specific, appropriate value of K. It should be noted again that dilution ventilation generally is not recommended for highly toxic contaminants, regardless of how well design parameters can be specified.

By looking at the simplest relationship in Table 1, i.e., for steady-state dilution, a high value of K, a high emission rate (G), and a low acceptable level (C) will combine to yield a high rate of dilution airflow and high cost for replacement air. When these conditions exist, and for other reasons, it is often more advantageous to provide local exhaust ventilation.

LOCAL EXHAUST VENTILATION (LEV)

Basic Concepts

LEV is designed to capture and remove airborne contaminants at their source of release. LEV should control, to an acceptable extent, the migration of contaminants within a workplace, particularly protecting employee breathing zones. LEV may be the only effective method to control highly toxic contaminants in some situations.

The exhaust volume flow rate needed for LEV will be significantly less than for dilution ventilation. This greatly reduces the volume and associated cost of replacement air; however, LEV does not eliminate the need for replacement air. A lack of replacement airflow is probably the most common reason, other than poor initial design, for failure of an LEV system to perform adequately.

LEV systems have four basic components: hood(s), duct, fan, and stack, which are illustrated in Figure 2. Hoods are the inlets to the system. They are designed to capture contaminants, and their design is usually highly specific for the operation of each contaminant source. Duct provides the pathway or conduit for airflow from the hood(s) to the exhaust fan. Fans are the devices that cause air to flow through the LEV system. The stack is the final section of duct from the fan to a discharge point outside the building.

Air cleaning devices are optional components of LEV systems. When used, they are usually installed close to and upstream (i.e., on the inlet side) of the fan, thereby reducing the flow of containments through the fan and to the atmosphere.

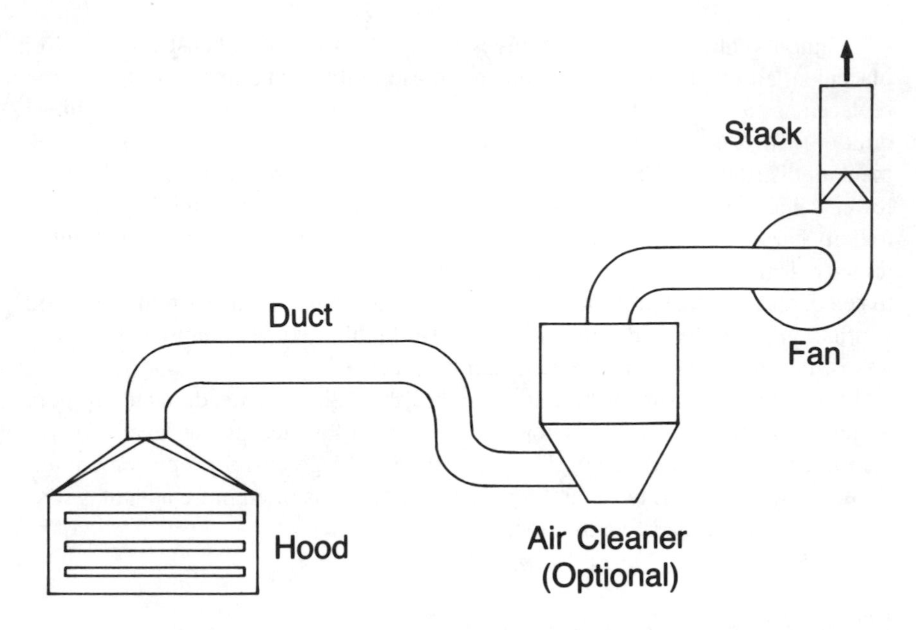

Figure 2. Basic components of an LEV system.

Exhaust Hoods

It can be argued that exhaust hoods are the most important components of LEV systems because they are the point at which contaminant control is either achieved or lost. They certainly are the part of the system for which the design is most specific to a particular application and for which an industrial hygienist should have major input in the design process.

There are three basic types of LEV hoods: capture, enclosure, and receiving. Figure 3 illustrates examples of the different hood types. A capture hood is designed to be located close to a contaminant source and to "reach out" to capture contaminants. An enclosure hood contains the contaminant source within itself. A receiving hood is intended to utilize the natural direction of contaminant movement and to receive contaminants and prevent their escape.

In some ways, capture hoods are the most difficult type to design. They offer the distinct advantage of minimizing interference with a process, intruding only close enough to affect airflow control; however, their performance can be severely inhibited by reduction in airflow rate, improper distance from the source, poor velocity distribution, and crossdrafts.

The exhaust Q for a capture hood is a function of the capture velocity needed for contaminant control and the inflowing velocity field generated by the hood. Design data and methods in both of these areas have significant limitations. Capture velocity may range from 50 to 2000 fpm depending upon contaminant release conditions, toxicity, surrounding airflow, and the size of the hood.[5] It will

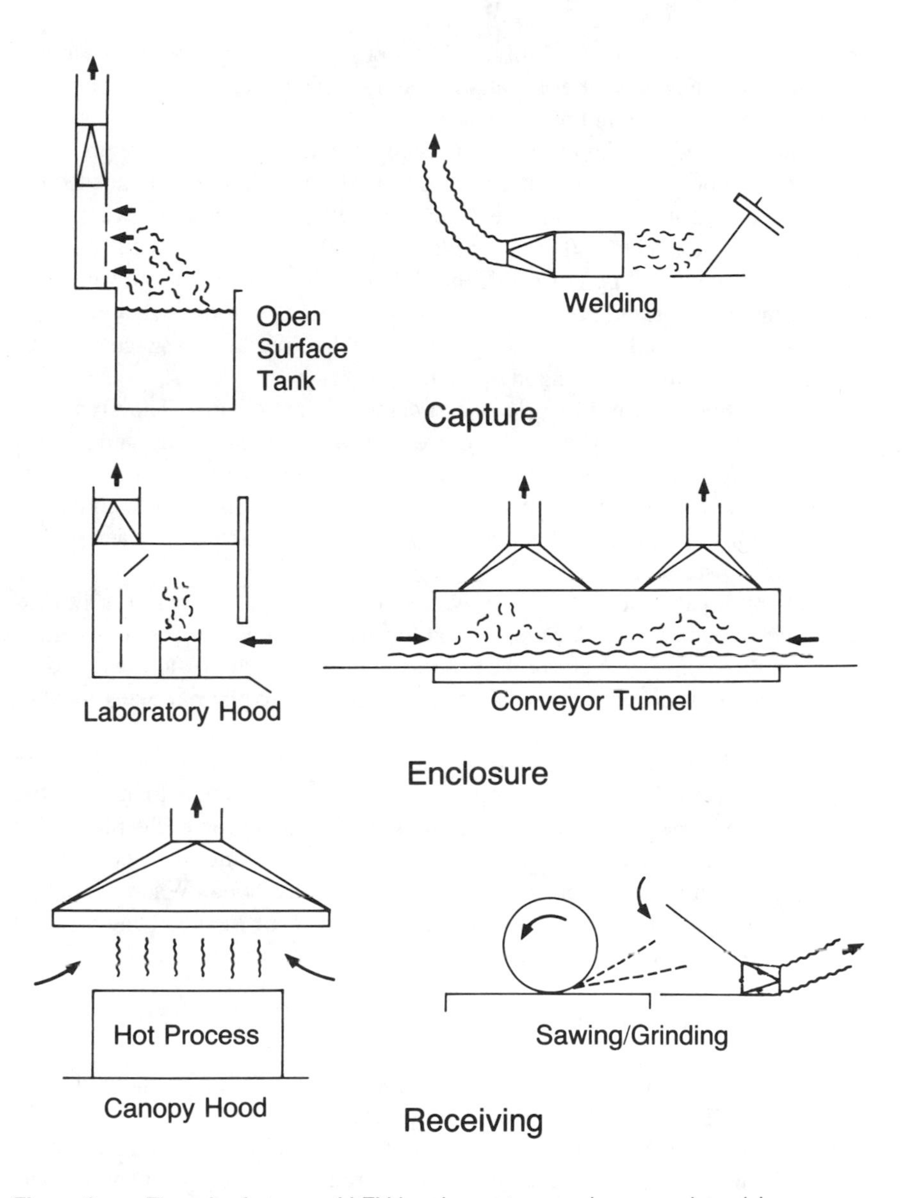

Figure 3. Three basic types of LEV hoods: capture, enclosure, and receiving.

rarely be known precisely. Velocity characteristics are typically described as simple functions of basic hood shape, usually expressed as the centerline (one-dimensional) velocity gradient (i.e., velocity change with centerline distance). Better three-dimensional models of capture hood velocity characteristics are under development, but have not yet been established for practical design applications.

Capture hoods can be quite effective despite difficulties in predictive design.

This often requires some flexibility in hood positioning and flow rate, careful evaluation and adjustment after installation, and strict adherence to work practices prescribed for hood use and maintenance.

An enclosure hood is, in effect, a box built around a contaminant source. Air flows into the hood through specific openings or "faces." The velocity needed for contaminant control at the hood face(s) must be sufficient to assure inward flow; this may range from 50 to 300 fpm, but typically will be between 75 and 150 fpm. A primary concern will be crossdraft interference. Tendencies for outward contaminant movement should be countered by face velocity and the size and design of the enclosure. Hood volume flow rate is equal to the average face velocity multiplied by the area of the hood opening(s).

Enclosure hoods do not reach out and capture contaminants. They typically offer the advantages of reduced Q, in comparison with capture hoods, and are less subject to interference caused by crossdrafts and poor work practices. Their principal disadvantages are limitation of access to process equipment contained inside (possibly complicating equipment operation and maintenance) and higher construction cost.

Receiving hoods can be effective when contaminants are released in limited and predictable directions. A canopy hood located over a hot process utilizes the natural upward convection above the process. A ventilated "trap" for dust thrown from a grinding wheel or circular saw can be quite effective in removing most of the airborne material.

Receiving hoods may be somewhat difficult to distinguish from capture or enclosing hoods. Baffles or other additional surfaces may be added to improve the performance of a receiving hood, thereby making it more like an enclosure. It also may be necessary to increase exhaust Q and decrease capture distance, thereby making a receiving hood perform more like a capture hood. When designing enclosure or capture hoods, it is always advisable to shape the hood and position duct take-offs to accommodate contaminant movement into the duct.

Hood Design Guidelines

There is a variety of general guidelines to consider and apply in hood design. Some have been addressed in the preceding discussion.

Utilize the "KISS" ("Keep It Simple, Stupid") principle. For hoods, this involves starting with a simple imaginary "box" around a contaminant source. The "sides" of the box should be removed and reshaped to the minimum possible extent. At least one opening is needed to allow airflow. In general, a six-sided box will give better control than a five-sided box, a five-sided box is usually preferable to a four-sided box, and so forth. The KISS principle also applies for other aspects of ventilation design (in fact, for many aspects of life in general).

Additional guidelines include (but are not limited to) the following:

- Know the process (i.e., good industrial hygiene) and make sure that the hood is usable.

•Consider other engineering control improvements (i.e., confirm that ventilation is needed).
•Protect the worker's breathing zone (contaminated air should flow away from it).
•Minimize conflicting air movement (doors, windows, fans, vehicular and personnel traffic, hot/cold processes).
•Utilize flanges, baffles, and/or partitions to help direct airflow in useful directions.
•Enclosure hoods are usually preferable to capture or receiving hoods (better control and lower flow rate).
•A canopy can work well over a hot process and probably will work poorly over a cold process (i.e., a good receiving hood, but a poor capture hood).
•Contaminants will be "heavier than air" only in very high concentrations and will behave like air in dilute mixture with air.
•Assure adequate replacement air to compensate the exhaust Q.

Duct

LEV duct functions to carry contaminated air from the hood to air cleaners, to air movers, and ultimately to atmospheric discharge. It is also used to distribute replacement air. The sizing and design of duct may be very important for establishing the distribution of airflow in multiple-hood LEV systems.

A variety of components comprise LEV duct. These include straight sections, elbows (e.g., turns of 90°, 45°), branch entries (two ducts joining into one), expansions/contractions, and transitions (e.g., circular to rectangular). Other components include clean-outs (openings into duct), guide vanes (to turn, straighten, or divide airflow), and blast-gates or dampers to regulate flow by creating adjustable resistance.

Duct will usually be of circular cross-section. Rectangular duct may be used to fit into walls or otherwise use space more efficiently. Most design data apply for circular duct, which is used to approximate characteristics for rectangular duct.

Duct may be fabricated from a variety of materials. These include galvanized steel sections (most common), spiral-wound steel, flexible duct (e.g., wire spiral covered with airtight fabric), fiberglass, various plastics, and concrete. Selection criteria include cost, ease of use, strength, corrosion and temperature characteristics, and need for mobility.

The discharge stack is a particularly important part of the duct. It must assure that contaminants do not remain near the building, where they could reenter and be recirculated. The shape of the stackhead (end) should favor vertical discharge at reasonably high velocity without excessive resistance (i.e., pressure loss). A "no loss," concentric stackhead is often recommended.[5] Rain caps covering the end of the stack tend to disperse contaminants across the roof and are not recommended. Stack height is also important in order to penetrate the building airflow "envelope." Industrial hygienists favor tall stacks, but architects often do not.

Duct velocity is an important design parameter. A minimum transport velocity must be maintained for particulate contaminants. This velocity can range from 2000 fpm for lightweight fumes and dusts to 4000 to 5000 fpm for large dust particles.[5] Transport velocity is not important for gas or vapor contaminants or clean (replacement) air. However, very large duct (for very low velocities) can be costly and consume valuable space. Minimum duct velocity for practical/economic considerations would typically be about 1000 fpm. Extremely high duct velocities can cause excessive pressure loss, high turbulence noise levels, and abrasive damage to duct.

Pressure Losses

Duct causes resistance to airflow principally by friction and turbulence. Friction results from the tendency of air close to a surface to adhere to the surface. Turbulence is a randomized, nondirectional pattern of flow that consumes energy and tends to obstruct and interfere with productive, directed air movement. Turbulence occurs when airflow is forced to change direction abruptly (e.g., elbows, expansions, branch entries, obstacles in the flow). Turbulent resistance occurs in hoods, duct, and air cleaners. The magnitude of this resistance is a function of velocity and the geometric shape of the component (i.e., the extent of directional change).

Friction is a function of airflow velocity, duct material (surface roughness), duct size (diameter), duct length, and properties of the air (density, viscosity). For a given duct size, friction will increase directly with the friction coefficient (surface roughness), the duct length, and the square of velocity, and inversely with the duct diameter. Diameter is a very important parameter. If Q is fixed, changes in diameter (e.g., reduction by 1/2) will cause large changes in cross-sectional area (reduce to 1/4) and velocity (increase of 4 times). The change in diameter (e.g., 1/2 reduction), combined with squaring the increase in velocity (e.g., 16 times) can cause drastic increases in frictional losses (e.g., 32 times higher, for 1/2 diameter with constant flow rate).

Resistance to airflow is commonly described as "pressure loss." In order for air to flow through hoods and duct, the SP must become increasingly lower in the direction of flow. It is necessary to reduce, or "lose," SP to cause air to flow against frictional and turbulence resistance in an LEV system.

SP must also be "lost" to cause acceleration of airflow from zero velocity (outside the hood) to duct velocity. This loss represents a conversion of energy from SP to VP.

Excessive pressure losses should be avoided. They waste energy (fan horsepower) and will cause reduction in overall Q. Turbulence and friction losses can be minimized by avoiding excessive hood and duct velocities, duct that is too small in diameter or unnecessarily long, and excessively abrupt changes in direction (e.g., sharp turns or abrupt branch entries or expansions/contractions/transitions).

System Design Method

The basic approach for designing an LEV system is to first establish the hood design(s), Q, and duct layout. It is then necessary to specify duct velocity criteria and select duct material. Duct is sized to meet velocity criteria and to achieve pressure balance within the system. Pressure losses are estimated for the hood(s) and duct. Losses are added throughout the system, including the stack, in order to estimate the total pressure differential needed to be created by the fan.

The procedure for estimating pressure losses can be quite tedious, especially for multiple-hood systems. It involves using a variety of empirical approximations for pressure losses of the system components. The method in general usage, the velocity pressure method, involves using loss factors (constants) that will approximate pressure loss when multiplied by the duct velocity pressure. Loss factors vary with the geometry of the hood ("entry" loss) and duct components (e.g., elbows, entries). Frictional losses for straight duct are described as a loss factor per foot of duct, to be multiplied by the duct length. Acceleration to duct velocity is figured as a loss factor of 1.0.

This may seem confusing. It is relatively easy, but the detailed procedure is beyond the scope of this chapter. The procedure can be complicated by the need (sometimes frequent) to redesign (usually change duct size) and make other adjustments to cause SP to balance at each junction where two airflows combine and each branch must maintain its own design flow rate.

In the end, the design procedure yields a total system flow rate and estimated fan SP differential. These data are necessary to select a fan (discussed subsequently). The procedure is, at best, an approximation. If the final system performs within 10 to 20% of the design, it should be considered a "win."

Types of LEV Systems

There are three basic variations of LEV systems: tapered-balanced, blast-gate, and plenum. These are illustrated in Figure 4. It may seem somewhat "horse after the cart" to introduce these systems at this point; however, some of the distinguishing characteristics require knowledge gained from the preceding discussion.

A tapered-balanced LEV system is designed to "balance" at the desired flow rates in each branch by virtue of duct design. It is designed as a fixed, relatively inflexible system. It should not need adjustment for airflow distribution and should not easily lose its inherent balance. These systems are usually designed to maintain a minimum duct velocity. The design procedure involves careful duct sizing to achieve SP balance, beginning at the junction farthest from the fan and proceeding to the fan, junction by junction. This type of system is preferred when flexibility is not required, transport velocity is important (for particulate contami-

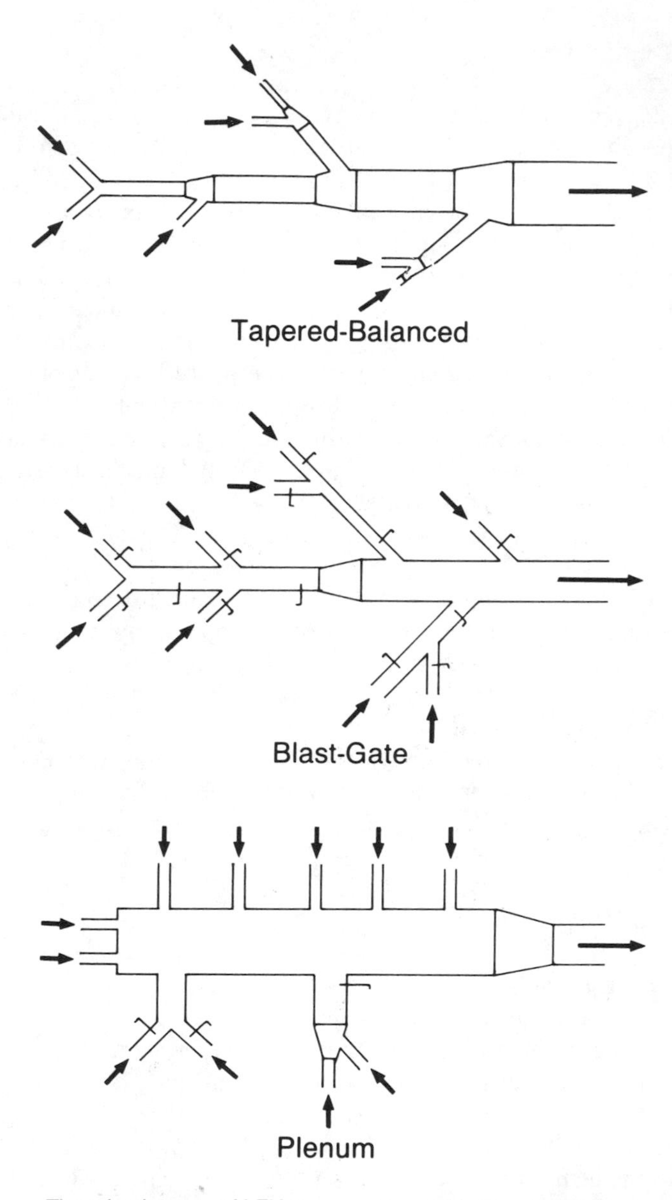

Figure 4. Three basic types of LEV systems: tapered-balanced, blast-gate, and plenum.

nants), and when contaminants are highly toxic (i.e., misadjustment or poor balance can be hazardous).

Blast-gate systems utilize variable resistance devices (blast gates) in each branch. The system is designed for the total desired flow rate and the SP loss of the "governing" branch (duct path having the greatest SP loss). Transport velocity may or may not be maintained throughout the system. The system is installed and balance (desired airflow distribution) is achieved by adjusting the blast-gates, measuring branch flow rates, and readjusting/remeasuring as needed. If the incorrect governing branch is selected, it may not be possible to achieve the desired balance. This type of system is more flexible than a tapered-balanced system; however, it can be thrown out of balance by damage to or tampering with the blast-gates. High velocities and turbulence can wear out the gates and lead to high SP losses, high noise, and material accumulation in the duct.

The plenum system consists of a large main duct (i.e., a plenum) in which velocities are quite low and SP is relatively uniform throughout. Branches connect to the plenum, with each having about the same SP available to cause airflow. Branches may or may not utilize blast-gates. Some branches can be added, moved, and removed with relatively little overall change in system performance. Materials can be expected to accumulate in the plenum; indeed, the plenum is sometimes designed to be a primary dust collector. Fires and explosions can occur in accumulated material if it is combustible and becomes ignited.

AIR MOVING DEVICES

General Considerations

All mechanical ventilation systems (dilution exhaust, LEV, or replacement air) require some sort of device to cause the desired air flow rate to move against the pressure losses inherent to the system. Fans comprise the vast majority of these devices and are of two basic types: axial and centrifugal (see Figure 5).

Fans can be manufactured from a variety of materials including steel, stainless steel, aluminum, fiberglass, and plastics. The choice of material depends upon the conditions under which a fan must operate. These may include highly corrosive contaminants (e.g., acids, caustics, organic solvents) and high temperatures (e.g., furnaces). It may also be important to avoid having fans that can be ignition sources for flammable air contaminants by preventing metal-to-metal contact (i.e., a spark) between fan blades and housing, and by keeping electric motors and wiring outside the duct.

Fans are selected for specific applications by deciding on the basic type best suited for an application, and by considering appropriate materials. There may be several manufacturers with fans meeting the basic criteria.

Fan manufacturers provide fan tables or fan charts for all of their products. These allow selection of a specific model and size of fan. The ventilation design process yields the desired Q (flow rate) and, for LEV, the associated system SP loss. Given these criteria, it is possible to determine the speed of rotation and horsepower required for a specific fan from the fan tables/charts. It is also possible to determine which fan will operate efficiently and quietly, and this will usually

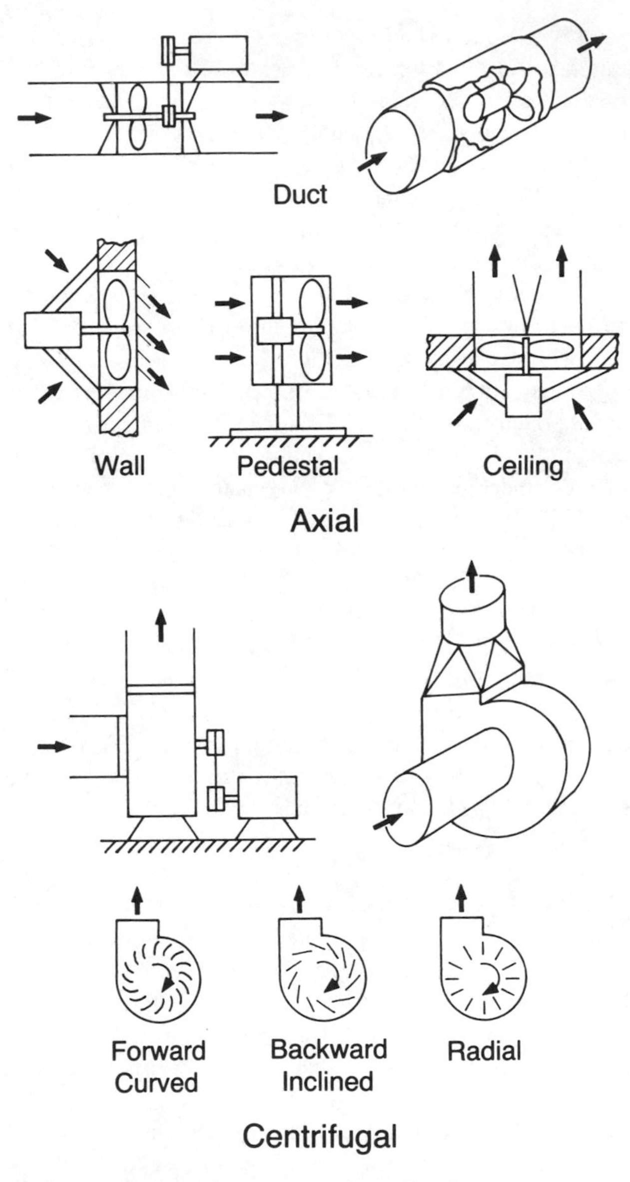

Figure 5. Two basic types of fans: axial and centrifugal.

be the fan of choice. Knowing the fan horsepower and speed, it is then possible to finalize the design by specifying the motor horsepower, rotational speed, and (if appropriate) pulley sizes for the fan and motor.

It is also possible to utilize fan tables/charts to evaluate existing fan performance. For example, Q could be estimated for a specific fan by measuring the fan speed, motor horsepower, and SP at the fan inlet.

Axial Fans

Axial fans are characterized by airflow entering and leaving along the axis of rotation (e.g., ceiling fans, pedestal fans). There are several basic types, differing in blade design. Disk blade fans are very common. They are used without duct and for clean air only. Propeller fans may be used with short sections of duct. They may be used with contaminated air and may be quite noisy. Vane-axial fans are the most energy-efficient and are typically used for clean (particulate-free) air.

Compared to centrifugal fans, axial fans are usually more compact (for a given flow rate), less expensive, and noisier. They generally perform well for moving large volumes of air against relatively small resistance. This makes them well suited for dilution ventilation.

Centrifugal Fans

Airflow enters a centrifugal fan along the axis of rotation and leaves perpendicular to the axis. The air accelerates through the rotating fan blades under centrifugal force. Centrifugal fans have several basic types of blade designs. Forward-curved blades operate well at low speed and against moderate pressure losses. Backward-inclined blades operate more efficiently and tend to be "nonoverloading" (a horsepower-limiting characteristic). Radial blades are "self-cleaning" and can work well for heavy dust loading.

Centrifugal fans are generally larger (for a given flow rate), more expensive, and quieter than axial fans. They can operate against significantly higher pressure losses and are much less sensitive to changes in SP. These characteristics make centrifugal fans the most common choice for operating LEV systems.

AIR CLEANING DEVICES

General Considerations

There are various reasons for removing air contaminants from exhaust airflow before discharging to the atmosphere. These include preventing pollution of outside air, compliance with air pollution standards, recovery of valuable materials (i.e., the "contaminant"), and maintaining good employee and public relations. In some situations, it may be necessary to clean and recirculate exhaust air to the workplace, rather than to discharge it into the atmosphere.

There are many ways by which contaminants can be removed from an airstream. Detailed discussion is beyond the scope of this chapter. It is usually necessary to meet with manufacturers of air cleaning equipment and carefully review the materials and operating conditions that apply for each situation.

It is useful to consider air cleaning in two principal categories, based upon the type of contaminants: particulates and gases/vapors. These are not necessarily exclusive categories; some contaminants are mixtures of these, and some air cleaning devices can remove both types of contaminants with variable effectiveness. It should also be understood that collected contaminants may be hazardous to personnel and the environment.

Removing Particulates

Particulates (e.g., dusts and mists) may be removed in settling chambers by slowing airflow and allowing larger particulates to fall under the influence of gravity. Baffles may be used to force abrupt changes in airflow direction, causing primarily larger particles to impact on the surfaces and be removed from the airstream. Some collectors, commonly called cyclones, cause contaminated air to spin around in circular paths. The centrifugal force resulting from this motion causes outward movement of particles. Again, the larger sizes are most affected, and particles impact on the walls of the collector and are removed from the airstream.

Smaller particles, which may be of greater significance for health concerns (i.e., are more "respirable" than larger particles), typically require other collection methods. Mechanical filtration is a primary means. There are many different types of filtration devices. All utilize some sort of fabric to trap the particulates and remove them from the air. Filter media may be disposable or reusable. If reusable, there must be some method, either manual or automatic, to remove the collected material and thereby recondition the filter media for further use.

Fine particles may be removed by electrostatic precipitators. These devices generate electrical charges on the particles. The charged particles migrate within electrical fields between collection plates (+ and - charge) where they remain until the collection plates are reconditioned (cleaned).

Particles can also be removed by introducing water droplets to the airstream. The water droplets attach to the particles, making them larger and more massive, and enabling collection by various means. These types of air cleaners are called wet scrubbers.

Removing Gases and Vapors

Wet scrubbers can also be used to remove airborne gases and vapors. Water can be used if the contaminants are water soluble. Other liquids can be used if they are better solvents for certain contaminants. It may be desirable to utilize chemical reaction to facilitate collection, such as collecting and neutralizing acid mist in a basic solution. Wet scrubbers may utilize spray nozzles, baffles/trays, beds packed with ceramic pellets, or other means of facilitating contact between the airstream and the collection liquid.

Contaminant gases may also be absorbed into solid chemicals or adsorbed onto solid surfaces. It is sometimes possible to remove the contaminant from the collection media and reuse the media.

Gases and vapors can be eliminated by incineration. This method is often used for substances of high toxicity and/or highly unpleasant odors. Incineration may be effective for contaminants of unknown composition and for emergency releases.

REPLACEMENT AIR VENTILATION

General Considerations

The replacement air flow rate in a workplace is always equal to the exhaust flow rate, unless significant amounts of air are generated or consumed. If replacement airflow is restricted, or nonexistent, so is the exhaust airflow. A dilution exhaust fan or an LEV system will fail to remove the design air flow rate if provisions are not made to replace the same flow rate.

One of the most common reasons for poor performance from contaminant control ventilation is limited replacement air. Replacement air is usually fresh air and it may have to be conditioned for comfort in the workplace. It is important to verify that replacement air is clean and uncontaminated. It may also be important to minimize exhaust flow rate in order to minimize the potentially high cost of regulating the temperature and humidity of replacement air.

A lack of replacement air can cause a variety of other problems. These include strong crossdrafts, improper operation of combustion stacks (possibly causing combustion products to enter the workplace), temperature extremes (e.g., overheated and underheated zones), and pressure differentials on doors (a possible safety hazard).

Recirculation of Exhaust Airflow

It is possible to recirculate exhaust airflow as a means of reducing the cost of replacement air. Recirculation of exhaust airflow is not a simple process. The systems must have the means for removing the contaminant and safeguards against failure of the air cleaning system. Safeguard measures include automatic contaminant monitoring systems, back-up air cleaning systems, alarms, and emergency bypass mechanisms to discharge contaminated air and intake fresh air. Recirculation of exhaust airflow is potentially hazardous and should be avoided whenever possible.

Recirculation is appropriate only when economic reasons are very strong, e.g., it would not otherwise be possible to have LEV. Air cleaners for recirculating systems do not make air "fresh" (i.e., replace oxygen, remove carbon dioxide); therefore, it is always necessary to have some fresh air brought to the workplace.

Recirculation systems require strict maintenance procedures. Recirculation should never be used for substances that have high acute (i.e., rapid) toxicity or are otherwise highly hazardous (e.g., carcinogens).

Ventilation Philosophy

Considerable amounts of air enter and leave through the ventilation systems of industrial facilities. It is advisable to make the best possible use of the airflow. Consideration should be given to what, for lack of a better term, may be called a ventilation philosophy.

Air handling units serving office areas typically move very little (if any) hazardous chemical contaminants, but may move relatively large flow rates to assure fresh air and to meet building codes. It is wasteful to discharge this relatively clean air, if it could adequately meet replacement air requirements for production or warehousing areas. It is similarly wasteful, if alternatives exist, to condition fresh air to serve solely as replacement air for LEV systems in production areas.

To whatever extent possible, the overall flow of air through a facility should focus on introducing fresh conditioned air where it will serve the greatest number of people with minimum contamination. Provisions should be made to direct this air to locations that require conditioned air, and finally to serve as replacement air and be removed by exhaust systems. This is a simple concept, often difficult to implement, but worth consideration in ventilation design.

VENTILATION TESTING

Ventilation systems should be tested routinely. This may be done after installation, on a periodic basis, and/or after changes in manufacturing processes and procedures or modification of ventilation systems. Testing is done to evaluate performance characteristics, such as airflow patterns, velocities (capture, transport, others), pressures (SP, VP, TP) and flow rates (hood, branch, fan). A variety of devices can be used.

The most useful devices for evaluating ventilation controls are those that generate a visible "smoke." These cause a particulate cloud (a noncombustive "smoke") to be generated in concentrations high enough to be visible. The "smoke" particles are fine enough to remain airborne and to move with the air. Small quantities (puffs, streams) of smoke can be made using small tubes (i.e., smoke tubes) with a hand-held squeeze bulb. Air movement into exhaust hoods, through doorways, and at points in a room can be observed in this way. Smoke "bombs" can be used to generate copious quantities of smoke in order to observe general airflow patterns and rates of clearance for dilution ventilation.

Pressures can be measured using fluid columns (i.e., manometers), mechanical pressure gauges, or electronic pressure transducers. Pressure measurements can be made at various locations near hoods, branch entries, and fan inlets and outlets of LEV systems. These data can be very helpful in establishing system performance criteria and diagnosing LEV system problems. SP measurements made in rooms relative to the outside (i.e., atmospheric pressure) or to other rooms can be useful for both LEV and dilution ventilation.

Velocities can be measured by several devices. A pitot tube is an L-shaped device used to measure relatively high velocities (over 1000 fpm), such as found in duct. It is inserted into the duct connected to a pressure gauge (e.g., manometer), and measures VP as the difference between SP and TP (these can also be measured separately). A pitot tube cannot normally be used to measure low velocities (below 500 fpm).

Lower velocities, such as for room air currents or hood face velocities, are measured with instruments called anemometers; they are also capable of measuring higher velocities. The most common instrument is a thermal anemometer, which utilizes a wand-like probe. The tip of the probe is heated by electricity from a battery; airflow exerts a cooling effect that alters the electrical signal from the probe, and this signal can be calibrated (e.g., against a pitot tube) to read out in feet per minute on analog or digital scales. Some probe tips are relatively sensitive to damage. A swinging-vane anemometer is a mechanical instrument comparable to a thermal anemometer, with some significant differences. All anemometers should have their calibration checked after heavy usage and on a regular (e.g., annual) basis.

Volume flow rate (Q) can be estimated by multiplying airflow cross-sectional areas (e.g., hood openings, duct, doorways) by airflow velocities. Velocities frequently are not distributed uniformly. It is usually necessary to make several velocity measurements, in equal-area subdivisions of the airflow cross-section, and to use the average of these measurements for calculating Q.

VENTILATION STANDARDS AND GUIDELINES

Some of the standards enforced by the Occupational Health and Safety Administration (OSHA) include specifications for ventilation controls. These apply for a relatively limited number of operations found in industry, including (but not limited to) abrasive blasting, grinding/buffing/polishing, open surface tanks, and welding/cutting/brazing. Details of these are beyond the scope of this chapter. Most of the provisions pertaining to these operations may be found in the Code of Federal Regulations 1910.94 and 1910.252.

The OSHA standards for ventilation have the primary purposes of protecting health and reducing the possibility of fire. OSHA has adopted performance standards, called permissible exposure levels (PELs), for several hundred airborne contaminants. Although not mentioned specifically, ventilation is understood to be a primary means of meeting these acceptable limits. If the PELs are exceeded, a citation may be issued and settlement of the citation will require a written plan and action to reduce airborne contaminant levels. A specific settlement agreement may include ventilation as an engineering control measure.

OSHA will not typically cite for failure to ventilate unless there are specific standards that apply, such as for the operations mentioned above. OSHA has established rather detailed standards for a relatively small number of chemicals, some of which (e.g., asbestos) have specific provisions for ventilation that could be a basis for violation. Provisions for fire protection include operations such as welding, spray finishing, storage of flammable/combustible liquids, and the use of such liquids for some types of operations (e.g., unit operations, bulk plants, service stations).

Many of the current OSHA standards were adapted in the early 1970s from what had been guidelines offered by consensus organizations such as the American National Standard Institute (ANSI) and the National Fire Protection Association (NFPA). These and other organizations still offer guidelines that may be useful in ventilation design.

Perhaps the most useful guidelines for LEV are offered by the ACGIH. They publish a book that represents the combined experience of many practitioners.[5] Unique to this text is the inclusion of conceptual drawings of LEV hoods for a wide range of operations. It provides useful design data and serves as an "idea book."

Guidelines pertaining to dilution ventilation for offices and other buildings are offered by the American Society of Heating, Refrigeration, and Air-Conditioning Engineers (ASHRAE). These recommend, for different types of building occupancies, minimum amounts of fresh (replacement) air that must be provided, usually on a volume flow rate (cubic feet per minute) per person basis. These guidelines are mainly intended to prevent odors, assure adequate oxygen, and control pollutants from tobacco smoking. These guidelines can take on the importance of standards if they are incorporated into municipal building codes.

OSHA standards are important because they are the law. As they pertain to certain situations, they should be known and complied with; however, compliance does not assure employee protection, even for the limited number of situations that address ventilation specifically. It is always necessary to apply good industrial hygiene practice and to design ventilation as needed to control airflow and protect employees.

CONCLUSION

Industrial ventilation, as applied for occupational health, strives to control the movement of air and airborne contaminants in order to protect people in their workplaces. It is a form of engineering control, not necessarily preferable to other measures that might be more effective in reducing a hazardous condition. When feasible, ventilation is preferable to administrative controls and/or personal protective equipment (PPE) controls, that rely upon the implementation of procedures to reduce risk, without changing the hazardous condition, and are subject to human error or failure to act.

Ventilation is an important tool for use in occupational health. Health protection, like many tasks, often requires more than one tool to get the job done. This chapter has briefly discussed a variety of topics to introduce ventilation for contaminant control. Much more can be learned from the cited references.

REFERENCES

1. Alden, J.L., and J.M. Kane. *Design of Industrial Ventilation Systems,* 5th ed. (New York: Industrial Press, 1982).
2. Burgess, W.A., M.J. Ellenbecker, and R.D. Treitman. *Ventilation for Control of the Work Environment* (New York: John Wiley & Sons, 1989).
3. Burton, D.J. *Industrial Ventilation Workbook* (Salt Lake City: DJBA, Inc., 1989).
4. *Controlling Airborne Contaminants in the Workplace* (Leeds, U.K.: British Occupational Hygiene Society, Technology Committee, 1987).
5. *Industrial Ventilation — A Manual of Recommended Practice*, 20th ed. (Cincinnati OH: American Conference of Industrial Hygienists, Committee on Industrial Ventilation, 1988).
6. McDermott, H.J. *Handbook of Ventilation for Contaminant Control* (Boston: Butterworths, 1985).

CHAPTER 9

Heat Stress

David R. Fulton

INTRODUCTION

The human body has only limited capacity to adjust to extremes of temperature and humidity. When these limits are exceeded, heat exhaustion, collapse, and heat stroke can occur. As heat stress progresses, mental functions become impaired and the worker can no longer be the best judge of his/her own condition.

The semiconscious firefighter who has his/her turn-out gear hastily thrown open and is being cooled with water after fighting an August brush fire; the construction worker who has sharp, painful leg cramps after 8 hr in the sun; and the asbestos removal contractor who collapses after several hours of work in the hot, humid environment of an asbestos abatement project all have something in common: **heat stress.**

Heat exhaustion or prostration, is also caused by salt depletion and is characterized by excessive sweating, but cold, pale, and clammy skin. Dizziness, blurred vision, and unconsciousness may accompany a rapid but weak pulse. Nausea and rapid, shallow panting may also be present.

Heat collapse can occur due to strain on the circulatory system as the skin blood vessels dilate to bring more blood from the core to the surface of the skin and as working muscles demand more blood. The increased vascular volume results in less blood being available to fully fill the heart between beats. In order to compensate for the reduced volume pumped by each beat, the heart rate must increase to maintain the same cardiac output.[1] Thus, as deep core temperature rises, so must the work done by the circulatory system. Oxygen deficit results when the demands for oxygen exceed the capacity to provide it. Dizziness, pale and sweaty skin, and headache occur. The onset of heat collapse is aggravated by dehydration since blood volume is further reduced when the body is dehydrated.

Heat stroke is the most serious heat disorder. It is a medical emergency requiring immediate attention. When a person works physically hard in a hot, humid environment, the body will attempt to discard heat through a number of mechanisms. If the duration of the exposure, the intensity of the work load, or environmental factors such as clothing and humidity result in more heat being stored than can be given off, the person's deep body core temperature will rise rapidly. Brain functions become impaired and the body's thermoregulation may break down entirely. Sweating ceases and deep core temperatures continue to rise. Symptoms of heat stroke are collapse, flushed face, hot and dry skin, and noisy breathing. There can be a loss of consciousness, convulsions, nausea, and vomiting. The pulse will be initially rapid and strong but may become weak in later stages of the illness. Deep body core temperature will increase to 40°C or higher. Death or permanent physical damage can result from heat stroke.

All heat-induced illness is preventable, however, through proper combinations of task modification, engineering controls, and the use of personal protective equipment. Heat strain from yard work or playing tennis under the sun is just as serious a problem as heat strain during work. It is the responsibility of everyone involved in thermally hazardous situations to gain the knowledge to evaluate the potential for and take steps to avoid heat stress.

The following pages give a brief review of the body's response to heat and the techniques used to evaluate the potential for heat stress.

BODY CORE TEMPERATURE

The concept of a body "core" temperature refers to blood and tissues that are located deep enough not to be affected by a temperature gradient through the surface tissues. However, temperature gradients do exist within the body core as the vascular blood flow varies. A single unique "core" temperature does not exist and therefore cannot be measured as such. The average temperature of the thermally isolated body core can, however, be inferred from measurements at various body points. These measurements can either represent the mean temperature of the body mass or the temperature of the arterial blood supply to the brain and hence to the hypothalmic thermoregulation centers. The temperature of the arterial blood irrigating the thermoregulatory centers in the hypothalamus is considered to be the true indicator of "core" temperature.

Unfortunately, the temperature of hypothalamic blood supply cannot be measured directly, except by surgical insertion of a catheter containing a thermistor. Since this practice is clearly unsuited for monitoring in an industrial setting, surrogate monitoring must be considered. Rectal, oral, aural, and skin temperatures are relatively easy to obtain and provide an indication of the body core temperature. The advantages and limitations of each measuring site are explained in the discussion of physiological monitoring of heat stress.

THERMOREGULATION

The maintenance of a deep body core temperature between about 36 to 38°C is essential for continued optimal functioning of the body. The human thermoregulatory system attempts to maintain a heat balance between the heat generated by the body, the heat lost or gained by means of convection and radiation, and the heat lost through evaporation of body fluids.

The hypothalamus, located at the base of the brain, is considered to be the primary control center for thermoregulation. The anterior hypothalamus is thought to measure the various physiological signals indicative of the thermal status. The anterior hypothalamus integrates the information sent from thermal receptors in the muscles, skin, and internal organs with information regarding the temperature of each arterial blood supply. The anterior hypothalamus is also the body's "thermostat," responding to this integrated thermal status, with reference to a deep body temperature "set-point" established by the posterior hypothalamus.

The posterior hypothalamus set-point is normally within the range from 36 to 38°C. Triggered by the anterior hypothalamus "thermostat," the posterior hypothalamus also initiates physiologic responses to generate or discard body heat in an attempt to maintain the deep body core temperature within the normal range. The blood vessels of the skin can dilate, increasing the blood flow to the skin when heat loss is required. If the heat loss is not sufficient to return the deep core temperature to the set-point, sweating is induced to increase heat removal through evaporative cooling of the skin. The skin cooling thus reduces the skin blood temperature, thus increasing the temperature differential between the body core and the skin surface. Since heat transfer is a function of the temperature difference between two materials, the increased core to skin temperature differential enhances the heat transfer from deep core to the skin layers. This improved heat loss is in addition to evaporative cooling of the skin tissue.

If heat gain or heat conservation is required to preserve the body core temperature within the normal range, the blood vessels of the skin can become constricted. The decreased blood flow to the skin surface reduces the body's heat loss. If heat conservation is not sufficient to return the deep core temperature to the set-point, shivering is induced to increase heat production in the muscles.

The body's control of deep core temperature can be described by a mathematical model that can be used to predict when heat stress might occur.

The metabolic rate is the rate at which energy is made available for use by the body. It is the rate at which energy is produced through enzymatically controlled exothermic (heat generating) oxidation of carbohydrates, fats, and proteins. About 70 to 90% of the energy produced is heat, with the balance applied to perform work. Since the quantity of heat produced within the body contributes to the overall heat balance, it is essential to be able to determine the metabolic rate, if direct measurement of the core temperature is not used to evaluate the presence of heat stress.

The metabolic rate can be determined by measurement of the worker's oxygen consumption or by estimation from reference tables of activity vs metabolic rate. The former technique relies on measurement of the worker's consumption of oxygen and the approximate equivalent of 5 kcal of energy produced per liter of oxygen consumed. Typical oxygen consumptions range from 0.22 L/min (66 kcal/hr) while sleeping to >3 L/min (900 kcal/hr) while marching double time. Such high levels can be maintained for only a few minutes. While this technique yields some of the most accurate metabolic rate measurements, it is not considered to be practical for use in a work environment.

Usually, metabolic rates are estimated from timed observations of activities and subsequent correlation with tables of the activities' metabolic requirements. The physical tasks being performed; the type and distance of reaches; whether objects are lifted, pushed, or pulled; the number of body bends and squats; and the body posture during task performance all have an impact on the metabolic requirements of the task. Ideally, the worker would be videotaped from several angles simultaneously, objects weighed, distances and heights measured, and the metabolic requirements of the task subsequently calculated; however, gross approximations of metabolic rate usually are made based on the predominant features of the task being observed. A basal metabolic rate of 60 kcal/hr is added to the task metabolic rate to arrive at the total metabolic requirements of the worker. Light metabolic rates (up to 200 kcal/hr) include such activities as sitting, light hand and arm work, drill press operation while standing, and casual walking at <2 MPH. Moderate metabolic rates (200 to 350 kcal/hr) include such activities as sustained hand-arm work driving nails, work with pneumatic hammers, pushing a lightweight wheelbarrow, and walking at 2 to 3.5 MPH. Heavy metabolic rates (350 to 500 kcal/hr) include such activities as chipping castings, sawing wood, sledgehammer work, intense arm and trunk work, and walking at >3.5 MPH.

Accurate estimation of the metabolic rate is critical to heat stress management if wet bulb globe temperature (WBGT) -based exposure control is used; however, if insufficient care is taken in this determination, the resultant metabolic rates will, at best, be accurate to only ±30 to 50%. In that case, correct classification of the task between light and moderate metabolic rates or between moderate and heavy metabolic rate becomes ambiguous. WBGT is discussed below.

HEAT BALANCE MODEL

The body heat balance equation is

$$S = M - W - R - C - E - C_{RES} - E_{RES} - K \qquad (1)$$

where

S	=	heat storage rate
M	=	metabolic rate
W	=	external work
R	=	heat loss by radiation from body
C	=	heat loss by convection from body
E	=	heat loss by evaporation of body fluids
C_{RES}	=	dry heat loss by respiration
E_{RES}	=	latent heat loss by respiration
K	=	heat loss by conduction

The units of Equation (1) are usually British thermal units per hour or kilocalories per hour.

The heat losses in respiration, the external work, and the heat loss by conduction to surrounding surfaces are all negligible compared to contributions from the other parameters and usually can be ignored in heat balance considerations. This assumption results in a conservative approach, i.e., erring on the side of safety for the worker. Therefore, if the heat content of the body is to remain at a constant level, the gain of heat from metabolic heat production and both radiant and convection heat transfer to the body must be offset by an equal loss of heat through evaporation and loss through both radiation and convection.

At thermoregulatory balance :

$$M \pm R \pm C = E \quad (2)$$

In other words, the combined metabolic, radiant, and convection heat loads determine the evaporative heat exchange rate required to maintain thermal balance. The body core temperature will rise when the evaporation of sweat is not enough to maintain heat balance. The heat balance may then be reestablished at a higher body core temperature. If the thermoregulatory capacity is exceeded, the body core temperature may increase uncontrollably, resulting in heat illness and possibly death.

The heat lost through evaporation of sweat, E, is limited either by the maximum sweat rate (about 0.14 g/sec) or by the relationship between air velocity and the water pressure difference between the ambient air and the perspiring skin given by:[1]

$$E = 14V_a^{0.6}(p_{sk} - p_a) \quad (3)$$

where:

E	=	evaporative hear exchange rate, kcal/h
V_a	=	air velocity, m/sec
p_a	=	water vapor pressure of ambient air, mmHg
p_{sk}	=	water vapor pressure on the skin, mmHg

At the assumed mean weighted skin temperature of 35°C the water vapor pressure of the sweat in the skin is equal to 42 mmHg.

MEASUREMENT OF HEAT STRESS BY WET BULB GLOBE TEMPERATURE (WBGT)

The WBGT combines the effect of humidity and air velocity (natural wet bulb temperature), ambient air temperature, velocity and radiant energy (globe temperature) and air temperature (dry bulb temperature) into a single index. The method for determining the WBGT is briefly described in the ACGIH publication, "Documentation of the Threshold Limit Values and Biological Exposure Indices," for the heat stress exposure standard[2]. The method is also detailed in the International Organization for Standardization standard number ISO 7243-1982 entitled,"Hot Environments — Estimation of the Heat Stress on Working Man, Based on the WBGT-Index (Wet Bulb Globe Temperature)[3].

Briefly, WBGT is defined by either of the following equations:

Outdoors with solar load:

$$\text{WBGT} = 0.7\ \text{NWB} + 0.2\ \text{GT} + 0.1\ \text{DB} \tag{4}$$

Indoors or outdoors with no solar load:

$$\text{WBGT} = 0.7\ \text{NWB} + 0.3\ \text{GT} \tag{5}$$

where
- WBGT = wet bulb globe temperature index, °C
- NWB = natural wet bulb temperature, °C
- GT = globe temperature, °C
- DB = dry bulb temperature, °C

Environmental measurements are made and the worker's metabolic work rate is estimated from tables of task work rates. The calculated WBGT index can then be compared to permissible heat exposure criteria given in units of the WBGT index for the applicable level of workload.

The often-quoted WBGT index is actually an environmental measurement from which the potential for heat stress is inferred. The true indicators of heat stress are physiological in nature. This is not to say that the WBGT is not useful information; on the contrary, without WBGT information regarding the source and intensity of the ambient environmental heat sources, industrial hygienists would not be able to direct their efforts to minimize the hazard.

ENVIRONMENTAL VS PHYSIOLOGICAL MEASUREMENTS

The use of environmental measurements to predict and manage heat stress is complicated by the difference in integration periods between the WBGT monitor and the body. While the modern WBGT instruments have temperature and humidity condition integration periods of about 5 min, the body takes 10 to 60 min to achieve equilibrium in a constant environment, depending upon the clothing worn, the air velocity, and the sources of the heat. When the worker is constantly in motion in a thermally nonhomogeneous environment, it is highly likely that the WBGT time weighted average (TWA) will at best approximate the average integrated heat stress. At worst, the WBGT could seriously underestimate the heat stress exposure.

WBGT measurements become relatively meaningless to workers, such as firefighters, chemical plant workers, nuclear power plant workers, asbestos abatement crews, pesticide applicators and foundry workers, who routinely wear vapor impermeable or semipermeable protective equipment. This is because the microenvironment within their protective equipment often can have a greater impact on their thermal balance than the macroenvironment in which they are working. Such protective clothing prevents heat loss by evaporation of perspiration. WBGT instrumentation does not take into account the physical condition of the individual as measured by parameters such as water and electrolyte balance or cardiovascular heat transport capability. It cannot account for the benefits or the liabilities imposed by protective clothing. Clearly physiological indices of heat stress would provide more valid information for managing heat stress.

PHYSIOLOGICAL MONITORING OF HEAT STRESS

The choice of monitoring site for the body temperature determines that interpretation will come from the information gathered, with respect to whether the person is exhibiting heat strain. These measurements can either represent the mean temperature of the body mass or the temperature of the arterial blood supply to the brain, and hence to the hypothalmic thermoregulation centers.

A variety of techniques exist for monitoring deep body core temperature but only a few are practical for use in an industrial setting. Some techniques such as esophageal thermometric measurements and/or determinations of urine temperatures are best left in the laboratory. A brief review of the more practical methods for measuring body core temperatures follows.

Rectal Temperature

This method belongs in the laboratory simply because user acceptance is universally low. The method is mentioned only because comparisons of temperatures determined rectally vs the other methods described in this section are inevitable. Many people hold the belief that rectal temperatures are equal to core temperatures and that rectal temperatures are not influenced by external factors. While it is true that the rectal mass is relatively insulated from the external temperature by virtue of its sheer mass, this area contains sufficient blood vessels influenced by heat generation in the leg muscles. As a result of warm blood returning from the leg muscles, when the majority of the work is performed by the leg muscles, rectal temperatures will rise before the core temperature. Conversely, when the majority of the work load is performed by the upper body, the rectal temperature will lag behind the core temperature.

Rectal temperatures are indicative of the mean temperature of the body mass for a subject at rest, however. When the body is subjected to heat stress, either through intense metabolic work rates or high ambient thermal loads, the rectal temperature rises slower than the core temperature. The rectal temperature continues to rise after exposure has stopped as warm pooled blood continues to transfer heat to the rectal mass. When the core temperature begins to decrease, the rectal temperature drops more slowly than the core temperature due to the low thermal conductivity of the rectal mass. This lead/lag problem means that (much to the industrial workers' relief) rectal temperatures are inappropriate to estimate heat stress.

Oral Temperature

The National Institutes for Occupational Safety and Health (NIOSH) concluded that physiologic monitoring, consisting of recovery heart rate and/or oral temperature, could help protect all workers, including the heat-intolerant worker at hot worksites.[1] Recovery heart rate is the heart rate measured at a prescribed time interval, usually 3 min after all physical work ceases.

Oral temperatures are determined by measurement of the tissue temperature at the base of the tongue. These tissues are in close proximity to the arterial branches of the lingual artery. This supplies a close measurement of the hypothalmic arterial blood temperature provided certain precautions are met. The worker must not have consumed food or beverages within the previous 15 min, the ambient temperature must be greater than 18°C, and the measurement must not be taken before the thermometer sensor has been in position for at least 5 min at ambient temperatures above 30°C and for 8 min below 30°C. Thermocouples and thermistors will have a shorter equilibrium period. Subjects shall, of course, keep their mouths closed during this entire time and breath only through the nose. Radiant heat and high air velocity conditions must be avoided during the measurement. The oral temperature measured differs from the hypothalmic blood temperature by about –0.4°C, depending upon the precautions taken during the measurement.

The development of accurate, low cost digital thermometers makes oral temperature measurements attractive for evaluating heat stress; however, continuous monitoring of the oral temperature is not possible for measuring occupational heat stress because work disruptions are inevitable.

Skin Temperature

The temperature of the skin is not the same at all locations, therefore, the temperature at a single point on the skin cannot be representative of the core temperature. However, the mean skin temperature over the entire skin surface can be determined by weighting a group of skin temperatures according to the areas that they represent. In the absence of asymmetrical radiant energy sources, the following weighting scheme has been used with reasonably accurate results:

Location	Weighting Factor
Back of neck	0.28 × skin temperature
Center of right scapula	0.28 × skin temperature
Back of left hand	0.16 × skin temperature
Front of right shin	0.28 × skin temperature

The mean skin temperature, t_{sk}, is the sum of the individual products of the localized skin temperatures times the weighting factor for that area of the body. In asymmetrical radiant energy fields, the number of test points must be increased to 8 or 12 in order to obtain an accurate mean skin temperature.

While this information is essential for evaluating the thermal exchange between the skin and the ambient environment, its usefulness for evaluating the thermal status of the body is limited and further diminished when the localized cooling of the skin through sweating is considered. When sweating occurs, the skin is cooled to a lower temperature, resulting in enhanced thermal transfer from the core. The temperature differential between the core and the skin increases with the metabolic rate. For example, if evaporative and convective heat transfer are not impeded by clothing, and if conditions of low air velocity and high humidity are not present, the core temperature is equal to T_c:

$$T_c = 36.7 + 0.004\ M\ °C \tag{6}$$

where M equals the metabolic heat production rate, in Watts. The mean skin temperature is lower than the core temperature by about 3.3 + 0.006 M°C when evaporation is not impeded. Thus, with increasing metabolic rates due to the performance of work, the core temperature rises by 0.004 M°C and the mean skin temperature drops by 0.006 M°C. The differential between the core and the mean skin temperature increases by approximately 1°C/100 W (86 kcal).[1] At a moderate work rate of 350 kcal/hr the core vs skin temperature differential is almost 4°C more than at rest. Any measurement system that depended upon a fixed correction factor to infer core temperature would provide grossly low core temperature results as metabolic rates increased. Therefore, skin temperature measurements suffer not only the problem of measuring the mean value, but also an uncertainty due to being a function of the metabolic and perspiration rates.

Aural (Ear) Temperature

A branch of the internal carotid artery supplies blood to the hypothalamus where thermoregulation occurs. The external carotid artery supplies blood to the tympanic membrane and adjacent auditory canal; thus, their temperature is affected by the core temperature and by variations in blood flow around the ear. Provided that the temperature measurement of the external auditory canal is made close to the tympanic membrane, and that the auditory canal is isolated thermally from exchange with the external environment, this site offers a reasonably precise measurement of the core temperature. Aural temperature measurements have been shown to closely parallel those made by esophageal probes, known to accurately reflect the temperature of the blood leaving the heart; thus, aural temperatures are considered to be representative of the blood supply to the hypothalamus.

The differential between core and aural temperatures varies among individuals as a function of the distance the temperature sensor is located from the tympanic membrane. Typical differentials of 0.5 to 2.5°C are observed. A fixed depth temperature probe enables reproducible reinsertion depth of the temperature sensor within the auditory canal and thus reproducible temperature differentials can be achieved. An ear canal insert, usually urethane foam or wax, provides a comfortable centering device that thermally isolates the small volume of air within the auditory canal from the external environment. Thus, the impact of the external environment upon the temperature differential between the core and aural measurements is minimized. The oral temperature can be used to calibrate the core vs aural differential since the oral temperature is a reflection of the core temperature, as previously discussed.

Aural temperature measurements have the advantage of being free from the shifting differentials experienced with skin measurement techniques because sweating and evaporation of sweat does not occur in the isolated external auditory canal. Aural techniques are also considered nonintrusive among industrial workers. Best of all, aural temperature measurement requires no special skills training. Workers who can insert a foam ear plug can be shown how to calibrate an aural thermometer in minutes.

Heat Stress Physiological Standards

The maximum permissible core temperature is limited by the requirement that there must be at least 1°C difference between the core and mean skin temperatures in order to effectively transfer heat energy out of the core.[1] Below this temperature differential only 1 to 2 kcal can be transferred per liter of blood flowing from the core to the skin. The body thus cannot eliminate heat when a core to skin temperature differential of at least 1°C cannot be maintained.

If skin temperature has risen because evaporative cooling is impaired or is simply insufficient to meet the body's requirements, heat transfer from the core will decrease and the core temperature will increase rapidly. When the mean skin temperature is 38°C and the core temperature is 39.2°C the risk of heat exhaustion or collapse is about 25%.[1] The risk of heat collapse at this stage increases dramatically with as small as 0.2°C increases in the core temperature. This is because the rate of heat transfer from the core and the elimination of heat by sweat evaporative cooling are already insufficient. At this point the body's compensation mechanisms have already been overwhelmed and no spare capacity is available.

The World Health Organization (WHO) has recommended that rectal body temperatures should not be permitted to exceed 38°C in prolonged daily work, but that in closely controlled conditions the core temperature could be permitted to rise to 39°C.[1] The ACGIH recommended a physiological exposure limit, stating that even acclimatized persons should cease work when deep body core temperatures exceed 38°C. [4]

The consensus is that the risk of heat stress increases when core temperature is elevated above 38°C for prolonged periods. That is not to say that everyone whose core temperature rises to 38°C or even 39°C will suffer heat strain; however, it recognizes that the region where thermoregulation is lost is being approached. There are some individuals who naturally have core temperatures close to 38°C in whom there is no evidence of disease or illness. Their thermoregulation set-point is simply a little higher, however, all workers should be certified by a physician as capable of withstanding prolonged core temperatures above 38°C before permitting such an exposure. The choice of core temperature permissible in the workplace will depend somewhat on the extent of the existing heat stress program. Closely monitored workers might be permitted by their medical advisors to use a core temperature criteria level up to 38.5°C.

CONTROLLING HEAT STRESS

Heat stress and strain are preventable by the use of administrative and engineering controls, supplemented by the wearing of appropriate personal protective clothing and the consumption of sufficient fluid to maintain water balance.

Administrative controls involve the management of the workload and duration to maintain the heat load at the point at which the heat lost through sweat evaporation can balance the combined metabolic, radiant, and convection heat loads. The metabolic heat load is further managed by controlling the worker's work-rest regimen. If workers are provided with core temperature monitoring instrumentation, they can regulate their own work-rest regimen without the need for complex WBGT measurements or estimates of metabolic rates, in order to maintain a core temperature below 38°C. Otherwise, the ACGIH threshold limit values (TLVs) for heat stress can provide guidance for managing the work-rest regimen. Having determined the WBGT and estimated the worker's metabolic rate, the following table is used to determine the permissible heat exposure TLVs. Note that the work-rest regimens are ratios not to be exceeded in each 1-hr period. The work-rest regimens are not daily average values and assume that the workers are dressed in light summer-weight clothing, are fully acclimatized, and have adequate supplies of water and electrolytes.

Permissible Heat Exposure Threshold Limit Values

(Values are given in °C WBGT)

	Work Load		
Work-Rest Regimen[a]	**Light**	**Moderate**	**Heavy**
Continuous work	30.0	26.7	25.0
75% Work/25% rest	30.6	28.0	25.9
50% Work/50% rest	31.4	29.4	27.9
25% Work/75% rest	32.2	31.1	30.0

[a] Each hour.

Reproduced with permission of the American Conference of Governmental Industrial Hygienists.

Engineering controls must be chosen with care. Materials that absorb heat make poor shields since they will heat up and re-radiate the heat from the "cool" side to the skin. The exception is heat exchanging thermal shield designed to transfer the absorbed heat to a liquid (water) for transport away from the workplace. Effective radiant heat shields can be made from reflective surfaces such as aluminum sheeting; however, as the surface becomes soiled or tarnished, the effectiveness of the reflective surface diminishes. Cooler and drier air can be blown, or less effectively, drawn, into the workplace to reduce the ambient temperature and humidity level. The cooler air will reduce the workplace heat load. Since the worker's evaporative cooling rate is partly dependent upon the vapor pressure of water in the workplace air, bringing in drier air will enhance the potential for evaporative cooling of the body. Increasing the workplace air velocity can cool the worker provided that a number of conditions are met. If the air is at a temperature above the worker's skin temperature (average 35°C), the air will actually transfer some heat energy to the worker; however, if the worker is not dehydrated so that sufficient sweat can secreted, and if the air is not saturated with water vapor, the heat gain to the worker through convection could still be lower than the heat lost through evaporation. If these conditions are not met, the increased air velocity could result in a net heat gain by the worker.

Often personal protective clothing is necessary to avoid heat stress. Reflective garments of aluminized materials, such as the fireproof Nomex (tradename of E. I. duPont de Nemours, Wilmington, DE), can reflect as much as 90% of the radiant heat; however, these types of garments provide little protection from conductive and convective heat loads. Heat transfer from high ambient temperatures can be reduced by wearing thermal insulating garments. Unfortunately, these protective garments are necessarily bulky and heavy, which adds to the worker's metabolic load. Another approach is to provide cooling for the worker, either by the use of coolant vests or compressed air-driven vortex cooling vests or suits. Any protective clothing that inhibits sweat evaporation will reduce heat loss from the body.

Cooling vests fitted with sealed packages of ice or gelled coolants can provide several hours of effective upper torso cooling. When selecting a cooling vest an employer should look for simplicity of design and some means to adjust the vest to closely fit to the worker's body. These devices work poorly when not in snug contact with the worker's outer clothing or skin. A recent improvement is that some vest manufacturers have started to offer their products in Nomex rather than flammable cotton.

Vortex-cooled coveralls are probably the best means of cooling a worker, provided that unrestrained mobility is not required. Vortex-cooled garments require a source of 20 to 30 ft^3/min of dry air at 50 to 125 psi. This means that the wearer is tied to an air line. A caution regarding the use of vortex-cooled garments: The cool, dry air enhances the removal of sweat through sustained evaporation. Since the worker can sweat and not develop a film of perspiration on their body, there is no visible evidence of the loss of moisture. In extreme cases, dehydration of the worker can occur without the worker's awareness. It is especially important to periodically have the worker consume water when using air or vortex-cooled garments.

Water balance is important in preventing heat stress injury. Water should be located close to the workplace and kept reasonably cool (10 to 15°C). The worker should be encouraged to frequently drink small quantities (100 to 150 cm^3) throughout the work period at 20 to 30 min intervals.

An acclimatized worker has an increased tolerance for heat, evidenced by a higher core temperature without accompanying heat stress symptoms. Their perspiration production is greater and its electrolyte concentration is less than that of an unacclimatized worker. Acclimatization is acquired by working in a hot environment for periods of 2 hr/day for approximately 1 to 3 weeks. Acclimatization is effectively lost 5 to 7 days after ceasing work in a hot environment, but is regained within 2 to 3 days after returning to work.

REFERENCES

1. Criteria for a Recommended Standard. Occupational Exposure To Hot Environments, Revised Criteria 1986, National Institute for Occupational Safety and Health (1986).
2. Documentation of the Threshold Limit Values and Biological Exposure Indices, American Conference of Governmental Industrial Hygienists (1986).
3. Hot Environments — Estimation of the Heat Stress on Working Man, Based on the WBGT-Index (Wet Bulb Globe Temperature), ISO 7243-1982, International Organization for Standardization (1982).
4. Threshold Limit Values and Biological Indices for 1988-1989, American Conference of Governmental Industrial Hygienists (1988).

CHAPTER 10

Industrial Plant Noise Abatement

Joseph McGuire

BASIC NOISE CONTROL CONCEPTS AND TERMS

There are a number of words and concepts that must be understood before beginning a discussion of noise control methods. Sound is produced when a sound source vibrates the air nearest to it in wave motion. This motion spreads to air particles surrounding the sound source, and the sound travels in air at a speed of about 1129 ft/sec. This rate of travel is greater in liquids and solids: 4925 ft/sec in water and 16,417 ft/sec in steel. The velocity of a sound wave therefore depends upon the medium through which it travels, and the frequency of a sound wave refers to the number of vibrations per second, measured in Hertz (Hz) or cycles per second (cps).

The human ear responds to sound waves in a frequency range from as low as 16 to as high as 20,000 Hz, with a great deal of individual variation. In general, perception of high frequencies is best in early childhood, with a gradual decrease in acuity throughout life. A normal adult may have difficulty hearing sounds pitched higher than 10,000 or 12,000 Hz. Anyone who hears over a range from 20 to 20,000 Hz has an unusually broad frequency response. Anyone who can hear tones in the range of 20 to 2000 Hz has a capacity for hearing that is adequate for speech communication.

The frequency of a sound wave has no relationship to the energy of the wave. The magnitude of the pressure variation constituting a sound wave provides a measure of the strength or intensity of the sound. Wavelengths in air at 70°F for

Table 1. Wavelength in Feet for Specific Frequencies

Frequency (Hz)	63	125	250	500	1000	2000	4000	8000
Wavelength (ft)	17.9	9.0	4.5	2.3	1.1	0.56	0.28	0.14

different frequencies are found in Table 1. In general, the longer the wavelength or lower the frequency, the more difficult the attenuation by any medium.

Sounds that one person may consider pleasant, another individual may consider to be noise. Unusual sounds coming from the engine of an automobile are usually defined as noise, because they are undesired sounds signaling trouble and expense. In the occupational health field, noises are defined as those sounds with an intensity of sufficient magnitude to cause biological damage, especially to the auditory organs.

The ear reacts to sound waves, and the auditory nerve carries the sound signals to the brain. The brain interprets the sound as either pleasant or unpleasant with the primary determinants being the frequency of the sound wave and the energy associated with the sound.

Sound Pressure Level

For practical noise measurement reasons, the decibel is employed to measure sound intensity. Decibels are expressed on a logarithmic scale in dimensionless units employed to express the ratio of a measured quantity and a reference quantity. The formula for expressing a sound pressure level (SPL) is as follows, where p is the measured root mean square (rms) sound pressure, and p_{ref} is the reference sound pressure:

$$\begin{aligned} SPL &= 10 \log_{10} (p^2/p^2_{ref}) \\ &= 20 \log_{10} (p/p_{ref}) \end{aligned} \quad (1)$$

The standard reference pressure of 20 $\mu N/m^2$ (or 0.0000004 lb/ft^2) is the parameter employed by nearly all acoustical instrumentation. The measurement unit employed is the decibel (dB). SPL may be specified for different frequency bands, such as the 500 or 2000 Hz octave band.

Sound Power Level

Sound power (W) is the rate of acoustic energy flow; the unit is the acoustic watt. For a sound source, W is a measure of acoustic output. Sound power level (PWL or L_w) is defined as:

$$PWL = 10 \log_{10} (W/W_{ref}) \quad (2)$$

The standard reference power W_{ref} is 10^{-12} Watts. The measurement unit is the

Table 2. Types of Noise

Types of Noise
Steady-state
Broadband
Steady noise with pure tones
Nonsteady and fluctuating
Intermittent
Impulsive

decibel, just as for SPL. In any given situation, the relationship between the SPL at a point and the PWL of the sound source present is such that

$$SPL = PWL + K \tag{3}$$

where K is a constant dependent upon geometry and other aspects of the situation. An important relationship between W and PWL is that every doubling of acoustic power increases the power level by 3 dB. Therefore, if an 85 dB noise source is doubled, the acoustical power increases to 88 dB, whereas, if the same noise source increases by a factor of 10, the total power level will increase by 10 dB, or to 95 dB.

Types of Noise

Noise may be classified into two major categories: steady and nonsteady. In addition, each of these categories may be subdivided further. The choosing of equipment and measurement procedures is dependent on the type of noise. Although they are not mutually exclusive, six types of noise are shown in Table 2.

Steady-state noise maintains a relatively stable level during a given time period and can normally be classified simply by listening. Changes in levels of <1 dB are difficult to detect while changes in level of 6 dB or more are quite apparent. A-weighted measurements, at slow meter response, which display variations over a range of less than 6 dB, may be considered to be measurements of steady-state noise. Steady noise may be categorized into two types: broadband noise, in which there are no perceptible pure tones, and noise that contains apparent pure tones.

Broadband noise is noise that exhibits a relatively constant level over a period of time, contains no apparent discrete frequency components (tones, whines, etc.), and is the most straightforward type to measure in the field. For surveys, A-weighted sound level measurements with a slow meter response (SLM) are adequate for evaluations of employee exposure, speech interference, and noise impact on communities. Data obtained with an octave band or a one third octave band analyzer normally provide ample information to describe the spectral content of the noise source for control purposes.

Steady noise may contain one or more audible discrete frequencies or pure tone components, and is known as steady noise with pure tones. In such a case, these components may be characterized by such descriptive terms as "hum," "whine," or "screech," depending upon the frequency. It is usually desirable to ascertain the frequency and relative strength of pure tone components in order to evaluate human effects and noise abatement treatments.

To precisely measure the amplitude and frequency of pure tone components requires a narrowband frequency analyzer or more sophisticated instrumentation. Tape recording the noise for subsequent analysis in the laboratory is an expedient procedure for field work. A prominent pure tone is one whose level is 10 dB or more above the level at which the pure tone would just be audible in the presence of the other noise components.

Nonsteady and fluctuating noise may be classified as fluctuating, intermittent, or impulsive, depending on its temporal characteristics. Fluctuating noise is characterized by time-variant sound levels, the pattern of which may be related to meter needle movements.

Intermittent noises are generally created by sources that cycle on and off. To be classified as intermittent, a noise must have average on-times >1 sec. For on-times <1 sec, the noise should be measured by impulsive-type noise methods.

Impulsive noises may be categorized into two separate types: isolated bursts and quasi-steady noise. Isolated bursts are short bursts of noise that are separated by a time duration long enough to allow the level of one burst to decay sufficiently so that a subsequent burst is definitely distinguishable. Quasi-steady noise is comprised of a train of noise bursts with a burst repetition rate which prevents detailed resolution of individual bursts when measured by SLM.

Sound Directivity Factor

The directivity factor is a measure of the degree to which sound is concentrated in a certain direction rather than radiated equally in a full spherical pattern. Factors are actually portions of spherical radiation pattern related to the surface of a sphere. Where there is no reflection, the directivity factor is Q=1. For hemispherical radiation, or reflection from a single plane reflective surface, Q=2, and the noise level could increase by 3 dB. With the intersection of two effective surfaces the directivity factor would increase to 4 and noise levels could increase by 6 dB. Near a corner formed by three plane surfaces, Q increases to 8, and the noise level could increase by 9 dB. The directivity factor for typical radiation patterns is shown in Figure 1.

Sound Absorption and Reflection

When a sound is incident upon a surface, its energy is partially absorbed, partially transmitted, and partially reflected. The absorptive quality of the surface is described by its sound absorption coefficient (α), which is the ratio of energy absorbed and the energy incident upon the surface. For a perfect reflector, $\alpha = 0$; for a perfect absorber, $\alpha = 1$. Absorption coefficients are strongly dependent on the frequency of incident sound. It is therefore customary to specify absorption

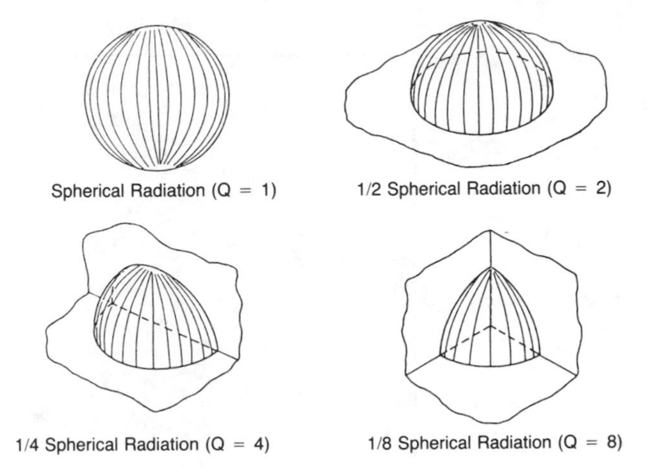

Figure 1. Directivity factors (Q) for typical noise radiation patterns.

coefficients for a given material at octave frequencies from 125 to 4000 Hz. Significant noise reductions can be achieved by room noise absorption. For example, highly reverberant areas, a room with hard, smooth, and impervious or "noise reflecting" surfaces could be changed to one with soft, rough, and porous or "sound absorbing" surfaces. The ability of the room to absorb sound is dependent on the ratio of sound energy absorbed to the sound energy incident upon the material and is influenced by the nature of the material, thickness, mounting, and the frequency of the sound. A theoretical noise reduction of an area should be determined before commencing this type of activity in order to decide if effective results can be met.

Sound Transmission Loss

There is an important distinction between sound absorption and sound isolation and the types of materials employed in each circumstance. Sound absorption is accomplished with materials that are usually porous and dissipate acoustic energy in the form of heat. Sound isolation is accomplished with materials that are poor acoustical energy transmitters. The two principles are usually employed concurrently. An important measure of the sound isolating capability of a panel is its transmission coefficient, which is the ratio of transmitted sound intensity to incident sound intensity. The noise transmission loss of a panel or partition is therefore related to the material's transmission coefficient. The noise transmission coefficient or loss for a partition is determined by its surface density, dynamic bending, stiffness, and internal damping. As the surface weight is doubled, noise transmission loss usually increases by 5 to 6 dB. The sound transmission class (STC) rating of a material is obtained by fitting a standard contour, similar to the A-weighting curve for sound level meters, to the plot of sound transmission losses

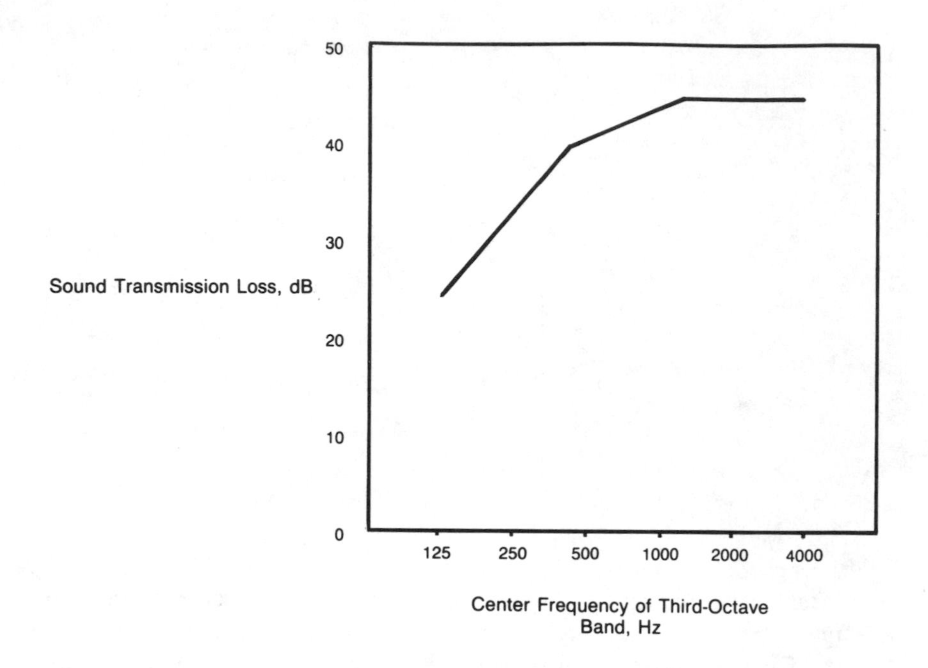

Figure 2. Typical STC contour (STC-40).

measured for a barrier over a speech frequency range of sounds. This STC rating sometimes involves complex calculations and estimates. An STC-40 contour curve is shown in Figure 2. The STC number 40 is the transmission loss in decibels at the specific octave band of 500 Hz point on the standard contour curve.

NOISE-MEASURING EQUIPMENT AND PRACTICES

Sound measurement falls into two broad categories: source measurement and ambient noise measurement. Source measurements involve the collection of acoustical data for the purpose of determining the characteristics of noise generated by a source. The source may be a single piece of equipment or a combination. Specific source measurements frequently are made in the presence of other, extraneous sources; the noise level created by the other sources is generally known as the background noise level. Although it is sometimes difficult to make a clear distinction between the two categories of measurement, it is important to understand that source measurements describe the characteristics of a particular sound source and ambient measurements describe the characteristics of a sound field due to unspecified, unknown, or ambient noise sources.

To acquire acoustical data by either source or ambient measurements, two different techniques, survey and field, may be employed. The survey technique requires the basic operation of a standard sound level meter to obtain A-weighted sound level data. Field measurements involve the collection of data that describe spectral and other specific characteristics of the noise; this often requires more sophisticated noise measuring instrumentation.

Standard Sound Level Meter

The standard sound level meter is the basic noise measuring instrument. It consists of a microphone, an amplifier with calibrated sound intensity control, and an indicating meter. It measures the rms sound pressure level in decibels proportional to intensity or sound energy flow. Basic standards for sound level meters specify performance characteristics that will require all conforming instruments to yield consistent readings under similar circumstances.

Three weighting networks, designated A, B, and C, are provided on standard noise level meters in an attempt to duplicate the response of the human ear to various sounds. These weighting networks cause the sensitivity of the meter to vary with frequency and the intensity of sound. The relative responses of the three networks are as follows. The A, B, and C weightings mock or imitate ear response to low, medium, and high intensity sounds, respectively. These A, B, and C meter response curves correspond to the 40, 70, and 100 phon equal loudness contours. For reference, the loudness level of a particular contour, in phons, is numerically equal to the standard pressure level of the 1000 Hz standard or base.

Acoustical Calibrator

The overall accuracy of all sound measuring equipment should invariably be checked by employing an acoustical calibrator. It consists of a small, stable sound source that fits over a microphone and generates a predetermined sound level within a fraction of a decibel. If the meter reading is found to vary from the known calibration level, the meter should be adjusted to eliminate this error.

Octave Band or Spectrum Analysis

When the sound to be measured is complex, consisting of a number of pure tones spread over many frequencies, it may be necessary to determine the sound pressure distribution according to each frequency, octave, or one third octave bands. Industrial noise is usually a mixture of complex sounds that combines low-frequency roars with high-frequency squeals. Employees react to these sounds in different ways, depending on the overall level and composition of the noise as a function of frequency. An octave or one third octave band analysis indicates how the sound energy is distributed over the audible range of frequencies and establishes which sounds should be reduced.

Figure 3 shows diverse straight-line octave band spectra.[1] Each line is a combination of various octave band frequencies and different pressure levels that

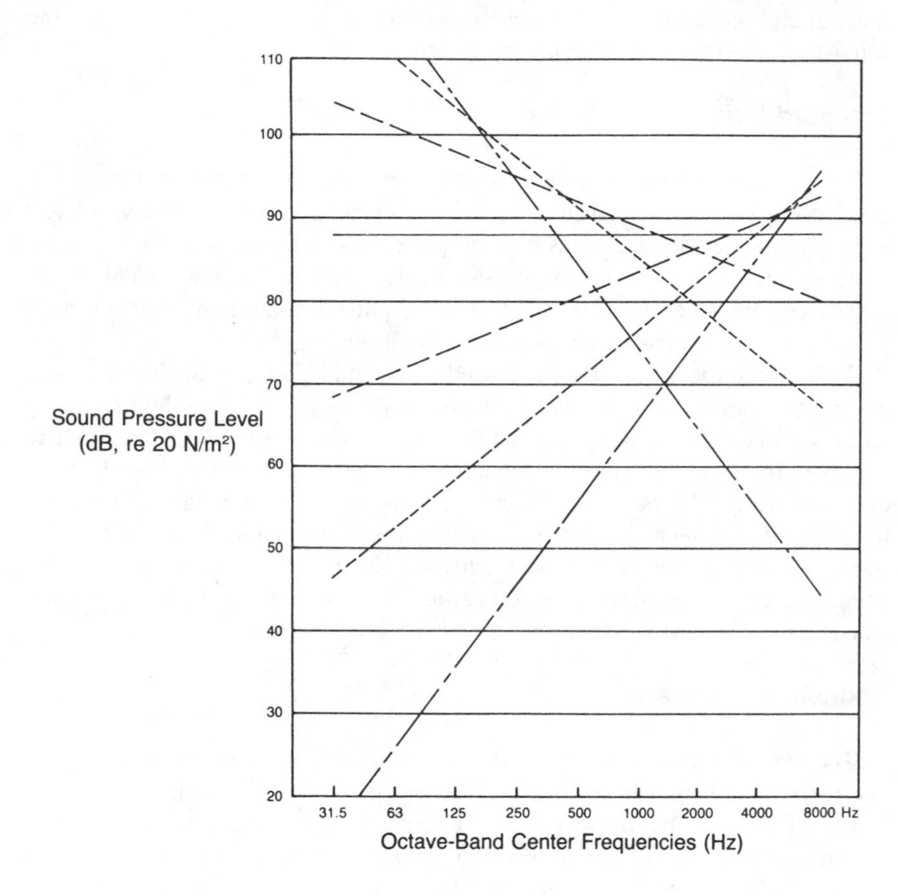

Figure 3. Diverse spectra having the same 95 dBA value.

results (strangely enough) in the same total sound pressure level of 95 dBA (decibels measured on the A scale). It is difficult to imagine that these various spectra can be assumed to be equal in auditory or nonauditory effects; however, the present Occupational Safety and Health Administration (OSHA) hearing conservation and noise regulations rely on this dBA equal damage criteria. The long-term adoption of this dBA criteria unfortunately discounts the utilization of octave band data that may be critical in field measurement, engineering noise prediction, and complicated noise problems where specific machinery or components must be singled out or analyzed. Even though there may be a great deal of dBA measurement data available, it may provide almost meaningless information in establishing noise abatement solutions and the reasons for some types of hearing loss.

Impulse Meter

The sound level meter is often too slow to indicate peak levels of transient noises lasting a fraction of a second, similar to punch press strokes. Such noises must be measured using a noise meter with a peak-hold feature or an impulse meter that indicates the peak level.

Statistical Analyzing Instruments

When sound levels vary erratically over a wide frequency range, it is difficult to describe the noise by standard meter readings; therefore, statistical noise analyzers have been developed to assist in this process, and many octave band or one third octave analyzers presently incorporate these analyzing features. These instruments indicate the percentage of time during which the sound level lies in certain predetermined ranges, determines mean levels, standard deviations, various statistical relationships, and provides various histogram-type printouts. Real-time analyzers can provide even more analytical sophistication with memory storage comparisons, transfer function analysis, and graphic relationships.

Vibration Measurements

Although vibration concepts or measurements are not discussed in this chapter, they have a significant relation to noise measurements. In vibration measurements, an accelerometer is employed to acquire vibrational data for determining the relative contribution of various vibrating surfaces to the overall noise field of interest. Output from the accelerometer or vibration attachment is usually connected to a sound level meter or noise analyzer via a control module. The module processes the pickup output signal voltage and relates it to the acceleration, velocity, or the displacement of the vibration being measured. An accelerometer for noise and vibration control analysis should have a uniform frequency response from about 10 to at least 10,000 Hz. Most new octave band noise analyzers can provide for the utilization of an accelerometer attachment.

Noise Dosimeters

Individual noise dosimeters that perform an integration of exposure in conformance with federal standards have been designed for noise exposure measurement, especially in circumstances where the employee visits several different noise environments on an unscheduled activity during the day. Whatever the activity, dosimeter microphones must be worn somewhere on the individual in designated parts of the ear or upper body.

Several models of dosimeters are on the market. The older instruments usually are designed to determine the employee's cumulative exposure dose or time above a preset sound level threshold, for example, total minutes above 85 dBA. Data such as these are obviously limited in usefulness. More recent instruments are usually designed to integrate noise exposure according to the 5 dB per doubling of time relationship implicit in current federal standards. Newer dosimeters allow for computer interfacing, memory storage of data, and statistical analysis of data with graphic and printed results.

INFORMATION REQUIRED PRIOR TO THE EQUIPMENT NOISE EVALUATION

A complete understanding of the nature of the noise situation is certainly one of the more important aspects influencing end results. Such information is essential in the selection of the equipment required and the employed noise abatement methods. Prior to commencement of actual noise abatement measurements, it is helpful to consider the question areas summarized in Table 3.

NOISE SOURCE MEASUREMENT PRACTICES

Source measurements are used primarily for acquiring acoustical data in support of noise control analyses. The location of the microphone and the number of positions required varies with source characteristics and the purpose of the measurements. The source may radiate sound uniformly in all directions or it may indicate distinct directional or multidirectional characteristics. For most noise sources in the chemical or other industries, the distribution of sound in the vicinity of the source is rather complex and significant directional and interference effects are usually present. Therefore, several microphone positions for each source measurement should be selected to provide adequate sampling of the noise field. Although various methods can be used for the selection of microphone positions, the approach illustrated in Figure 4 is recommended for plant equipment. An imaginary rectangular parallelepiped (box) around the source exterior provides reference planes for locating measuring points. It is good practice to have one vertical side of the parallelepiped to be parallel to one of the major vertical surfaces of the source. Microphone positions are selected at a distance of 3 ft or 1 m from the reference planes for most sources. For large sources, where the maximum levels may occur further away than 3 ft, a larger separation should be employed and specified in the recorded data. The elevation of the microphone above the floor or ground should be the dimension "h" shown in Figure 4, with

Table 3. Information Required Prior to the Equipment Noise Evaluation

Objective of the measurements
Hearing conservation
Community annoyance
Speech interference
Noise control
Equipment testing
Location and time of measurements
Environmental conditions
Temperature
Humidity
Wind speed and direction
Barometric pressure
Types of present noise sources
Noise source technical data
Name, manufacturer, and model number
Equipment function
Physical dimensions
Design performance or load
Operating performance or load
Rotational speed
Plant production rate
Noise source operation cycles
Operational periods
Operational times with respect to other equipment
Presence of workers during operation cycle
Noise characteristics
Broadband
Pure tone
Continuous
Nonsteady
Intermittent
Impulsive

a minimum height of 1 ft^2. Where measurements at this elevation are not practical, the microphone should be located at a height of 4.5 ft above floor level. If the noise source exhibits a relatively uniform directional pattern, measurements at the key measuring points are generally sufficient to describe source characteristics. The key measurement points should include a point opposite each major item in the equipment train. For highly directional noise sources, additional measuring points should be used. If significant background noise levels exist, background noise correction should be made as shown in Table 4.

Where the noise source under consideration is distant from interfering sources and obstructions, it is recommended that supplementary measurements be made at greater distances, such as 25 and 50 ft from the center of the source, in the direction of the significant points. These data provide a more accurate characterization of

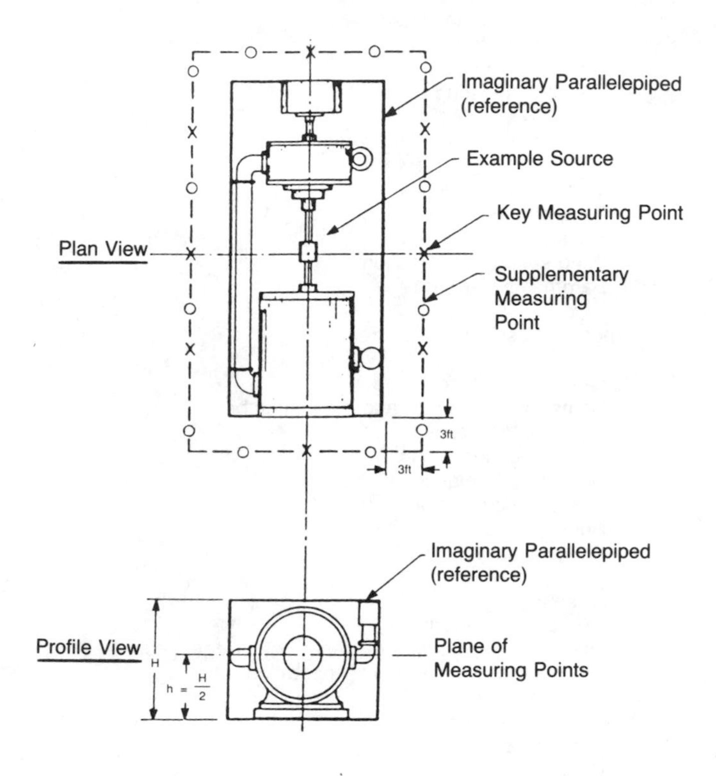

Figure 4. Location of noise measurement points. (Adapted from Guidelines on Noise, American Petroleum Institute Medical Research Report No. EA7301, prepared by Tracor, Inc., 1973.)

a noise source in terms of sound power level and SPLs. In most cases involving field measurements, other noise sources or objects will be located in the vicinity of the source being measured. Such situations are unavoidable and become a part of the measurement problem. Even when the sound field is not disturbed by other noise sources, objects, and reflections, sound reverberation from the ground can affect measurements, especially if the sound has strong pure tone content.

Point source contributions should be determined at all locations, and if possible, the machine or the noise source being tested should be shut down to determine ambient or the noise contributions by other sources. The on/off technique is always more reliable than any analytical or computer projection procedure.

Table 4. Background Noise Correction for Source Level Measurements

Difference between Source SPL and Background SPL (Source Not Operating) (dB)	Correction for Background Noise to be Subtracted From Source SPL (dB)
>9	0
6—9	1
4—5	2
3	3
<3	(Not reliable)

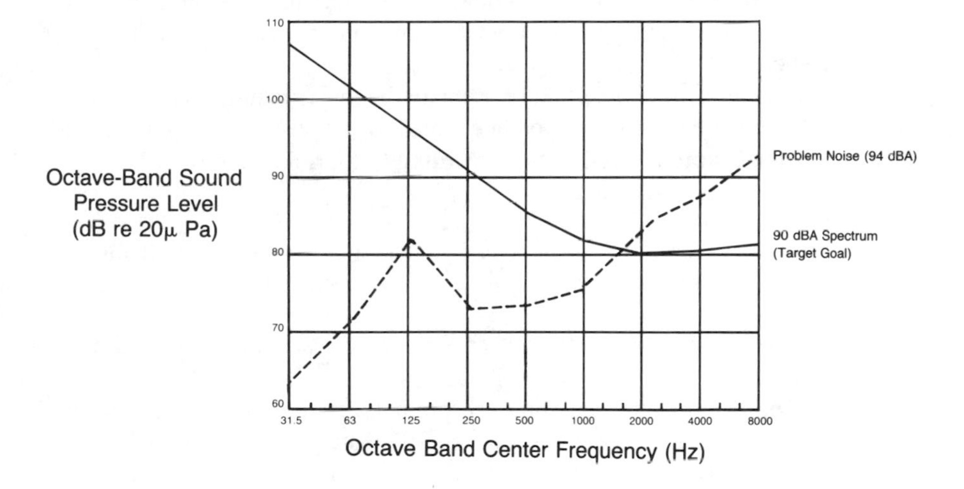

Figure 5. Determination of required noise reduction.

If source measurements are conducted with a background noise that is not at least 10 dB below the source level at each frequency of interest, corrections should be made to the observed sound pressure level or sound level readings in accordance with Table 4. It is, therefore, useful to apply a frequency or octave-band analysis to the measurement of existing noise conditions. The usefulness of frequency analysis in evaluating the severity of a noise problem is apparent when the frequency of characteristics of a significant noise source can be determined.

Figure 5 illustrates a particular target goal spectrum often employed for noise problems in order to reach a 90 dBA sound level and a plot of a noise problem with a sound level of 94 dBA is shown. It should be noted that the target goal line is exceeded only in the 2000, 4000, and 8000 Hz octave bands. If the octave band

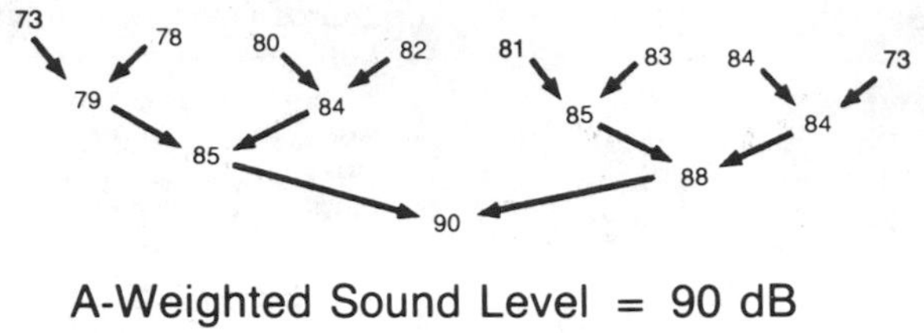

73 78 80 82 81 83 84 73

79 84 85 84

85 88

90

A-Weighted Sound Level = 90 dB

Figure 6. Noise source addition.

levels at these three frequencies were reduced to levels that fall below the target spectrum, there would be assurance that the total noise levels for the total spectrum would be below 90 dBA. The advantage of this approach is that there is no need to concentrate on lower frequency noise sources; the concentration should be made on dealing with the easier and more economical to attenuate higher frequency noise sources. This type of target spectrum automatically pinpoints problem frequencies.

To perform even the simplest noise problem diagnosis requires the process of adding individual noise sources to obtain a total noise level. The weighted sound levels are added successively in pairs according to the following rule:

If the decibel difference in sound levels is	0 or 1	2 or 3	4 to 9	>10 dB
Add to the higher level	3	2	1	0 dB

The process of addition is carried out until a single resultant is obtained as in Figure 6.

Another basic principle that should be remembered is related to the SPL formula and involves the loss of noise energy with distance. SPLs beyond the near field and traveling without disturbance drop in accordance with the inverse square law and equation, or with a 6-dB drop with the doubling of the distance from the noise source relative to the initial point of measurement. Or, in other words, the SPL diminishes as the square of a spherical distance, r, from the point source, since the sound disperses over the entire area, $4\pi r^2$, of the theoretical sphere. This means that the SPL is reduced by 6 dB with each doubling of the radius, r, of the sphere. Even at closer distances or in the near field the attenuation is not significantly different than the 6 dB rule. This rule can, of course, be modified to lower values if the noise path is disturbed in its transmission path and/or a researcher wishes to be conservative in projections.

NOISE CONTROL ENGINEERING PRINCIPLES

The primary means of compliance with all federal standards relates to the specification that noise control should initially be approached from the engineer-

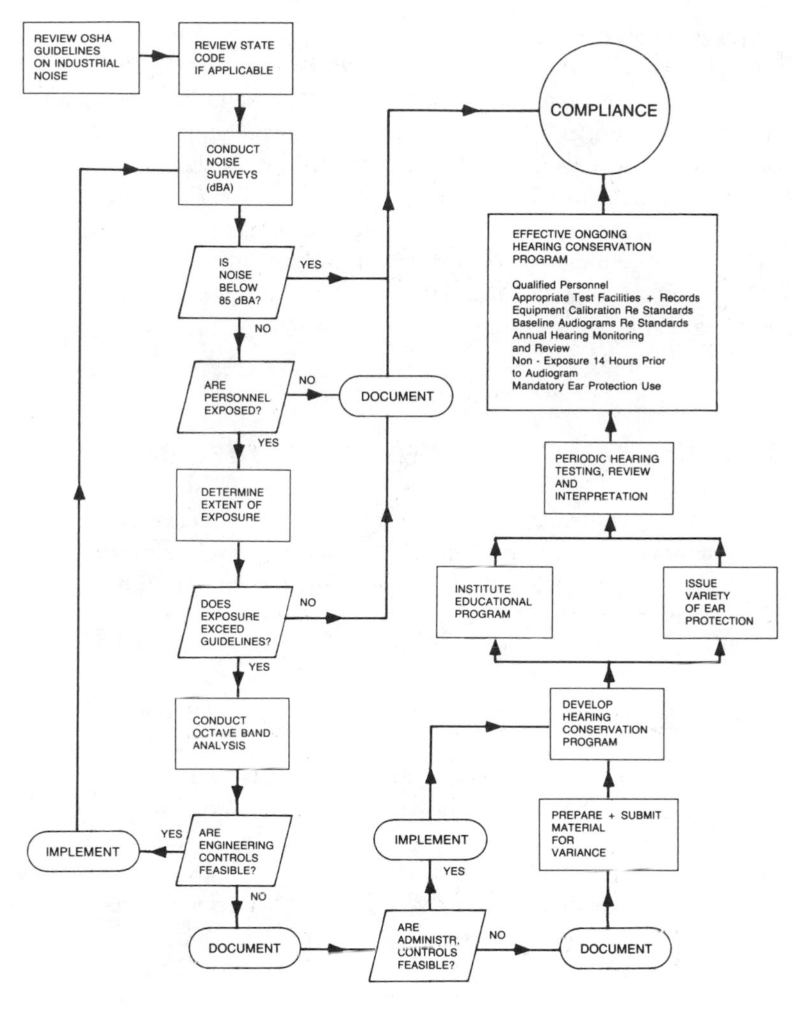

Figure 7. Industrial hearing conservation and noise abatement program flow chart. (From Henderson, D. et al. *Effects of Noise on Hearing,* New York: Raven Press, 1976. With permission.)

ing standpoint. Simply put, when engineering technology is feasible, it should be employed. The concept of "economic feasibility" with regard to the noise standard is often disputed in industry and the employment of hearing protection has unfortunately advanced as a more cost effective option, rather than as a temporary measure. Hearing conservation programs are an integral part of the occupational noise exposure regulations; however, as shown by the flow chart in Figure 7,[3] when engineering controls "are feasible" they should be implemented to reduce worker noise exposure, even if it is projected that the OSHA regulations or guidelines cannot be met. If engineering controls still result in potentially damaging noise exposure, this material must be documented and the feasibility of administrative controls such as work schedule modifications and plant operations must be reviewed and, if possible, implemented. Whether engineering and administrative controls are feasible, these facts must be documented in order for a company to prepare an application for a variance to the OSHA regulations.

Noise control can often be designed into equipment so that there are no plant or process design goal compromises. Careful plant and equipment acoustical design will result in a quieter, more economical plant. The substitution of quieter machines, the employment of vibration-isolation mountings, operational speed reductions, and the elimination of structural vibration paths that by-pass intended barriers, are only a few of the great number of noise control measures that should be considered. Table 5 lists examples of engineering principles that can be applied to reduce noise levels.[4]

SPECIFIC NOISE ABATEMENT DESIGNS AND TECHNIQUES

The selection of a particular noise reduction design is dictated by the characteristics of the specific noise problem. Care should be exercised to adopt the most effective technique for each problem. In this section are descriptions of noise control methods frequently applied to problems in process plants. A list of comments is also provided to enhance the noise abatement knowledge.

Enclosures

Noise from certain sources may be most effectively reduced by acoustically enclosing the source. Sound transmission loss characteristics of a composite partition composed of panels of different materials is a function of the total percentage of area occupied by each material and its transmission loss factor. Transmission loss characteristics of various commonly used materials for enclosures are available in most noise control handbooks.

Leaks or open areas in the enclosure create significant reduction in the overall transmission loss of a panel, as illustrated, in Figure 8. For example, if an enclosure with a 60-dB expected noise transmission loss or noise reduction has an opening that represents 1.0% of the total area, the effective net noise transmission loss is reduced from 60 to 15 dB. The values shown in Figure 8 are only an indication of loss of effectiveness. To contain equipment requiring cooling air, an enclosure must have openings for air intake and exhaust. Such openings must be acoustically treated, as shown in the example of Figure 9.[2]

For maintenance purposes disassembly of an enclosure should be given careful design consideration. Lift-out panels, panels mounted on overhead tracks, or merely large, well-positioned enclosure doors with chain hoist accessibility often solve equipment removal and maintenance problems.

Quite often, thinner panels may be employed to absorb high frequency noises; however, for frequencies below 250 Hz, 4-in. thickness panels with a heavy type material septum is recommended. Panels should also be vibration isolation

Table 5. Noise Control Engineering Principles

1. Plant equipment layout and planning
2. Maintenance
   Replacement or adjustment or worn, loose, or unbalanced parts of machines
   Lubrication of machine parts
   Use of properly shaped and sharpened cutting tools
3. Substitution of machines
   Quieter equipment, processes, materials
   Larger, slower machines
   Presses for hammers
   Hydraulic presses for mechanical presses
   Belt drives for gears
4. Substitution of processes
   Compression riveting for impact riveting
   Welding for riveting
   Hot working for cold working
   Pressing for rolling or forging
5. Driving force of vibrating surfaces may be reduced by
   Reducing the forces
   Rotational speed
   Isolating
6. Response of vibrating surfaces may be reduced by
   Damping
   Adding support
   Increasing the stiffness and mass structure or bracing material
   Increasing vibrating member mass
   Changing resonance frequency
7. Sound radiation from the vibrating surfaces can be reduced by
   Reducing the radiating area
   Reducing overall size
   Perforating surfaces
   Reduce the sound transmission through solids by using
   Flexible mountings
   Flexible sections in pipe runs
   Flexible-shaft couplings
   Fabric sections in ducts
   Resilient flooring
   Coupling guard damping and isolating
8. Reduce the sound produced by gas flow by
   Intake and exhaust mufflers
   Fan blades designed to reduce turbulence
   Large, low-speed fans instead of smaller, high-speed fans
   Reducing the velocity of air flow
   Increasing the stream flow area
   Reducing the pressure
   Reducing air turbulence
9. Reduce noise transmission through air by
   Employing sound absorptive material on walls and ceilings
   Employing sound barriers and sound absorption along transmission paths
   Complete enclosure of individual machines
   Employing baffles
   Confining high-noise machines to insulated rooms
10. Isolate the operator by providing a soundproof booth
11. Employ source directionality

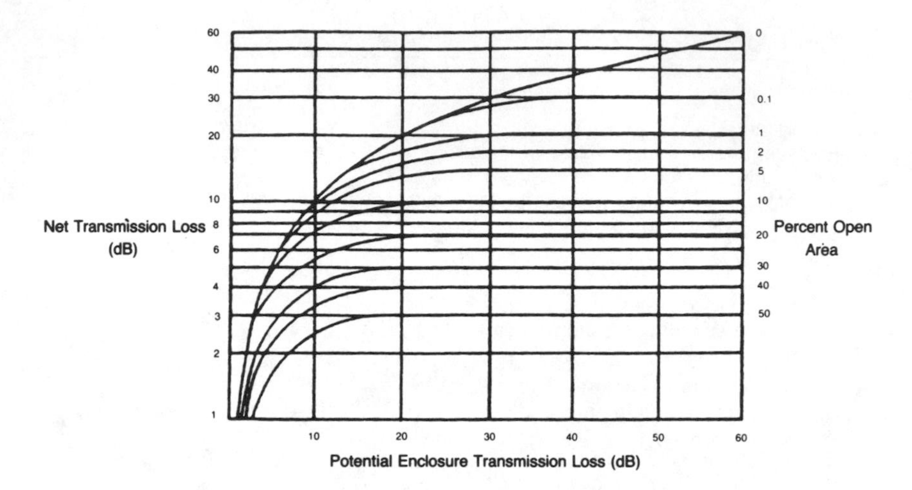

Figure 8. Effects of open areas on potential noise reduction of an enclosure.

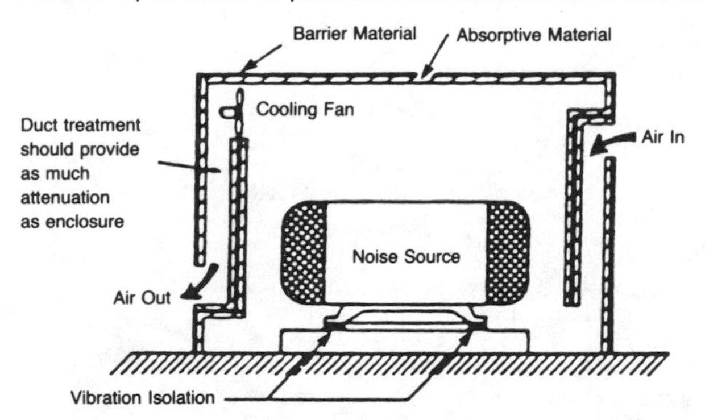

Figure 9. Concept design of enclosure for a noise source requiring cooling air. (Adapted from Guidelines on Noise, American Petroleum Institute Medical Research Report No. EA7301, prepared by Tracor, Inc., 1973.)

mounted to avoid the transmission of low frequency vibration to the floor or attachment structures.

Room Treatments

For equipment housed in buildings or rooms, consideration should be given to the noise reverberation reduction through the addition of absorptive treatments to the room, particularly if it has hard, acoustically reflective surfaces. The choice of an effective treatment is dictated primarily by the available area and noise absorptivity of room surfaces.

Room treatment will not be significant in the proximity of the noise source; however, it can be beneficial in reducing total area noise levels, especially rooms containing several noise sources. In most cases, the resulting noise reduction is usually not >3 dB.

Quite often in industry, area noise reverberations are increased by 3 to 4 dB merely by painting the cement block walls with a heavy, pore sealing paint, which significantly decreases the noise absorption of the walls. If area cosmetics are desired, color-tinted cement blocks and water-based nonpore sealing paint are available.

Reactive and Dissipative Silencers

Silencers or mufflers can be considered simply as a duct or pipe that has been acoustically treated or shaped specifically to reduce noise transmission in the contained medium. The noise may originate from a machine source or it may be flow-generated. Sources include blowdowns to atmosphere, draft fans, vacuum pumps, pelletizers, chillers, blowers, compressors, piping systems, pressure reduction valves, turbines, reciprocating engines, and other equipment. The characteristics of gas flow noise vary widely; therefore, a thorough analysis of the spectral content of gas flow noise is an important first step in the choice and application of a silencing mechanism.

There are two basic types of silencers, reactive and dissipative. These are categorized by the manner in which noise reduction is achieved. The acoustical properties of a reactive silencer are governed primarily by its internal configuration and the reduction of flow velocity by providing an expansion chamber. Reactive silencers are designed to take advantage of sound reflections from abrupt changes in shape and resonances of added branches or cavities to a pipe or duct. These reactive mechanisms obstruct the acoustical passage by impedance mismatch of acoustic energy flow within the pipe or duct. Reactive silencers are most effective at low frequency and limited spectral bandwidth applications.

The second basic type of silencer for noise is the dissipative silencer. Its acoustical properties are governed primarily by the presence of sound-absorbing material that dissipates acoustic energy. Materials such as rock wool, fiberglass, and felt, when deployed within a duct, form a dissipative silencer. Maximum absorption of such materials usually occurs at the higher frequencies, yet dissipative silencers usually have a relatively wideband noise reduction capability. A potentially undesirable feature of this type of silencer is that if improperly designed for a given service, bits of the absorptive material may be drawn into the gas stream. This not only eventually degrades silencer performance but also may endanger downstream equipment. Thus, care should be exercised to ensure that silencer design is compatible with service requirements.

The distinction between reactive and dissipative silencers is conceptual. In practice, all silencers achieve some noise reduction by both reactive and dissipative means. Certain silencers, however, are designed as combination reactive-dissipative devices for a specialized application.

In all muffler employment considerations, the back pressure increase in the duct must be considered and its effect on the total plant and equipment design specifications.

Piping Noise Silencers

Pulsating flow created by the intake and discharge of reciprocating compressors and pumps is a frequent and serious source of noise and vibration. Devices called "snubbers" or "pulse traps" are used to buffer this pulsating flow by providing both an expansion chamber and dissipative elements in the associated piping system. The performance of such devices is a function of the system in which they are installed; therefore, an analysis of a complete piping system should be performed to ensure operational compatibility and acoustical performance.

In-line silencers are useful in reducing gas stream noise within the body of the silencer and thus reducing the level of downstream noise propagation. These silencers are available in a wide variety of designs. Figure 10 presents a general description of this type of silencer and Table 6 lists the primary design characteristics of an in-line silencer.[2]

Acoustical Lagging

Acoustical lagging of noise sources, principally pipelines, valves, and ducts, consists of encapsulating the source with treatments that provide a combination of sound barrier and sound absorption mechanisms to obtain the maximum noise reduction. The optimum lagging design is dependent upon the spectral content of the radiated noise as well as the level of noise reduction required. All designs should avoid any mechanical coupling between pipe or duct surface and outer shell treatment and should have a layer of resilient absorptive material between the pipe, duct, or valve surface and the outer shell treatment. Increased thickness of material or a heavier composite material is required for low frequency performance. For thin-shelled pipes or ducts, a vibration damping material with adhesive backing should be directly applied on the surface. Figure 11 illustrates typical acoustical lagging designs.[2] For a 20-dB reduction, two 2-in. layers of 6 lb/ft^3 absorptive material with at least a 1.0 lb/ft^3 septum interposed between layers is an average requirement.

Vibration Isolation and Damping

The application of vibration isolation and damping techniques are generally required to silence structure-borne noise. A relatively small vibrating machine, pipe, or other mechanism, when closely coupled to a floor or panel and then radiates the vibration acoustically, can often produce objectionable noise levels. Structure-borne noise refers to the transmission through structures of mechanical vibrations that produce airborne noise when a panel or other structure is set into motion and radiates sound.

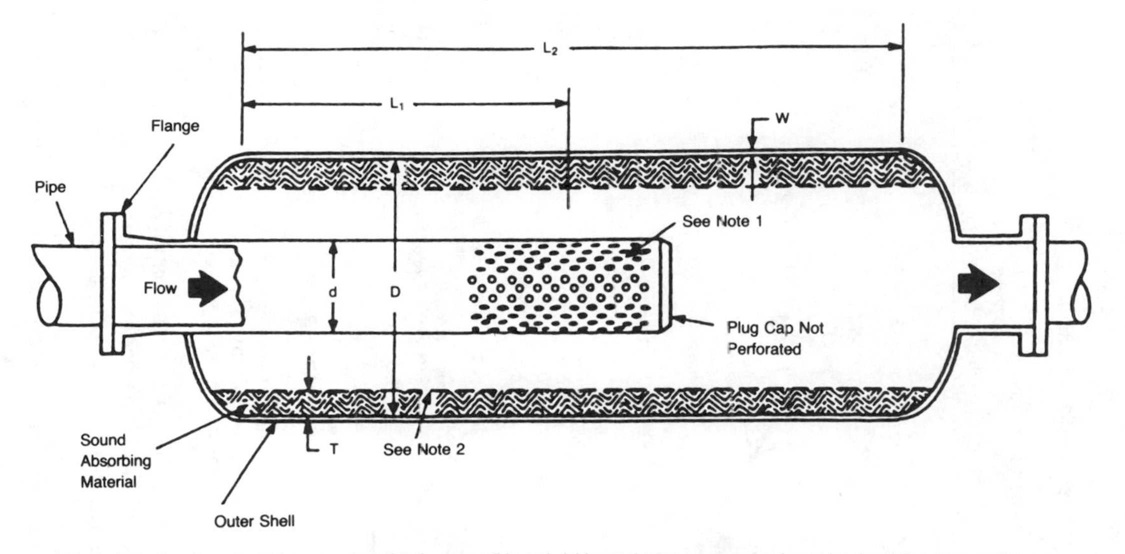

Note 1: Hole diameters should be spaced as closely as possible and yield a total open area greater than twice the piping cross-section area.

Note 2: Sixteen (16) gauge (or heavier) noncorrosive steel with at least 50% perforated open area and backed by 16 × 16 or 18 × 14 noncorrosive wire mesh.

Figure 10. General design features of an In-line silencer. (Adapted from Guidelines on Noise, American Petroleum Institute Medical Research Report No. EA7301, prepared by Tracor, Inc., 1973.)

Table 6. Primary Design Characteristics of an In-Line Silencer

1. Sound absorption — probably the single most important characteristic; the absorptive material should cover as much area as practical
2. Limitation of turbulence re-excitation as the gas leaves the silencer — can be achieved by avoiding sharp edges or area restrictions
3. Suppression or avoidance of resonances
4. Wide-band attenuation
5. Construction suitable to avoid external radiation of sound
6. No serious pressure loss in net flow
7. Construction suitable to withstand operating pressures
8. Simplicity of installation

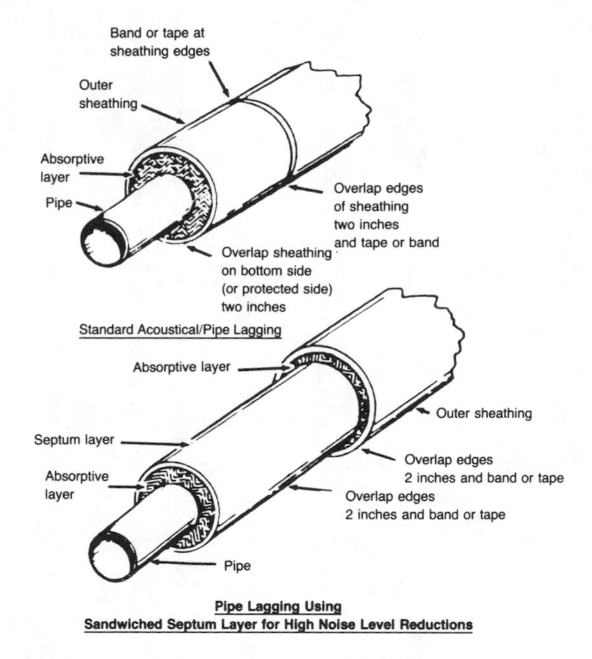

Figure 11. Typical acoustical lagging designs for noise radiating pipelines.(Adapted from Guidelines on Noise, American Petroleum Institute Medical Research Report No. EA7301, prepared by Tracor, Inc., 1973.)

Vibration suppression or reduction is generally accomplished by the installation of vibration mounts that combine the properties of resilience and vibration damping to provide two fundamental mechanisms for control:

1. Dissipation and reduction of vibrational energy generated within the system by conversion of that energy into heat.
2. Mechanical decoupling or removing the vibration paths of the system from its mounting structure and floor.

The process for selecting a particular vibration isolation or damping design is

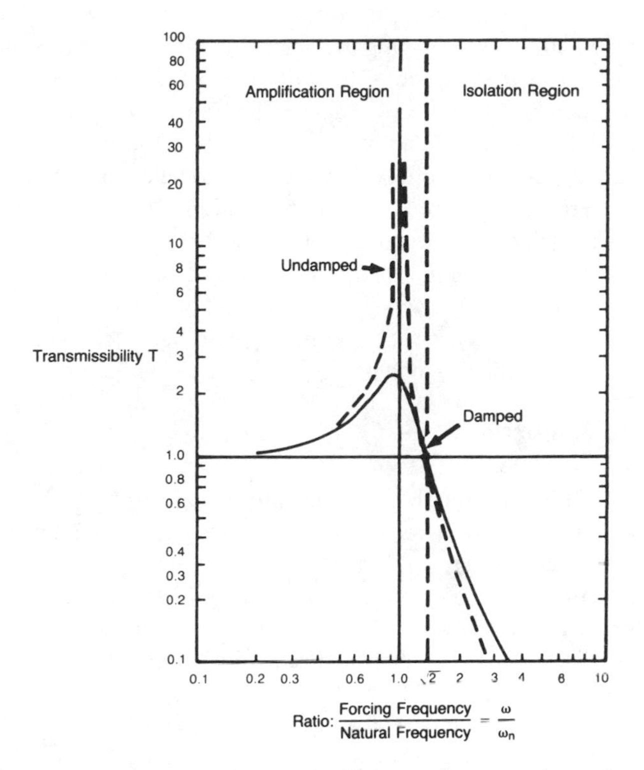

Figure 12. Vibration response of a mechanical system mounted on viscous vibration dampers.

described in detail in the literature supplied by manufacturers of such vibration isolation devices.

Figure 12 illustrates typical characteristics of the response of a mechanical system mounted on viscous dampers.[2] The ordinate represents the transmissibility of the mounting (T) which is the ratio of transmitted force to driving force and the abscissa is the ratio of the forcing frequency to the resonance or natural frequency of the system. The magnitude of the transmissibility in the figure at the ratio of $\omega/\omega_n = 1$ approaches $T = 1$ with increasing damping and a higher ω/ω_n ratio. Note that if the frequency ratio ω/ω_n is less than 2, amplification of the transmitted force will result and the mounts will do more harm than good. In summary, transmissibility is reduced when the forcing frequency, ω, exceeds the natural frequency, ω_n, in the isolation region of Figure 12. Equipment vibration consultants should be employed for effective vibration isolation and damping.

Viscoelastic materials typically are the most versatile and effective materials in providing the desired vibration damping and ultimate noise reduction, especially for thin structures. To determine what surfaces may require treatment by application of viscoelastic materials, it is necessary to measure vibration levels. For those surfaces having high vibration levels relative to other surfaces, decisions should be made as to the type(s) of appropriate treatment. If the surfaces are easily accessible and continuous, a lagging treatment may be employed; if the surfaces are irregular or not easily accessible, the application of damping materials may be the most expedient noise reduction approach. The characteristics of applied damping materials vary significantly with temperature and frequency; therefore, the manufacturer's data should be consulted prior to selection and application.

Coil springs are another form of vibration isolation. Coil springs are employed primarily for the isolaton of vibrating motion that have a low forcing frequency and where elastomeric mounting pads are not very effective. Coil springs are a transmission path for high-frequency vibrations and have large static deflections, therefore, design precautions must be taken in the proper selection of spring-type vibration isolation equipment mounts.

Motor Noise Reduction

Noise from electric motors is predominantly generated by cooling fans. The cooling fan is the dominant noise source in totally enclosed fan-cooled-type motors up to about 200 hp and thus the radiated noise level does not vary significantly with load. Motor noise levels vary with manufacturers. Personal computer fan noise has probably been recognized by many individuals, and the application of discussed noise abatement techniques would reduce these noise levels for most computers.

Noise reduction may be achieved by installing an acoustically treated air intake shroud. Such devices are commercially available for many motor sizes and configurations, or may be designed and fabricated to satisfy unique requirements. Substantial noise reduction can also be achieved by installing an enclosure around the motor using previously discussed enclosure principles. With a motor-driven pump the loudest noise quite often is in the coupling guard area. This guard is usually made of sheet metal, so with vibration damping of the coupling guard or the installation of a guard acoustical lining, the noise level would be significantly lowered.

Control Valve Noise Reduction

Valves are generally the primary sources of noise radiated by piping systems. Noise generation by valves is caused by at least one of three mechanisms:

1. Turbulent eddy interaction with solid surfaces
2. Turbulent mixing and cavitation
3. Shock-turbulence interaction and vibration

Regardless of the generating mechanism, the in-pipe valve-generated noise field is propagated downstream and decays very slowly with distance from the valve. Because of the mass of the valve wall, little noise is radiated directly from the valve. As a result, the piping system itself becomes the prime source of externally radiated noise. Effective noise control of valve generated noise may be approached by one or more of three methods:

1. Change the dynamics of fluid flow through the valve
2. Change the dynamics of fluid flow downstream of the valve and absorb acoustic energy by installing an in-line silencer
3. Intercept and absorb acoustic energy at the outside pipe wall surface with acoustical lagging

The first method is the most effective and involves the employment of special design type valves that limit the localized flow velocities by forcing the flow to travel through a plurality of passages. Quiet valves are available, however, cost tradeoffs should be examined with comparisons to low cost acoustical lagging and the employment of in-line silencers.

Miscellaneous Noise and Vibration Design Principles and Comments

Even with a great deal of education in the field, it appears that it is only after years of industrial noise abatement experience that a feeling for cost-effective noise abatement is developed. Formal education exposes the individual to theoretical and laboratory principles; however, extensive field experience is required to obtain cost-effective noise abatement solutions. Table 7 offers a summary of some common sense noise abatement "nuggets of knowledge."

NEW PLANT DESIGN CRITERIA

The preparations necessary to ensure that the design of a new plant meets criteria for allowable noise limits must begin in the early stages of planning. Noise specifications must be established for selection of equipment, equipment layout, and site selection. The initial step in planning for an acoustically acceptable plant is to specify an acceptable noise environment for the plant and its surroundings. For example, if a maximum plant noise level is set at 85 dBA in all areas, all equipment must meet at least an 80 dBA criteria at 3 ft or 1 m. This is considered to be an effective criteria since the physical spacing required for equipment seldom results in a composite or total noise level of >5 dB above individual source levels. Unless more detailed analysis or spacing is available or known, these limits should be employed. It is also recommended that a maximum total energy 90 dB criteria be set for all equipment to discourage equipment design practices that result in equipment with significant low frequency (below 1000 Hz) noise levels.

Employing the meeting of the OSHA 85 dBA time weighted average (TWA) criteria does not eliminate the existence of significant noise levels below 1000 Hz, which the dBA regulation discounts. If in the engineering abatement process high frequency noise is shifted to low frequency (below 1000 Hz) rather than converted to heat or eliminated, the low frequency levels will increase. If low frequency

Table 7. "Miscellaneous Nuggets" of Noise Abatement and Vibration Design Principles

1. One of the most common noise problems associated with fluid flow is that of venting high pressure gas to the atmosphere. When the ratio of absolute pressure up stream of the pressure reducing (PR) valve is 1.9 or greater, the port velocity in the valve will be sonic and the sound power level will vary directly with the quantity of gas flow. Such valves are called choked valves. Calculations for choked and nonchoked valves provide basic data for specifying mufflers to control this noise.

2. In order to reduce turbulent flow and subsequent high noise levels, the flow velocity should be reduced to a minimum. Additional noise reduction may be obtained by streamlining the equipment design. For example, the use of multiple opening nozzles to replace a single opening nozzle results in the reduction of the exit noise. Noise in the higher frequency octave bands are effectively reduced with substitution of multiple opening nozzles.

3. Surfaces radiating low frequency sounds can sometimes be made less efficient radiators by dividing them into smaller segments or otherwise reducing the total area. The employment of perforated or expanded metal can often result in less efficient sound radiation from sheet metal guards and cover pieces.

4. The response of a vibrating member to a driving force can be reduced by damping the member, increasing its stiffness, or increasing its mass. When the driving force is equal to the natural frequency of the member (including floors) being vibrated, a large surface displacement results with highly amplified noise levels. This condition is known as "resonance".

5. Changing the stiffness of the mount is the simplest method of altering the natural frequency of a system. The second parameter that must be known in designing an isolation system is the frequency of the vibration being isolated. If many frequencies exist, the lowest frequency is the one which must be considered. This is known as the forcing frequency.

6. Structural steel can transmit noise sources greater distances than air. Structure-borne noise sources or vibrating surfaces should be taken strongly into design considerations. It should also be remembered that a flexible pipe or hose for a hydraulic pump may expand and contract at the piston or gear frequency of the pump to radiate sound.

7. A noise source may radiate sound energy into air; this may cause a plate or sheet metal panel to vibrate, which in turn then reradiates sound energy.

8. At 1000 Hz, the wavelength of sound is approximately 1 ft. At lower frequencies the wavelengths become longer while at higher frequencies the wavelengths become shorter. At 200 Hz, a wavelength is 5.65 ft, while at 2000 Hz, a wavelength is 0.565 ft. Therefore, when the dimensions of a noise-radiating surface are comparable to the wavelength of the sound being emitted, the surface becomes a relatively efficient radiator. If the radiator is greater in size than 3 or more wavelengths, it becomes a very efficient radiator. On the other hand, if the dimensions of the radiating surface are reduced to one third or less of a wavelength, this radiating surface becomes a relatively inefficient radiator.

9. A good example of sound radiation: a tuning fork vibrating at 256 Hz (middle C) radiates a sound wave having a wavelength of approximately 4 ft. Because the radiating surface of the tuning fork prong is approximately 1/4 in. wide or 1/100 of the wavelength, it does not radiate sound very efficiently or effectively; however, if the base of the tuning fork is held in contact with a relatively large surface, such as a table top, the sound level is increased or amplified and the tone heard loud and clear. The reason, of course, is that the greater surface area of the table is more able to radiate

Table 7 (continued). "Miscellaneous Nuggets" of Noise Abatement and Vibration Design Principles

sound having a 4-ft wavelength. A piano, for instance, is an excellent radiator of sound energy. Likewise an extruder or compressor with low forcing frequency noise can have its sound radiated effectively through a supporting floor area. Vibration isolation of an extruder or compressor or large pieces of vibrating equipment is therefore necessary.

10. Surfaces that are small in comparison to the wavelength of the emitted sound radiate the sound relatively uniformly in all directions, whereas surfaces that are large in comparison to the wavelength are very directive sound sources. An example is to compare a "tweeter" (small loudspeaker) which produces high frequency sounds in a high-fidelity sound system to a bass speaker called a "woofer." The bass speaker will be very directive if used to reproduce higher frequency sounds.

11. In general, vibrating surfaces are very directive at high frequencies and less directive at low frequencies, since a wavelength of a sound wave is inversely proportional to its frequency. When studying equipment design it is important to keep in mind the directivity properties of the sound emitting surface and its relationship to the wavelength of the sound being produced. For example, a manufacturer may build a machine and then install covers or side panels to dress up the product. The manufacturer is then surprised that the sound is higher than before the side panels were placed on the machine. The panels in this case not only increased the radiating surface, but because the size of the panels were comparable to the wavelength of the sound being produced, they radiated the sound more efficiently than the structural members of the machine. This amplifying effect may be desirable with pianos, where the sounding board of a piano is employed to amplify sound; however, it is not a desirable design technique for industrial equipment. A large percentage of equipment manufacturers are not familiar with these very basic principles. In cases of a metal cover, the panel should be perforated with holes to reduce the radiating surface, or the panel can be acoustically damped or structurally modified. The sound level produced by a radiating panel may be reduced by 3 or more dB each time the radiating surface is halved. Disconnecting a significant vibrating part from the radiating area would, of course, significantly reduce the noise levels produced.

12. If a machine moving member has less surface to compress the air in front of it, creating sound, there will be less mechanical energy converted into sound.

13. Fan speed reduction significantly reduces noise levels. A 2:1 reduction in fan speed can result in a 16dB reduction in noise level. If the velocity of an airstream can be reduced by 50%, the noise level can drop by 20 dB.

14. Laminated construction is likely to have high internal damping which is helpful in controlling resonant vibrations. "Damping" is used to describe the conversion of resonant vibration energy into heat energy. This is an effective mechanism for noise control because once converted to heat, vibration energy is no longer available for generation of airborne noise.

15. Pumps operating at 3600 RPM are likely to produce more noise than similar larger pumps operating at 1800 RPM for a given rating.

16. A preventive equipment maintenance program, in addition to preventing equipment breakdown and maintaining product quality control, has a desirable by-product — the prevention of increased noise levels that would result from excessive wear of moving parts. Periodic machine vibration checks and the employing of a vibration accelerometer might be worthwhile during equipment maintenance checks.

17. The application of periodic forces or impulses may cause a machine to vibrate with a

Table 7 (continued). "Miscellaneous Nuggets" of Noise Abatement and Vibration Design Principles

frequency that may or may not be the natural frequency of the vibrating body. When the period of the forced vibration is the same as that of the free vibration, the two effects reinforce each other and large amplitudes of the vibrating body result in "resonance." Most mechanical structures resonate at a series of frequencies that become closely spaced at high frequencies. There are instances in which machines cannot be operated at certain speeds because of resonance; this on occasion results in structural failure.

18. When the frequency of the driving force coincides with the natural frequency, transmission of vibration force is increased rather than reduced by isolators. To counteract resonant vibration, damping is added to many isolators to increase energy dissipation. This is done with some sacrifice in the effectiveness of the isolator at other frequencies.

19. Acoustical damping is more effective at high frequencies than at low frequencies because damping is accomplished by internal friction of the material as it is stretched or compressed by the vibrating motion.

20. A technique for determining the presence of resonant vibrations and need for damping is to vary the speed (RPM) of equipment to above and below normal speed; listen for increased noise or changes in pitch, which may indicate that resonances are occurring and that damping treatment may be a useful measure for noise reduction. Instrument techniques may also be used to determine vibration levels and node patterns.

21. The response of a vibrating part to a driving force above the resonant frequency can be reduced by damping the member, improving its support, increasing its mass or stiffness, or in general, reducing its resonant frequency.

22. The noise transmission loss from 2 or more walls separated by air spaces is usually significantly greater than predicted on the basis of mass law attenuation. It is important that walls should not be solidly tied together in order to gain advantage of air space. Staggered wall studs are also recommended.

23. Partial noise enclosures are most effective in reducing high-frequency noise and least effective at low frequencies.

24. Barrier walls provide a noise shadow effect for high frequencies but are less effective for low frequency, long wavelength types of noise. Barriers are most effective when either noise source or the receiver or both are close to the barrier wall. The attenuation and advantages of a barrier should always be compared to those of an enclosure with regards to ventilation, noise reduction requirements, and cost.

25. Mass theory states that noise transmission loss should increase by 6 dB for each doubling of partition weight.

26. Relatively low noise transmission losses result when the frequency of the sound energy is coincident with resonant partition vibrations.

27. Small openings in sound barriers or enclosures greatly reduce their effectiveness. For example, a 1-in.^{2} hole will transmit slightly more sound energy than the entire surface of a 4×12 ft sheet lead barrier rated for a 40-dB transmission loss. Door crack openings and ventilation ducts are often noise passages. Commercially available special rubber seals for doors may be used to avoid excessive noise leaks.

28. The reverberant noise level in a room decreases 3 dB for each doubling of the total absorption. In most applications, installations of absorptive materials provide <10 dB

Table 7 (continued). "Miscellaneous Nuggets" of Noise Abatement and Vibration Design Principles

of noise reduction, and significantly <10 dB with regard to worker noise exposure levels.

29. It is impractical to acoustically treat more than 50% of a room surface area because the noise reduction gained by going beyond this percentage is slight. If room treatment is necessary, treatment of 20 to 50% of the boundary surface area is considered a practical approach.

30. A good muffler for a hydraulic system is a flexible hose or an accumulator inserted in the line at the pump discharge. The flexible hose can then expand and contract in response to the hydraulic pulsations. The hose diameter is small compared to the wavelength of the sound generated by pressure fluctuations in the hydraulic fluid; therefore, the hose does not radiate noise efficiently.

31. A double wall noise enclosure can provide improved noise reduction over a single wall enclosure of the same weight.

32. Vibration isolators must be placed correctly with respect to the center of gravity of the machine. In cases of instability, i.e., "rocking" the effective center of gravity may be lowered by mounting the machine on a heavy mass and isolating the mass and the machine on a so-called "floating floor."

33. Coil springs are employed in vibration isolators primarily for the isolation of vibrating motion having a low forcing frequency. Consequently, coil springs must usually have relatively large static deflection. This introduces the danger of instability, with the possibility that the mounted equipment may fall sideways unless precautions are taken to assure stability of the equipment and isolator assembly. Instability is likely to result if the lateral stiffness of the isolator is too small or the static deflection is too large. Coil springs possess practically no damping; therefore, transmission at resonance is extremely high. Coil springs also allow high frequency surges to pass through the equipment being protected. Springs also have a transmission path for high frequency vibration, resulting in excessive noise levels. Rubber pads are employed at the bottom of spring isolators to overcome some of these deficiencies. Side-restrained metal spring isolators are available to avoid difficulties.

34. Since there is little inherent damping of resonant vibrations in most structural elements, external damping must be made to reduce vibrations. External damping can be applied in several ways: (a) by interface damping (letting two surfaces slide on each other under pressure, the dry friction produces the damping effect); (b) by application of a layer of material with high internal losses over the surfaces of the vibrating element; (c) by designing the critical elements as "sandwich" structures. (Damping through the use of sandwich structures refers to a layer of viscoelastic material placed between 2 equally thick plates or to a thin metal sheet placed over the viscoelastic material that covers the panel.)

energy causes more speech frequency hearing loss than high frequency noise (greater than 1000 Hz), which some experts believe, regulations can be met by employing some engineering noise abatement practices with a resultant increase in employee hearing losses. This could be the unfortunate result of honest engineering intentions. A limit on equipment total energy noise levels should therefore be set, and 90 dB has been determined to be a reasonable limit.

Noise projections should be made in the early plant design stages with reliable equipment noise data, and this should be observed field data and not data supplied

Table 8. Factors to be Considered in Setting Noise Abatement Tolerances

Most manufactured equipment is built to operate within a noise range; however, some units will exceed a specification based on the median value
Equipment noise levels tend to increase with age, wear and tear, lack of maintenance and increased load, etc
The measured equipment composite noise level may be increased by reflection from nearby objects
Equipment specified to meet an overall noise power level limit may produce a noise level that exceeds the specification when measured in a certain direction
An engineering allowance may be necessary where overall design procedures are not precise
Noise limits for individual sources must be set at a lower level to insure that the composite noise level for an area does not exceed the total noise level criteria
It should be realized that noise levels quoted by equipment manufacturers are usually lower than actual field experience noise levels

Table 9. General Guidelines for Site Selection and Plant Layout

Locate process areas or known noise sources at maximum distances from more sensitive plant areas; for example, large furnaces can often be placed near plant boundaries away from administrative offices and adjacent community areas
Utilize acoustical shielding situations offered by large structures such as storage tanks
Take advantage of hills, large neighboring structures, wooded areas, etc, to screen noise
In determining site location, the following factors should be considered:
Proximity to existing housing or light industries and general topography
Proximity to unoccupied land, to possible development, and to industrial neighbors
Local codes, ordinances, and standards
Transport of raw and finished products
These guidelines and factors should, of course, be integrated with other factors, considerations, future expansions, and regulations

by the equipment manufacturer. The task of combining noise sources and levels is a critical activity whether it is performed manually or by computer programs. In this type of activity the employment of computer assisted engineering drawings (CAD) has been employed very effectively with color-coded noise projection interfaces. The reduction of these engineering drawings with noise projections allows for quick reference in summary reports.

Noise limits should apply to typical or maximum operating modes. If there is more than one mode of normal operation, the limits should apply to the noisier condition. The resulting specifications to equipment manufacturers and engineering companies should be very clear as to the conditions assumed. Noise specifications must be designed with tolerances or built-in margins for unavoidable variations in noise levels. Several factors to be considered in setting tolerances are listed in Table 8.

General guidelines for site selection and plant layout are listed in Table 9. In determining plant layout, the general guidelines in Table 9 should be observed when possible.

Table 10. Noise Specifications to the Equipment Supplier

Equipment noise level limitations, noise testing procedures, and noise data documentation requirements. These limitations and requirements should apply to all stationary and mobile equipment and machinery that produce continuous, intermittent, and impulse noise.
Provisions for a uniform method of conducting and recording noise tests to be made on equipment.
Requirements that the equipment manufacturer and the engineering contractor guarantee to meet the noise limits set forth in the specification.
Statements indicating that if the noise survey of a completed plant indicates that an item of equipment is producing noise levels that exceed equipment specifications, the equipment manufacturer, subcontractor, or engineering contractor will be responsible for the extra cost of treating the equipment to bring noise levels within the equipment and plant specification requirements.
Agreements that all equipment manufacturers and engineering companies will be penalized if the equipment and plant and/or site noise specifications are not met.
Pre-bid and final test noise measurements shall be made on the purchased equipment and test data will be made available and determined acceptable by the buyer or their representative prior to shipment with authorized signatures.
Reservation of the right to send qualified representatives to the equipment manufacturing plant to observe or conduct noise tests.
Maximum acceptable noise levels for the plant site perimeter, plant area perimeters, and interior plant areas including production areas, control rooms, offices, laboratories, etc.
Maximum acceptable vibration levels for all equipment, and noise reverberation levels for all work and process areas.
Instrumentation and measurement techniques.
Pre-bid equipment noise level data sheets requiring equipment noise specification guarantee signature and buyer approval signature.
Final test noise level data sheets requiring equipment noise specification guarantee signatures and witness/or noise data acceptance signatures.

PLANT AND EQUIPMENT NOISE SPECIFICATIONS

The most effective and economical approach to noise control is to include noise control features as an integral part of the plant design. Such an approach is most efficiently handled by proper use of equipment performance and design specifications. Performance specifications require that the proposed equipment will satisfy the selected criterion; design specifications indicate to the supplier specific noise control features known to be effective and compatible with plant operations. The equipment noise specification package should be similar to those shown in Table 10.

In selecting equipment for a new plant or for equipment replacement or additions to an existing facility, satisfactory noise limits often can be obtained by proper attention to specific design features of the equipment. The prescribed equipment specifications may sound stringent, however, successful activities in all plants that have followed the criteria have not only been rewarding but have been cost-effective. A summary of some of these design features are shown in Table 11.[2]

Table 11. Desirable Equipment Design Features for Noise Reduction

Equipment	Source of Noise	Design Features
Heaters	Combustion at burners	Acoustic air intake plenum
	Inspiring of premix air at burners	Inspirating air intake silencer Acoustic air intake plenum
	Draft fans	Air intake silencer or acoustic plenum
	Ducts	Lagging
Motors	TEFC cooling air fan WP II cooling air openings	Acoustic fan shroud, unidirectional fan, and/or intake silencer
	Mechanical and electrical	Enclosure
Airfin coolers	Fan	Lower rpm (increased pitch) Tip and hub seals Increased number of blades Decreased static pressure drop More fin tubes
	Speed changer	Belts in place of gears
	Fan shroud	Streamlined air flow Damping and stiffening
Centrifugal compressors	Discharge piping and expansion joints	In-line silencer and/or lagging
	Antisurge bypass system	Quiet valves, reduced velocity, and streamlining Lagged valves and piping In-line silencers
	Intake piping and suction drum	Lagging
	Air intake/air discharge	Silencer
Screw compressors (axial)	Intake and discharge piping	Silencers and lagging
Speed changers	Gear meshing	Enclosure, constrained damping on case, or lagging
Engines	Exhaust	Silencer (muffler)
	Air intake	Silencer
	Cooling fan	Enclosed intake and/or quieter discharge
Condensing turbine	Expansion joint on steam discharge line	Lagging
Atmospheric exhausts and intakes	Discharge jet	Discharge silencer
	Upstream valves	Quiet valve or silencer
Piping	Leading pipe	Lagging
	Excess velocities	Limited velocities Smooth, gradual changes in size and direction Lagging
	Valves	Limited velocities Constant velocity or a quiet valve Divided pressure drop

Table 11 (continued). Desirable Equipment Design Features for Noise Reduction

Equipment	Source of Noise	Design Features
Pumps	Cavitation of fluid	Enclosure
Flares	Steam jets	Multiport nozzles on air injectors

COMMUNITY NOISE

Community noise surveys usually involve sophisticated instrumentation. Contour and plotting techniques are often useful for defining "noise impact" areas in the community. There are many potential sources of community noise; however, there are only six general classes of noise sources: (1) power generating units, (2) fluid control systems, (3) process equipment, (4) atmospheric inlets and discharges, (5) materials handling, and (6) plant traffic. Although not a source, plant architectural and engineering deficiencies also contribute to the community noise problem. The most significant community noise problems are usually related to forge hammers and punch presses and other impact noises, escaping air or steam; dynamic imbalance of high speed machinery, furnaces and heaters, fluid control systems, process equipment, saws, cooling towers, flare stacks, mechanical power transmission, the rapid pressure fluctuations at the discharge of diesel engines, or compressor inlets and discharges. One significant point to consider is that the lower the frequency, the longer the wavelength and the further sound travels.

The complexity of a community noise survey program depends upon the noise environment itself and the purpose of the program. The latter may include (1) investigation of complaints, (2) site evaluation for new construction, and (3) current assessment of community noise. In general, enough measurement locations in the community should be visited to obtain a consistent noise pattern. At least some measurement points should be selected well away from road traffic, since traffic noise tends to dominate in the vicinity of a roadway. Unless the sound can be related to the plant, and a specific frequency sound reduction with distance can be established, plant noise cannot be separated from ambient noise to determine the source of the problem.

If the predominant noise in a community is related to a single, steady source, the only temporal pattern that may require investigation is that due to wind or other factors affecting the propagation of sound. If the level of noise from either the primary source or other sources fluctuates with time, however, more detailed measurements are necessary. These may be obtained, again depending upon the complexity of the temporal pattern, by measurement for a few minutes every hour using a sound level meter, by a short tape recorded sample taken at regular intervals automatically, or with the use of a graphic, statistical recorder or more sophisticated instrumentation over several 24-hr periods. Care should be taken that

Table 12. Required Community Noise Data

Description of instrumentation
Location of sound sources (a scale site layout indicating trees, structures and other reflecting objects is most useful)
Meteorological conditions (wind speed and direction, relative humidity and temperature, barometric pressure)
Corrections for each measured value (cable length, temperature, etc.)
Operational mode of sound sources and special noise characteristics

unusually loud noise events are included in the required sample data to be recorded as shown in Table 12.

Community Noise Standards

Typical ranges of municipal ordinances in U.S. cities are 43 to 56 dBA for residential areas; 53 to 65 dBA for commercial/retail zones; and 58 to 70 dBA for industrial zones. These values are applicable to continuous steady broadband noise occurring in the daytime. Outdoors, a noise can be expected to prove annoying if it exceeds the ambient or background level by 10 dBA or more. Some ordinances require nighttime levels as much as 10 dB lower than specified daytime levels; this is the case mainly where the permissible daytime levels are relatively high. In general, enough measurement points should be taken to obtain a consistent delineation of the noise pattern. At least some points should be selected well away from road traffic, since traffic noise tends to dominate in the vicinity of a roadway and sometimes produces misleading results. Specific attention must be given to relate octave band frequencies to the plant and specific plant noise sources. Many regulations specify day-night noise levels. New Jersey, as an example, specifies a 65-dB daytime maximum for residential environments. Other regulations, such as EPA/HUD, specify day-night noise levels that involve more sophisticated data. Penalties vary with communities.

Complaints regarding noise do arise, even when efforts have been made to maintain a satisfactory community noise environment. Every effort should be made to determine the precise nature of the annoyance, enlisting the cooperation of the complainant as much as possible. Often the real object of concern is a particular aspect or time of the noise that might be overlooked in the course of measurement and evaluation. Good community relations should be maintained, and attempts made to constructively deal with complaints, but promises should not be made regarding any type of corrective action due to the complexity of the problem. Efforts should be publicized with care taken not to lose credibility by exaggeration or promises.

The available noise levels for industrial class property lines adjacent to residential class property can be obtained in the *Standard Land Use Coding Manual* (SLUCM).

For most purposes, the A-weighted sound level serves as an adequate descriptor of community noise. Furthermore, it is the descriptor most useful in community noise regulations and on which most community noise ordinances are based. It should be realized that the generally accepted low community noise standard for the U.S. is 55 dBA. In general, more affluent communities have lower than 55 dBA noise standards. The object of a community noise survey is to gather enough data to determine the noise sources, and its radiators, and to determine the most cost-effective method to reduce the noise to meet the desired criteria. This means that the amount of data required for various noise sources can vary widely and depend on the complexity of the noise radiating sources. For a single source with no or few connecting radiation points, very little data might be required to establish a control measure or treatment. Whereas for some large manufacturing sites, extensive testing may be required to investigate many noise sources before particular elements are cited and control procedures determined.

For detailed noise control studies, the A-weighted level measurements are supplemented or replaced by octave-band or one third octave band spectral noise analysis. It is rarely necessary to employ finer spectral analysis. In general, temporal and spatial variations in the outdoor noise environment are so large that placing emphasis on minor spectral variations should be avoided.

Figures 13 to 20 illustrate some findings from various community noise level studies. Figure 13 illustrates the average octave-band sound pressure level of ambient noise measured in a large number of residential areas in Chicago.[5] High frequency noise generated by a plant does not travel significant distances. Figure 14 shows the range of outdoor daytime A-weighted sound levels, and Figure 15 presents corresponding night average sound levels.[6]

Extensive records of community noise levels over long periods are basically unavailable. Figure 16 shows a comparison of urban noise surveys taken between 1937 and 1971, after adjustment for differences in method, objective, and instrumentation.[6] Figure 16 shows that the 50th-percentile exceeded the 50 dBA sound level for all the surveys, excluding 1954 data. Taking into account differences in locations, these data show that where land use does not change, there was no strong increase in the average suburban or urban residential noise levels over the 40 years studied. It is believed that similar conclusions can be drawn for later periods. The point of emphasis with these data is that an A-weighted median sound level of 52 dBA is a representative noise level for a residential community in the U.S. This also supports a low community noise standard of 55 dBA, at an industrial property line next to a residential area. Figures 17[7] and 18[8] show a relationship between noise exposure and the proportion of the community that will be annoyed. This relationship has to be used with caution, however, since it is based on a conglomeration of data from dissimilar studies, it is unlikely to be directly applicable to any specific case. Figure 19 is an example of community statistical noise spectra obtained from daytime and nighttime surveys.[7] The 90th percentile values are considered to be background ambient or indicating that the noise levels are above this value 90% of the time, while the 10th percentile values

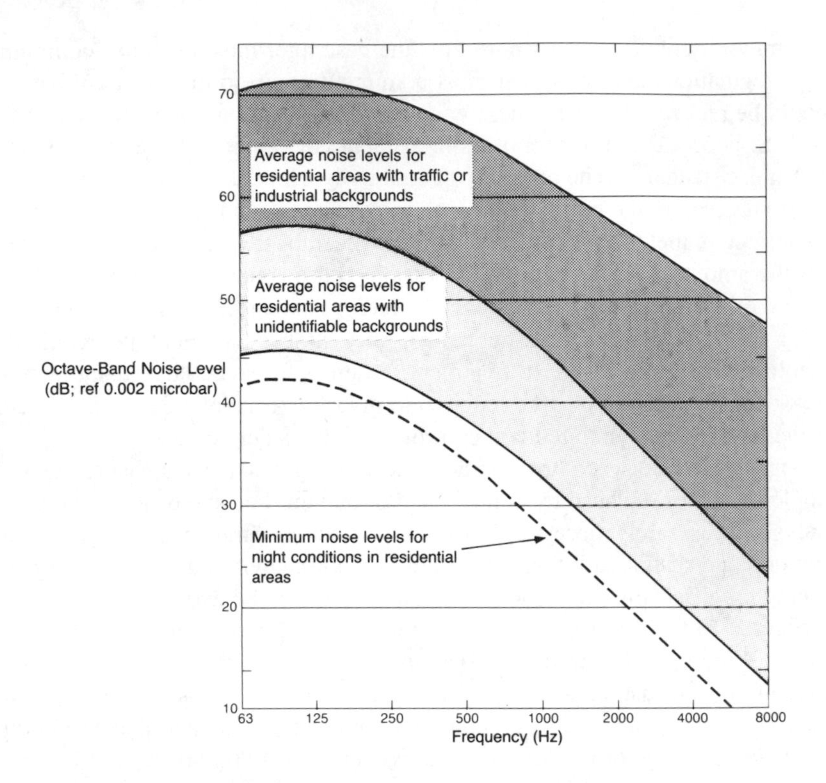

Figure 13. Average octave-band spectra of ambient noise measured in residential areas.

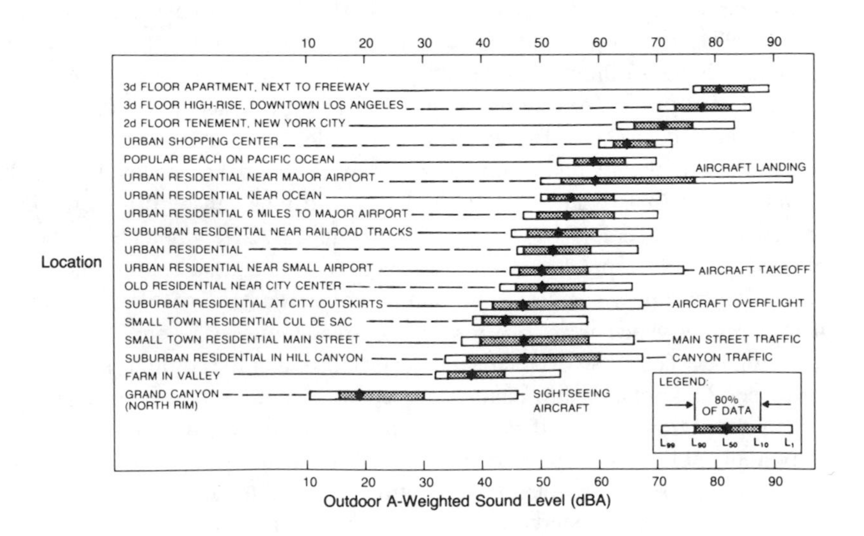

Figure 14. Daytime outdoor A-weighted sound level (7 a.m.—7 p.m.).

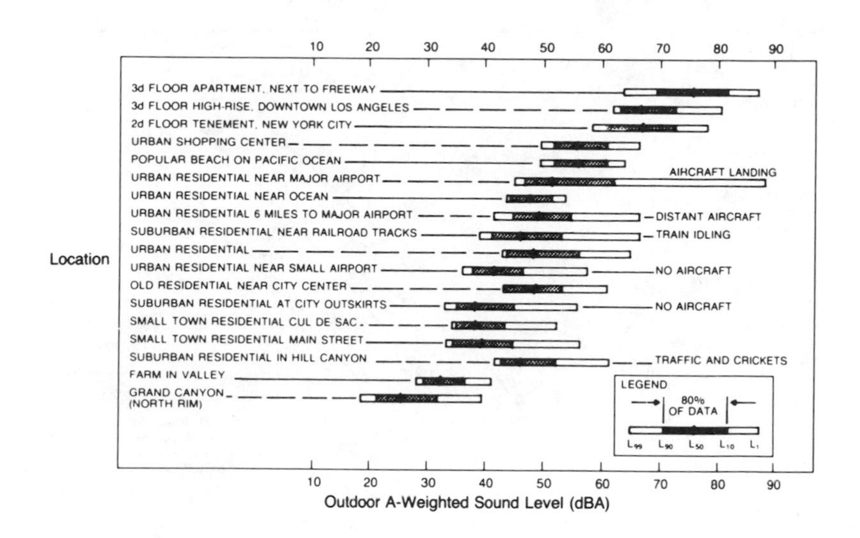

Figure 15. Nighttime outdoor A-weighted sound level (10 p.m.—7 a.m.).

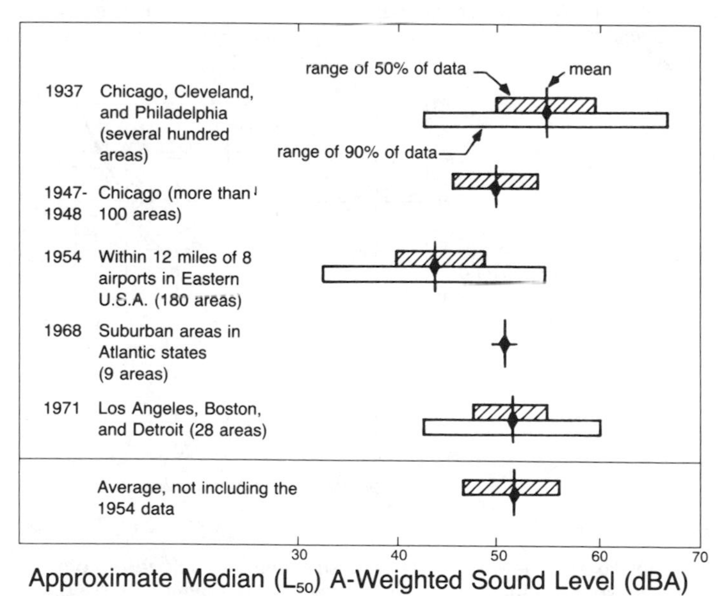

Figure 16. Urban noise survey undertaken between 1937 and 1971.

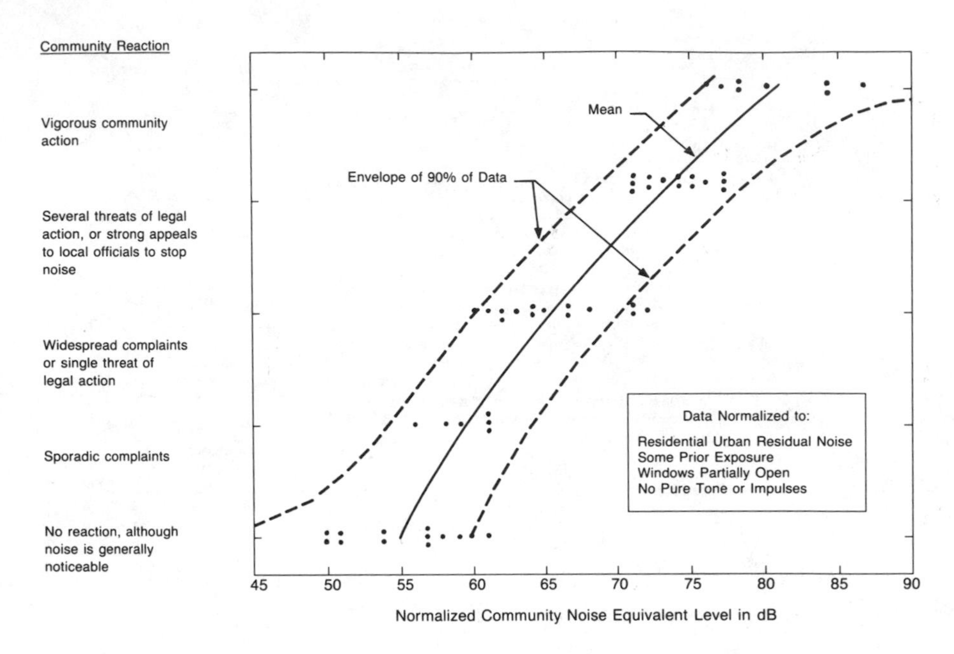

Figure 17. Relationship between normalized community noise levels and community reaction. (Adapted from "Noise from Industrial Plants," EPA Rep. NTID 300.2.)

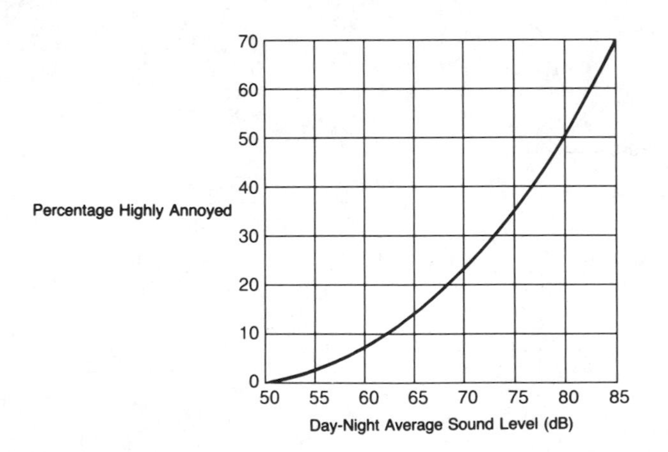

Figure 18. Relationship between noise exposure and percentage of community highly annoyed. (Adapted from "Noise from Industrial Plants," EPA Rep. NTID 300.2.)

are those of the intrusions. Figure 20 is an example of a noise survey around an industrial plant with noise levels measured directly with a sound level meter.

Although community noise statements are relatively simple, it is often very difficult to determine cost effectiveness and locate and attenuate the industrial

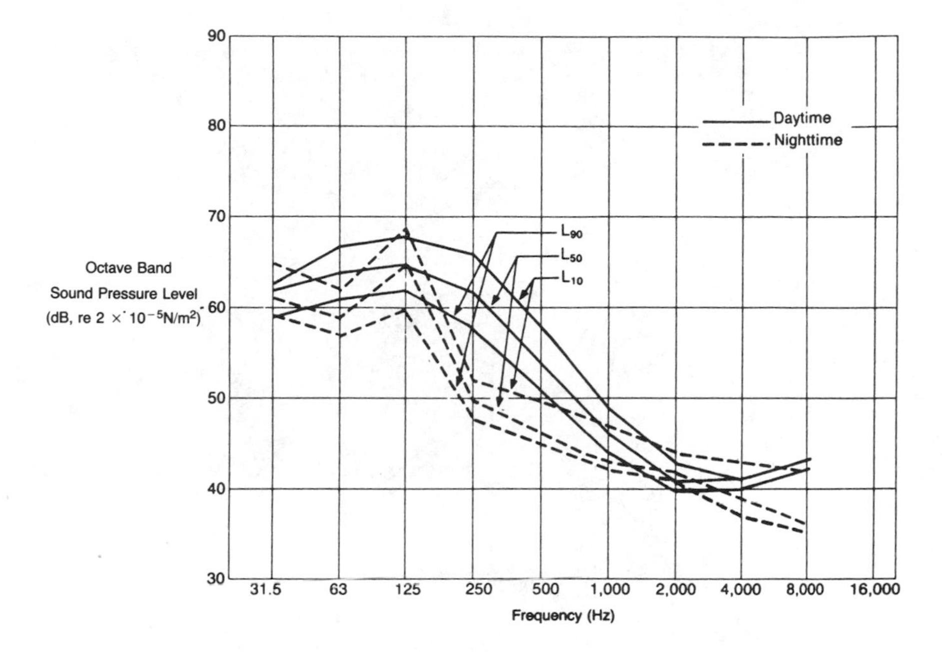

Figure 19. Example of community statistical noise spectra from daytime and nighttime survey. (Adapted from "Noise from Industrial Plants," EPA Rep. NTID 300.2.)

noise sources that can lower the community noise levels to acceptable levels. However, since many community noise problems can be resolved through common sense, they can be cost effective to abate.

SUMMARY OF KEY POINTS

1. A normal adult may have difficulty hearing sounds pitched higher than 12,000 Hz.
2. Every doubling of acoustic power increases the power level by 3 dB.
3. Near a corner formed by three plane surfaces, the noise level can increase by 9 dB.
4. Sound isolation is realized by materials that are poor transmitters of acoustic energy.
5. As a surface weight is doubled, transmission loss increases typically by 5 to 6 dB.
6. Octave or one third octave band data has been found to be very useful in engineering noise prediction and design.
7. Microphone positions are selected at a distance of 3 ft or 1 m from most noise sources if located in a free field.
8. A target spectrum can be useful in identifying problem noise frequencies.
9. Sound pressure levels drop by 6 dB in accordance with the inverse square law with the doubling of distance in nonreverberant fields.
10. Leaks and open areas create significant degradation in enclosure performance.
11. Noise levels can be significantly increased merely by painting cement block walls.

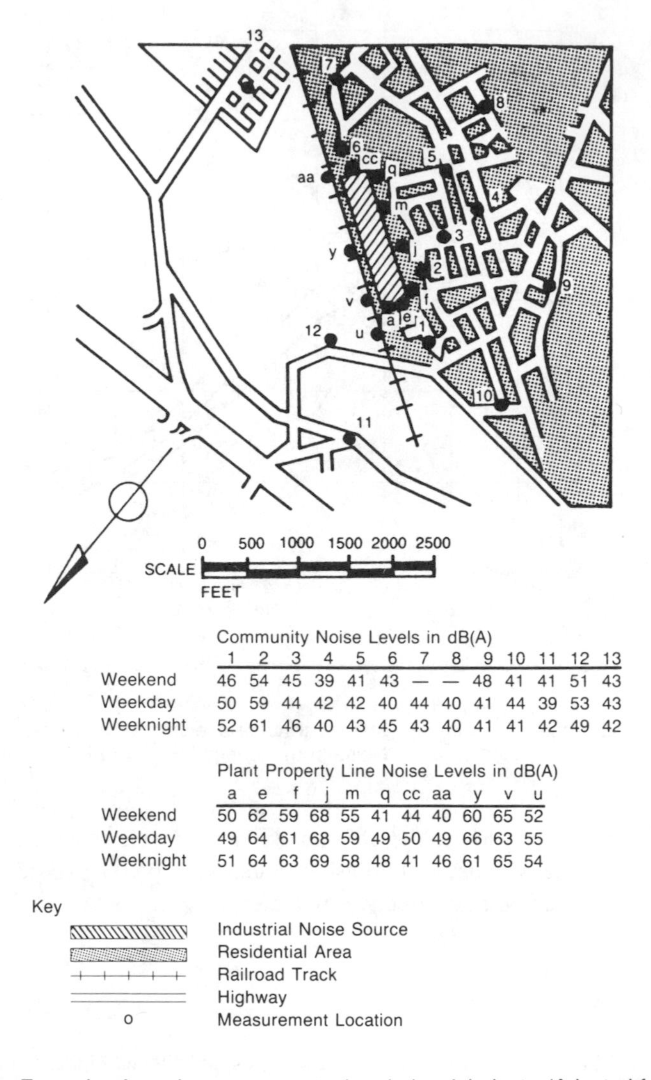

Community Noise Levels in dB(A)

	1	2	3	4	5	6	7	8	9	10	11	12	13
Weekend	46	54	45	39	41	43	—	—	48	41	41	51	43
Weekday	50	59	44	42	42	40	44	40	41	44	39	53	43
Weeknight	52	61	46	40	43	45	43	40	41	41	42	49	42

Plant Property Line Noise Levels in dB(A)

	a	e	f	j	m	q	cc	aa	y	v	u
Weekend	50	62	59	68	55	41	44	40	60	65	52
Weekday	49	64	61	68	59	49	50	49	66	63	55
Weeknight	51	64	63	69	58	48	41	46	61	65	54

Figure 20. Example of a noise survey around an industrial plant. (Adapted from "Noise from Industrial Plants," EPA Rep. NTID 300.2.)

12. Reactive silencers are generally suited for low frequency noise sources and dissipative silencers for high frequency sources.
13. The effectiveness of acoustical lagging is very dependent upon the spectral content of the noise.
14. Some isolation mounts can do more harm than good.
15. The most intense noise from a motor-driven pump is quite often in the coupling guard area.
16. The most effective valve noise control involves changing fluid flow dynamics.
17. Calculations for choked and nonchoked valves provide basic data for specifying noise control mufflers.

18. Changing the stiffness of the mount is the simplest method of altering the natural frequency of a system.
19. At 1000 cps, the wavelength of sound is approximately 1 ft.
20. Vibrating surfaces are very directive at high frequencies and less directive at low frequencies.
21. The sound level produced by a radiating panel may be reduced 3 dB by halving the radiating surface.
22. A 2:1 reduction in fan speed can result in a 16-dB reduction in noise level.
23. Many equipment manufacturers are not familiar with basic noise control principles.
24. When the frequency of the driving force coincides with the natural frequency, transmission of vibration increases rather than decreases.
25. It is impractical to acoustically treat more than 50% of a room surface area.
26. In the design of a new plant, the preparation necessary to meet noise limit criteria must begin in the early planning stages.
27. An attempt should be made to design equipment to meet a maximum noise level of 80 dBA and 90 dB at 3 ft or 1 m.
28. Equipment noise levels quoted by equipment manufacturers are quite often based on laboratory testing results and usually are significantly lower than actual field experience noise levels for various reasons.
29. Agreements should be made with all equipment manufacturers and engineering companies that they will be penalized and held responsible for the extra noise abatement cost if equipment and plant noise specifications are not met.
30. A generally accepted community noise standard for the U.S. is 55 dBA and even lower for many suburban communities.

REFERENCES

1. Barnett, N. E. *Noise Control in Mechanical Systems,* course notes, Chapter 8, p. 13 (1976).
2. Guidelines on Noise, American Petroleum Institute Medical Research Report No. EA7301 prepared by Tracor, Inc. (1973).
3. Feldman, A. S. "Industrial Hearing Conservation Programs,"in *Effects of Noise on Hearing,* D. Henderson, R. Hamernik, D. Dosangh, J. Mills, Eds. (New York: Raven Press, 1976).
4. Fundamentals of Industrial Hygiene (Chicago: National Safety Council, 1988), p. 178.
5. Bonvallet, G. L. *J. Acoust. Soc. Am.,* 23:435 (1951).
6. "Community Noise," EPA Report NTID 300.3, prepared by Wyle Laboratories (December 1971).
7. "Noise from Industrial Plants," EPA Report NTID 300.2.
8. Schultz, T. J. *Sound Vibr.*, 6(2):18 (1972).

SELECTED BIBLIOGRAPHY

1. Occupational Safety and Health Administration, Occupational Noise Exposure and Hearing Conservation Amendment. Fed. Regist. 46(11)(1981): 4,078-4,181; 46(162)(1981): 42,622-42,639; 48(46)(1983): 9,738-9,783.
2. Olishifski, J. B., and Harford, E. R. Industrial Noise and Hearing Conservation (Chicago: National Safety Council, 1975).
3. McGuire, J. L. The Effects and Dynamics of Worker Exposure to Industrial Noise, PhD thesis, University of Michigan, Ann Arbor (1983).
4. "Criteria for a Recommended Standard for Occupational Exposure to Noise," National Institutes for Occupational Safety and Health, U.S. Government Printing Office (1972).
5. Air Force Regulation No. 160-3 (October 29, 1956) and Air Force Regulation No. 161-35 (July 27, 1973) (Washington, D.C.: Medical Service Department of the U.S. Air Force, 1956 and 1973).
6. "Compendium of Materials for Noise Control," National Institute for Occupational Safety and Health, U.S. Government Printing Office (May 1980).
7. "Industrial Noise Control Manual," National Institute for Occupational Safety and Health,U.S. Government Printing Office (December 1978).
8. Harris, C. M., Ed. *Handbook of Noise Control,* 2nd ed. (New York: McGraw-Hill, 1979).
9. *Noise and Hearing Conservation Manual,* 4th ed. (Akron, OH: American Industrial Hygiene Association, 1986).
10. Noise Control, A Guide for Workers and Employers, U.S. Department of Labor, Occupational Safety and Health Administration, Office of Information, OSHA 3048 (1980).

CHAPTER 11

Hearing Conservation and Employee Protection

Joseph McGuire

INTRODUCTION

In the U.S. there are nearly 30 million people with hearing loss, and the compensation costs for occupational hearing loss approach $20 billion. It is hoped that the Occupational Safety and Health Administration (OSHA) hearing conservation regulations will reduce occupational hearing loss incidence. Noise exposure is a continuing health problem in the U.S. Excessive noise exposure can cause temporary loss of hearing, and if the exposure persists, the loss will become permanent. Recent research has even linked noise to changes in the cardiovascular, endocrine and immune systems, disturbances in the gastrointestinal tract, physiological and psychological stress, and fetal abnormalities. Given the often-excessive levels of noise present in both the workplace and the living environment, these detrimental effects of noise are becoming important issues that individuals, employers, and the government must address.

Table 1. EPA Study Conclusions on Auditory and Nonauditory Effects of Noise

1. Hearing impairment is a widespread problem in the U. S. today. Up to 13.2 million adults have some degree of impairment in their "better" ear and up to 27.5 million have some degree of impairment in their "worst" ear.
2. Occupational noise was identified as a major risk factor associated with the prevalence of hearing impairment among men.
3. Men whose current jobs entail exposure to high levels of noise have significantly poorer hearing than men employed in quieter environments.
4. Occupational noise exposure was found to have a weak, but nevertheless significant, association with hypertension for both men and women.
5. A direct relationship with elevated diastolic blood pressure and noise exposure was observed, especially for women.
6. Among men, occupational noise exposure was associated with overall physical health, whereas among women it was associated with psychological health.
7. Men in higher noise exposure occupations were more likely to be diagnosed as having some physical ailment or abnormality.
8. A significant decrease in psychological well-being was found among females in high noise levels occupations.

Under authority of the Noise Control Act of 1972, the Environmental Protection Agency (EPA) was charged with conducting research on the auditory and nonauditory effects of noise. As part of an EPA research program, a report was submitted on March 12, 1982. The analyses presented in the report were based on a national probability subsample of 6913 adults, aged 25 to 74 years, in 417 separate occupational categories, and 12 major noise exposure levels which ranged from below 70 to above 96 dBA. The major findings from this study are summarized in Table 1.

This data, summarized in Table 1, allowed for the generation of Figures 1 and 2 which show the relationship of worker hearing level/noise exposure and a correlation between hearing level and hypertension.[1] These and other findings strongly indicate the need for very effective hearing conservation in the U.S.

SUMMARY OF THE CURRENT OCCUPATIONAL NOISE EXPOSURE STANDARD AND HEARING CONSERVATION AMENDMENT

A summary of the hearing conservation amendment is presented to familiarize the reader with the amendment. Additional details can be found by obtaining a copy of the amendment in the *Federal Register*.[2]

Occupational Noise Exposure Standard

Relative to the current OSHA noise exposure standard, protection against the effects of noise exposure shall be provided when the sound levels exceed those shown in Table 2 when measured on the dBA scale of a standard sound level meter at slow response. In the *Federal Register*, Table 2 is noted as Table G-16. When

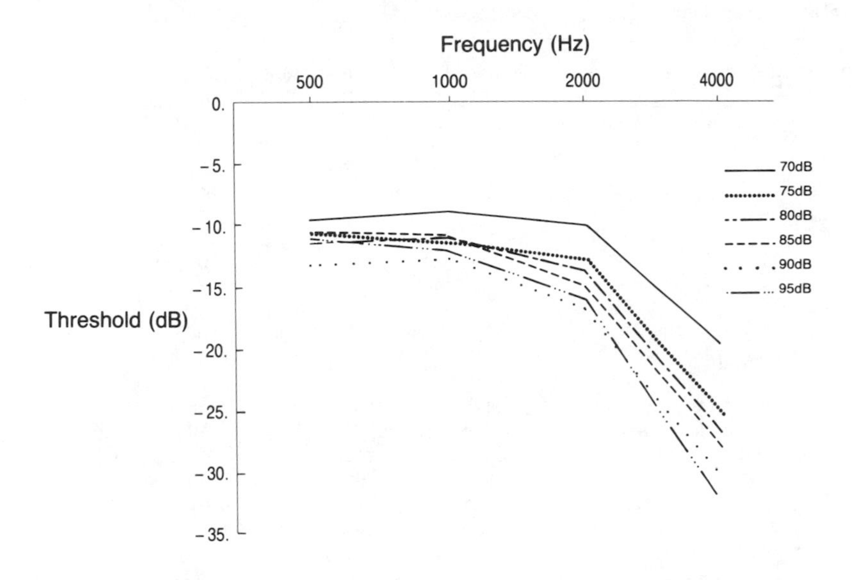

Figure 1. Mean air conduction hearing level in right ear relation to occupational noise exposure.

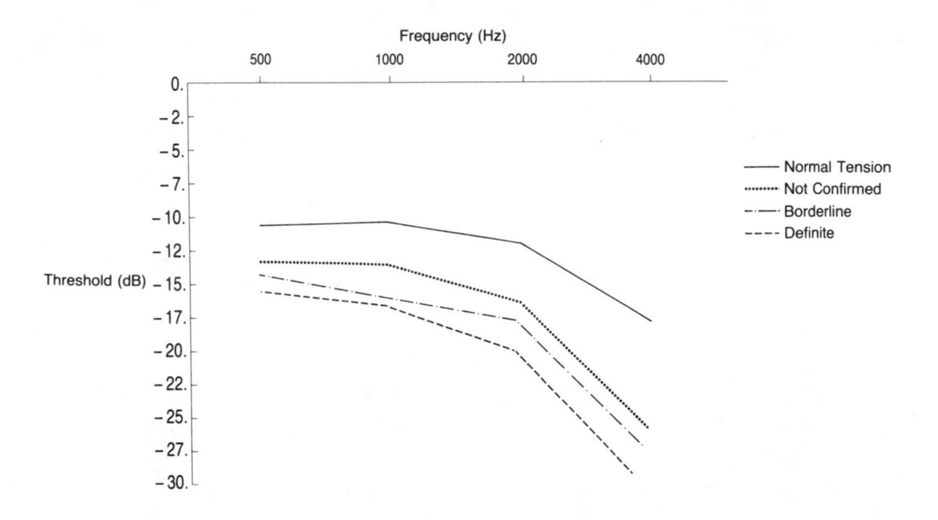

Figure 2. Mean air conduction hearing level in right ear relation to hypertension.

Table 2. Permissible Noise Exposures, OSHA

Duration/P Day, (hr)	Sound Level (dBA, slow response)
16	85
8	90
6	92
4	95
3	97
2	100
11/2	102
1	105
1/2	110
1/4 or less	115

Note: When the daily noise exposure is composed of two or more periods of noise exposure of different levels, the combined effect should be considered rather than the individual effect of each. If the sum of the following fractions: $C_1/T_1 + C_2/T_2 + \ldots + C_n/T_n$ exceeds unity, the mixed exposure should be considered to exceed the limit value. C_n indicates the total time of exposure at a specified noise level, and T_n indicates the total time of exposure permitted at that level. Exposure to impulsive or impact noise should not exceed 140 dB peak sound pressure level.

the noise levels are determined by octave band analysis, the equivalent A-weighted sound level may be determined by Figure 3.

When employees are subjected to sound exceeding those listed in Table 2, feasible administrative or engineering practices are utilized. If such control fails to reduce sound levels within the levels of the table, personal protective equipment shall be provided and used to reduce sound levels within those levels. If the variations in the measured noise level involve maxima at intervals of 1 sec or less, it is to be considered continuous sound.

In all cases where the sound levels exceed the values shown in Table 2, a continuing, effective hearing conservation program shall be administered. This complies with the current OSHA noise exposure standard.

Hearing Conservation Amendment to the Occupational Noise Exposure Standard

The following briefly summarizes the present hearing conservation amendment to the occupational noise exposure standard. For more details, the Federal Regulations Code, *Federal Register,* Volume 48, Pages 9738-9783, March 8, 1983, should be consulted.[2]

To comply with the latest OSHA hearing conservation amendment, rules, and regulations, the employer shall administer a continuing, effective hearing conservation program whenever employee noise exposures equal or exceed an 8-hr time-weighted average (TWA) sound level of 85 dBA (decibels measured on the A scale, slow response) or, equivalently, a dose of 50%. For purposes of the hearing conservation program, employee noise exposures shall be computed without

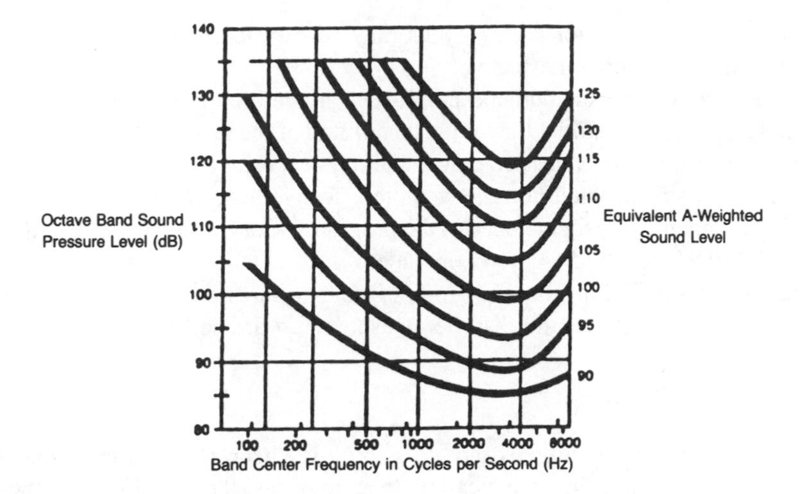

Figure 3. Equivalent sound level contours (Federal Regulations, Figure G-9). Octave band sound pressure levels may be converted to the equivalent A-weighted sound level by plotting them on this graph and noting the A-weighted sound level corresponding to the point of highest penetration into the sound level contours. This equivalent A-weighted sound level, which may differ from the actual A-weighted sound level of the noise, is used to determine exposure limits from Table 2.

regard to any attenuation provided by the use of personal protective equipment. It is stated in the Department of Labor Occupational Noise Exposure Hearing Conservation Amendment that an 8-hr TWA of 85 dBA or a dose of 50% shall be referred to as the "action level" as discussed in the employee noise limit section of this chapter.

Monitoring

The hearing conservation amendment requires employers to monitor noise exposure levels in a manner that will accurately identify employees who are exposed at or above an 85-dBA 8-hr TWA. The exposure measurement must include all noise levels between 80 and 130 dBA.

When information indicates that any exposure may equal or exceed an 8-hr exposure of 85 dB, the employer shall develop and implement a monitoring program for employees. The sampling strategy shall be designed to identify employees for inclusion in the hearing conservation program and to enable the proper selection of hearing protectors.

Where circumstances such as high worker mobility, significant variations in sound level, or a significant component of impulse noise make area monitoring generally inappropriate, the employer shall use representative personal sampling to comply with the monitoring requirements unless the employer can show that area sampling produces equivalent results. Monitoring shall be repeated whenever a change in production process, equipment, or controls increases noise exposures to the extent that additional employees may be exposed and employed hearing protection may become ineffective.

Under the revised amendment, employees are entitled to observe monitoring procedures and, in addition, they must be notified of the results of noise exposure monitoring. Employers must remonitor worker's exposures whenever noise exposure changes are sufficient to require more effective hearing protection or cause employees not previously included in the program (because they were not exposed to an 8-hr TWA of 85 dBA) to be included.

Instruments used for monitoring employee exposures must be OSHA acceptable and calibrated to ensure that the measurements are accurate.

Audiometric Testing

Audiometric testing not only monitors employee hearing acuity over time, but provides hearing protection education to employees. The audiometric testing program includes baseline audiograms, annual audiograms, training, and follow-up procedures. The audiometric testing program should indicate whether hearing loss is being prevented by the employer's hearing conservation program. Audiometric testing must be made available to all employees who have an average 8-hr noise exposure level of 85 dBA or above. An audiologist, otolaryngologist, or physician must be responsible for the program. This responsibility includes overseeing, reviewing, and determining necessary referrals. The program shall be provided at no cost to employees.

There are two types of audiograms required in the hearing conservation program: baseline and annual audiograms. All audiograms obtained must meet audiometric instrument requirements. The baseline audiogram is the reference audiogram against which future audiograms are compared. Valid baseline audiograms must be provided within 6 months or in some circumstances 1 year after an employee's first exposure to workplace noise at or above a TWA of 85 dBA. Under any circumstance, after 6 months of noise exposure at or above 85 dBA, employees must be issued and fitted with effective hearing protection. Testing to establish a baseline audiogram shall be preceded by at least 14 hr without exposure to workplace noise. Hearing protectors may be used as a substitute for the requirement that baseline audiograms be preceded by 14 hr without exposure to workplace noise. The employer shall notify employees of the need to avoid high levels of nonoccupational noise exposure during the 14-hr period immediately preceding the audiometric examination.

Annual audiograms must be routinely compared to baseline audiograms to determine whether the audiogram is accurate and to determine whether the employee has lost hearing ability, or if a significant threshold shift (STS) has occurred. An effective program carefully monitors employee STS variations. Audiometric tests shall be pure tone, air conduction, hearing threshold examinations, with test frequencies including as a minimum 500, 1000, 2000, 3000, 4000, and 6000 Hz. Tests at each frequency shall be taken separately for each ear. An

STS is defined by the hearing conservation amendment as an average hearing shift in either ear of 10 or more decibels at 2000, 3000, and 4000 Hz. In determining whether an STS has occurred, allowance may be made for the contribution of aging (presbycusis) to the change in hearing level by correcting the annual audiogram according to a specific OSHA recommended procedure.

If an STS is identified, employees must be fitted or refitted with effective hearing protection, shown how to employ the protection, and be required to wear it. In addition, employees must be notified within 21 days from the time the determination is made that their audiometric test results indicated an STS. Some employees with an STS may require further testing and physician referrals. An employer may obtain a retest in 30 days to determine STS validity.

Hearing Protectors

Hearing protection must be made available to all workers exposed at or above the 85 dBA action level. Hearing protection is also mandatory for employees who have previously incurred STSs since these workers have demonstrated that they are susceptible to a noise-induced hearing loss.

Employees should decide with trained assistance which size and type of hearing protection is most suitable for them and their working environment. Employers shall provide a variety of hearing protectors and insure proper fit. Hearing protection must provide adequate attenuation for each employee's work environment. In some severe noise situations, finding effective protection presents some difficulty. Employed hearing protection must attenuate worker exposure to a 90-dBA noise exposure level or below. For employees with a significant hearing level shift, the required attenuation level is 85 dBA TWA for 8 hr. The effectiveness of the hearing protection shall be re-evaluated if employee noise exposures increase.

Training

Employee training is important because employees must understand the reasons for the hearing conservation program requirements and the need to protect their hearing and shelter them from the nonauditory effects of noise. Employees exposed to TWAs of 85 dBA and above must be trained at least annually in the effects of noise; the purpose, advantages, disadvantages, and attenuation of various types of hearing protectors; the selection, fitting, and care of protectors; and the purpose and procedures of audiometric testing. The training must emphasize on a yearly basis that noise is hazardous to hearing, and that hearing loss can be prevented by wearing hearing protection; participation in audiometric testing; and their assistance in abatement methods to reduce noise exposures.

The employer shall provide to affected employees any informational materials pertaining to the occupational health standard that are supplied to the employer by the Assistant Secretary of Labor. The employer shall provide, upon request, all materials related to the employer's training and education program pertaining to

Table 3. Expanded Permissible Noise Exposure Limits

T-Duration/ Day (hr)	Sound Level (dBA)	T-Duration/ Day (hr)	Sound Level (dBA)
16.00	85	1.73	101
13.90	86	1.52	102
12.10	87	1.32	103
10.60	88	1.15	104
9.18	89	1.00	105
8.00	90	0.86	106
6.96	91	0.76	107
6.06	92	0.66	108
5.28	93	0.56	109
4.60	94	0.50	110
4.00	95	0.43	111
3.48	96	0.38	112
3.03	97	0.33	113
2.63	98	0.28	114
2.30	99	0.25	115
2.00	100		

the noise exposure standard to the Assistant Secretary of Labor or Director of OSHA. The employer shall maintain an accurate record of all employee noise exposure measurements and audiometric test records for prescribed specific periods of time with employee access to records and transfer to successor employers.

Employee Noise Exposure Limits and Hearing Protection Evaluations

Employee Noise Exposure Limits

When employees are subjected to sound levels exceeding those listed in Table 3, which is an expansion of Table 2, feasible engineering or administrative controls should be utilized. It should be recognized that only 5-dB exchange rate instruments may be used for OSHA compliance measurements (the doubling of noise exposure time with a 5-dBA decrease in sound levels). The Department of Defense employees use a 4-dB exchange rate, while the EPA and most foreign governments use a 3-dB exchange rate.

The accuracy of noise-measuring instruments is considered by OSHA in regulation compliance determinations. American National Standards Institute (ANSI) Type 2 sound level meters are considered to have an accuracy of ±2 dBA, a reading with an ANSI Type 1 sound level meter is considered to have an accuracy of ±1 dBA, and readings with a noise dosimeter are considered to have an accuracy of ±2 dBA. Therefore, the enforcement level with reference to percent exposure level shown in Table 4, which relates TWA in dBA to employee dose or percent noise exposure, would be equal to a noise dosimeter exposure of 87 dBA (TWA). This is equal to a percent noise exposure enforcement level of 66% and not 50% with respect to the ±2 dBA accuracy of a Type 2 noise dosimeter. These types of

Table 4. Conversion from Percent Noise Exposure to 8-hr TWA Sound Level

Percent Noise Exposure	TWA (dBA)
25	80.0
50	85.0
100	90.0
150	92.5
200	95.0
250	96.6
300	97.9
350	99.0
400	100.0
450	100.8
500	101.6
600	102.9
700	104.0
800	105.0
900	105.8
1000	106.6
1500	109.5
2000	111.6

facts are not well known but are contained in OSHA compliance manuals and reference documents.

In addition to a noise control program, a hearing conservation program must be initiated if employees have noise exposure levels equal to or exceeding an 8-hr TWA of 85 dBA action level. It is not required that engineering or administrative controls be initiated until exposure levels, as indicated by Table 3, are exceeded. In many industrial operations, all employees are required to wear hearing protection when noise levels in an area exceed 85 or 90 dBA. This eliminates the dependency of the program and establishment of employee noise exposure monitoring and many other hearing conservation activities; however, the resulting number of employees involved in the hearing conservation activities will increase significantly in most operations with this requirement. A few aggressive companies have been successful in the design or reduction of noise levels in industrial facilities to <85 dBA, and have therefore eliminated the need for a hearing conservation program and significant employee noise exposure monitoring. This is a wise cost-effective decision for some operations and an almost impossible task for others.

Employee Noise Exposure Calculations

When a daily noise exposure consists of two or more periods of different levels of noise exposure, the combined rather than the individual effect of each noise exposure should be considered. The total noise exposure for a given circumstance can be calculated with Table 3 and the described noise exposure equation.

From the equation $D = 100 (C_1/T_1 + C_2/T_2 + \ldots C_n/T_n)$, it should be realized that D is the daily noise dose in percent; C_1, C_2, and C_n are the actual durations of employee exposure expressed in hours; and T_1, T_2, and T_n are the respective duration limits obtained from Table 2.

As an example, suppose an operator is exposed to 100 dBA for 15 min, 95 dBA for 1 hr, 90 dBA for 1.5 hr, and 85 dBA for 5.25 hr (permissible noise exposure durations are from Table 2).

C_1 (15 min @ 100 dBA)	=	0.25 hr
C_2 (1.00 hr @ 95 dBA)	=	1.00 hr
C_3 (1.5 hr @ 90 dBA)	=	1.50 hr
C_4 (5.25 hr @ 85 dBA)	=	5.25 hr
T_1 (2.00 hr max @ 100 dBA)	=	2.0 hr
T_2 (4.00 hr max @ 95 dBA)	=	4.0 hr
T_3 (8.00 hr max @ 90 dBA)	=	8.0 hr
T_4 (16.00 hr max @ 85 dBA)	=	16.0 hr

Inserting these values into the equation:

$$D = 100 \times \frac{0.25}{2.0} + \frac{1.0}{4.0} + \frac{1.5}{8.0} + \frac{5.25}{16.0}$$

$$100 \times 0.125 + 0.25 + 0.19 + 0.33$$

$$= 89\%$$

The operator's exposure is below 100% and is in compliance with OSHA's noise abatement standard of 90 dBA for 8 hr; however, the worker should wear effective hearing protection to comply with OSHA's hearing conservation regulations. Since this daily noise dose is above 50%, or 0.5, OSHA requires that hearing conservation regulations be met.

Hearing Protection Evaluation

OSHA standards place primary emphasis on engineering and administrative noise controls in light of the inherent deficiencies of hearing protection. It is assumed, however, that the plant industrial hygienist will determine the effectiveness of the hearing protection when used as an interim measure until engineering or administrative controls have corrected the noise hazard, or where controls have been determined to be infeasible. Generally, the actual effectiveness of any individual hearing protector is not obtainable under field conditions; however, the following guidelines are offered by OSHA to determine attenuation of the hearing protection:

OSHA Guidelines

1. Because hearing protection requires individual fitting and different types have various characteristics, they are prone to acoustical leaks and generally do not have a field attenuation equal to the printed or laboratory attenuation factors.

2. Since most earmuffs do not have to be specially fitted to the individual, they may be prone to have more acoustical leaks than the ear plug. Ear plugs may become loose, require reseating, and are somewhat more difficult to monitor due to the fact that they are less visible.
3. When the hearing protection attenuation required is greater than 12 dBA for plugs, or 20 dBA for muffs, an octave band should be obtained and calculations should be made similar to the described sample in the following (Figure 4 is a calculation example):

Employee Hearing Protection Calculation Example (see Figure 4)

1. List on line 1 the measured (or estimated) octave band noises that are representative of the employee noise level in decibels.
2. Enter on line 2, the A-weighted adjusted, defined by ANSI.
3. Arithmetically subtract line 2 from line 1. The result is the A-weighted octave band level. Logarithmic addition of these values will result in a total A-weighted sound level.
4. Enter on line 4, the mean attenuation of the protector to be employed (see NIOSH Publication 76-120).
5. Enter on line 5 the manufacturer's (or NIOSH's) standard deviation × 2.
6. Arithmetically add lines 3, 4, and 5. The result (line 6) is the "protected" A-weighted, octave band level. The logarithmic addition of these levels yields the protected total A-weighted sound level.
7. Arithmetically subtract the total level (line 6) from unprotected level (line 3) to obtain the calculated noise reduction or hearing protector reduction factor in dBA (line 7).
8. The calculated dBA reduction in this example is 23, which results in the employee having an actual unattenuated exposure of 96 dBA. Therefore, this hearing protection has been determined to be ineffective for an exposure duration of more than 3.5 hr.

For OSHA compliance purposes a minimally effective hearing conservation program consists of the following items:

1. A baseline audiogram for all employees exposed to noise levels equal to or in excess of the standard
2. Periodic audiograms for each overexposed employee
3. Analysis of audiometric results with retesting and referrals, and recording on OSHA Form 200 when necessary
4. Providing and fitting of effective hearing protection and an OSHA recommended hearing conservation employee training program when required
5. Commencement of an administrative and/or engineering noise abatement program when necessary to reduce noise exposures to <90 dBA

Method for the Calculation of the Hearing Protection Factor:

Octave Band Frequency	125	250	500	1000	2000	4000	8000Hz	dB
1. Octave band level (measured)	100	102	106	110	113	114	104	118
2. A-weighted adjustment	-16	-9	-3	0	+1	+1	-1	---
3. A-weighted octave band level	84	93	103	110	114	115	103	119
4. Mean attenuation of hearing protector	-25	-25	-25	-26	-35	-40	-37	----
5. Standard deviation X 2	+ 8	+ 9	+ 12	+ 9	+ 7	+ 9	+ 11	----
6. "Protected" octave band level, A-weighted	67	77	90	93	86	84	77	96
7. Calculated dBA reduction								23

Figure 4. Sample evaluation of hearing protection.

AUDITORY AND NONAUDITORY EFFECTS OF NOISE

Although in many cases it takes an otologist to diagnose the cause of a hearing loss, it is advisable that individuals responsible for the management of an industrial hearing-conservation program have a working knowledge of the difference between a conductive- and a sensori-neural hearing loss.

Conductive Hearing Loss

If the site of the damage causing the hearing loss is located in the middle or outer ear, the hearing loss is called "conductive." This implies that part of the sound conducting mechanism of the ear is impaired. Conductive hearing loss is generally correctable, whereas sensori-neural hearing loss (or damage to the inner ear) is rarely correctable. Impacted wax or foreign material in the ear canal, perforated eardrums, fluid in the middle ear, dislocation or deterioration of the middle ear bones (ossicles), fixation of the stapes bone, and infection of the outer ear or middle ear, can often cause a conductive hearing loss. Anything that obstructs the passage of sound waves through the ear canal may cause a conduc-

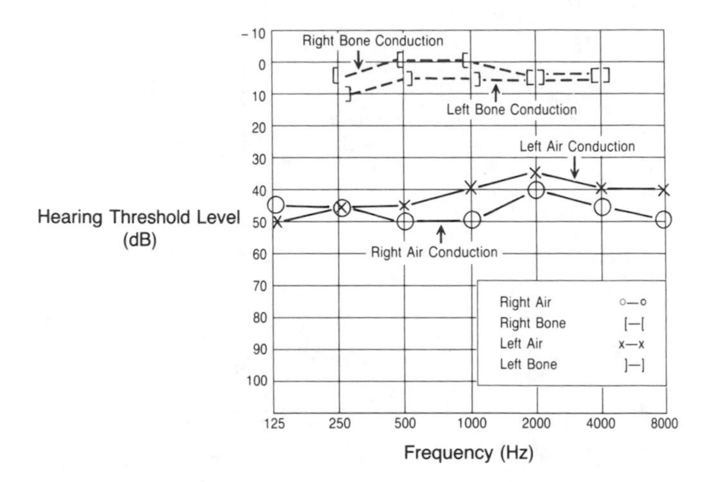

Figure 5. Conductive hearing loss with bone conduction level and air-bone gap. An audiogram indicates conductive loss. Note that there are standard symbols to record thresholds on an audiogram. Circles are used to record air conduction thresholds for the right ear; "X's" are used for the left ear. Brackets opening to the right indicate bone conduction thresholds for the right ear, and brackets opening to the left indicate thresholds for the left ear. (Adapted from EPA Hearing Status Study Report 1982.)

tive hearing loss; for example, an inflammation of the middle ear (otitis media) or fluid in the middle ear can cause a conductive hearing loss. Figure 5 is an audiogram of an individual with a conductive hearing loss where the audiometric slope is generally flat or ascending, indicating greater hearing loss at low rather than higher frequencies.

The test for conductive hearing loss is performed by applying sound energy directly to the bones of the skull, where it is transmitted to the inner ear or cochlea, thus bypassing the structures in the middle ear. The dotted line in Figure 5 shows the bone conduction level (or how much the individual could actually hear if the obstruction or condition causing the hearing loss was alleviated). The solid line shows the amount that the patient hears through air conduction, or through the middle ear. The difference is called the "air-bone gap." This is an example of the kind of audiogram that might be seen for a person with otosclerosis or a middle ear problem, which is surgically correctable.

Sensori-Neural Type Hearing Loss

A hearing impairment that involves only the inner ear or auditory nerve is classified as a sensori-neural impairment. "Sensori" refers to the sensing mechanism in the inner ear and "neural" refers to the nerve fibers. A sensori-neural hearing loss can involve either impairment to the cochlea, the auditory nerve, or both. An audiogram can generally not be employed to separate inner ear or cochlea or sensing mechanism damage from auditory nerve impairment, resulting in the term sensori-neural hearing loss.

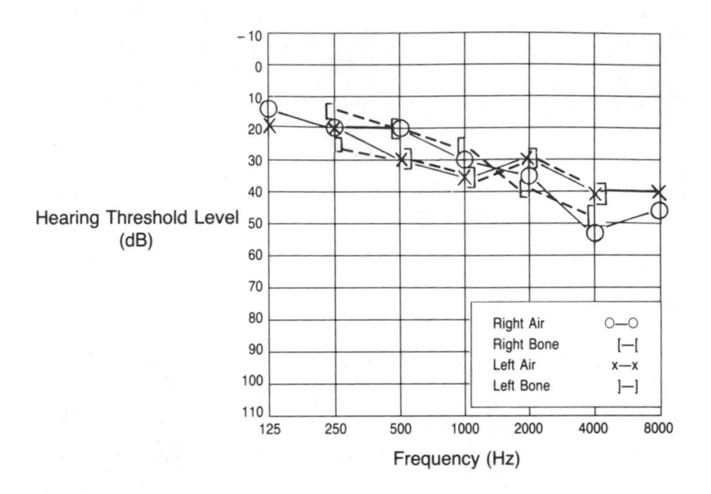

Figure 6. Sensori-neural hearing loss. These results are typical of audiograms for individuals having a mild sensori-neural type hearing loss. The individual hears just as poorly by bone conduction as he does by air conduction. (Adapted from EPA Hearing Status Study Report 1982.)

Significant long exposure to excessive noise causes sensori-neural, not conductive, hearing impairment; however, noise is not the only cause of sensori-neural impairment. Other common causes are congenital conditions, use of certain drugs (such as streptomycin, neomycin, kanamycin, and quinine), severe head trauma, explosions, viral infection, blood disease, vascular spasms, Ménière's disease, aging, and others. Any of these conditions can produce a hearing profile similar to that caused by a significant period of exposure to intense noise.

The typical audiometric pattern of a sensori-neural hearing impairment is a high frequency hearing loss with a typical audiogram as shown in Figure 6. Although it is unusual, the same audiometric pattern can be found for a conductive hearing loss. It is apparent, then, that a hearing loss diagnosis cannot always be based on the shape of the audiogram. It is often necessary to refer the employee to an otologist for a diagnosis.

One major reason to obtain a diagnosis for all present and future employees is to provide effective hearing protection and noise environments for the employee having a progressive type of nerve deafness.

Mixed Hearing Loss

A third type of hearing loss is a combination of conductive and sensori-neural, mixed hearing loss, and the employee will show a combination of both types of loss. In this case, hearing loss is usually significant at both high and low frequencies with a less pronounced descending slope as seen in a sensori-neural loss. It should always be remembered that with audiometric slopes, ascending slopes usually indicate possible conductive hearing loss, while descending slopes usually indicate inner ear or sensori-neural hearing loss.

Noise-Induced Hearing Loss

Noise-induced hearing loss is a cumulative and permanent sensori-neural hearing loss that develops over weeks or years of hazardous noise exposure. When a normal ear is exposed to noise at damaging intensities for a sufficient period of time, a "temporary" depression of hearing results. This "temporary hearing loss" or "fatigue" is a physiological phenomenon referred to as temporary threshold shift (TTS). This temporary shift is believed to occur in the hair cells of the organ of corti of the inner ear as a result of vigorous stimulation of hair cell structures, mechanical stresses that jar the hair cells loose from supporting cells, or changes in the metabolic processes essential for cellular life. Recovery of hearing level after TTS usually occurs within the first 2 hr after noise exposure has ended.

As a general rule, a noise capable of causing significant TTS with brief noise exposures is probably capable of causing permanent threshold shifts or hearing losses, given prolonged or recurrent exposures. In fact, some limited evidence from animal studies suggests the presence of minor but permanent hair cell damage even in those ears showing "complete" recovery from noise-induced TTS.

In general, when the noise exposure is continued and is of significant duration, at a higher level, or both, a stage is reached where the hearing threshold level "does not recover," and the hearing loss is called noise-induced permanent threshold shift (NIPTS) and is designated as noise-induced hearing loss. There appears to be a distinct relationship between temporary and permanent threshold shifts. This is complicated even further with experimental studies on the causes of a TTS that appears to demonstrate that "continuous" noise exposure produces different effects in the ear than "interrupted" noise exposure. Quiet employee rest periods away from the noise exposure have consequently been determined to produce lower level threshold shifts.

Most significant noise-induced hearing losses occur initially in the high frequency range, most prominently at 3000 to 4000 Hz. There does not appear to be a consensus of the various theories to explain this specific high damage area.

As individual noise exposures continue, the hearing threshold dip broadens and spreads to higher and lower frequencies, as shown in Figure 7. Impairment of hearing is usually not noticed by the employee until threshold levels of important speech frequencies (500, 1000, 2000 Hz) average more than 25 dB. This 25 dB average loss is the general basis for accepted procedures for rating auditory handicaps. At the present time, 3000 Hz is included in most handicap judgments since the recognition of speech consonants may depend on a hearing sensitivity for sounds higher than 2000 Hz. This consequently increases the number of individuals who are regarded to have hearing loss handicaps.

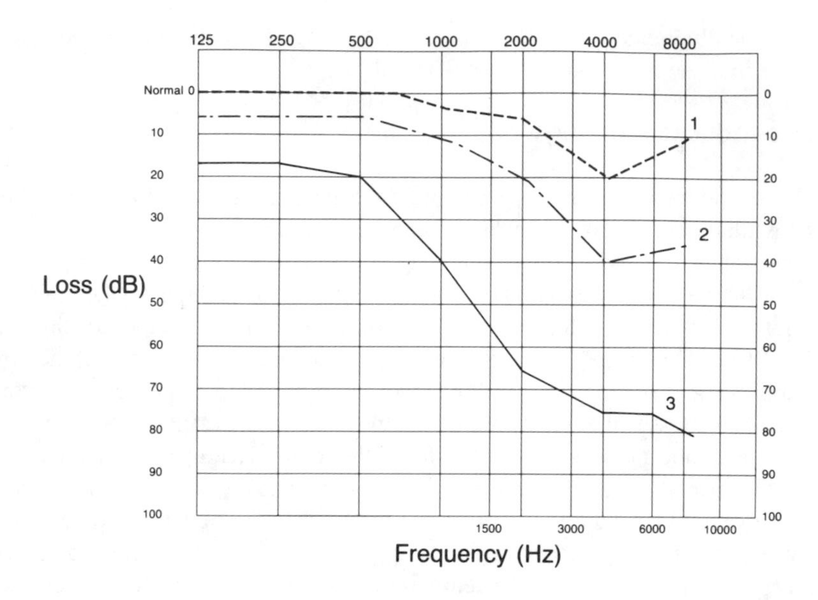

Figure 7. Hearing threshold shift broadening and spreading to higher and lower frequencies. (Adapted from EPA Hearing Status Study Report 1982.)

Hearing Loss Factors

When evaluating hearing loss due to industrial noise, a number of factors must be considered. Even in the absence of occupational noise exposures, hearing sensitivity normally decreases as individuals grow older. This decrease in hearing sensitivity is known as "presbycusis." Hearing losses due to aging are similar to those caused by excessive noise, but they are differentially greater at higher frequencies. Hearing data from various age and sex groups with minimum noise exposure have been employed in correction factors to reduce the aging component from audiograms collected on noise-exposed employees. These corrections are incorporated in workmen's compensation formulas used by various states in rating hearing loss disability from occupational noise exposure.

Other causes of hearing loss include use of drugs, illness and disease processes, and blows to the head. Off-the-job noise conditions, particularly in recreation, can pose some risk of hearing change or can aggravate hearing losses associated with the job situation. Such complicating factors should, however, not minimize the seriousness of the noise, vibration, and hearing loss problems in industry. Noise surveys in various manufacturing, construction, mining, transportation, and farm operations have noise conditions that are definitely potentially harmful to millions of workers. The previously mentioned 1982 EPA hearing loss study indicated, surprisingly, that farm workers have the highest percentage of hearing loss per working population.

The ear, owing to its sensitivity to acoustic energy, is most vulnerable to damage from overexposure to noise. If permissible noise levels are set properly, other physiological functions less sensitive to sound stimuli would not be as susceptible to noise-induced alterations or damage, and the individual's nonauditory health disturbance would be reduced. Unfortunately, the fact that those who work in high noise areas show greater medical difficulties than those who work under quieter conditions is not conclusive evidence that noise is the primary "causal" factor. Research on noise-induced biologic effects unfortunately remains inconclusive. In many studied cases, it is quite possible that the differences in the specified health parameters may be explained by other factors such as age, other environmental contaminants, workload, job habits, etc.

Presbycusis and Age Correction

The hearing conservation amendment permits the application of an age (presbycusis) correction to audiometric test results in order to discount the results of aging. In determining whether a significant threshold shift has occurred, allowance may be made for the contribution of aging to the change in hearing level by adjusting the most recent audiogram. If the employer chooses to adjust the audiogram, procedures are described in Appendix F of the hearing conservation amendment.[2] This procedure was developed by NIOSH in the criteria document entitled, "Criteria for a Recommended . . . Occupational Exposure to Noise".[3] Although the Amendment Appendix Tables should be employed for presbycusis correction, representative presbycusis threshold shifts for males and females are illustrated in Figure 8.

HEARING PROTECTION DEVICES

Hearing protection devices may be broadly categorized into *earmuffs*, which fit over and around the ears to provide an acoustic seal against the head; *earplugs*, which are placed in the ear canal to form a seal; and *semiaural devices*; which are held against the ear canal entrance with a headband to provide an acoustical seal at that point. Other hearing protection devices also exist for specialized activities. In the selection of hearing protection, a wide range of head and ear anatomical characteristics must be considered in the fitting of an effective hearing protection device.

Earmuffs

Earmuffs normally consist of rigid, molded plastic earcups that seal around the ear using foam or fluid-filled cushions and are held in place with metal or plastic headbands, or by a spring-loaded assembly attached to a hard hat. The cups are lined with an acoustical material, typically an open-cell polyurethane foam, to absorb high-frequency energy. Earmuffs are relatively easy to distribute since one size fits most adults. Earmuffs must, however, be evaluated for fit when initially

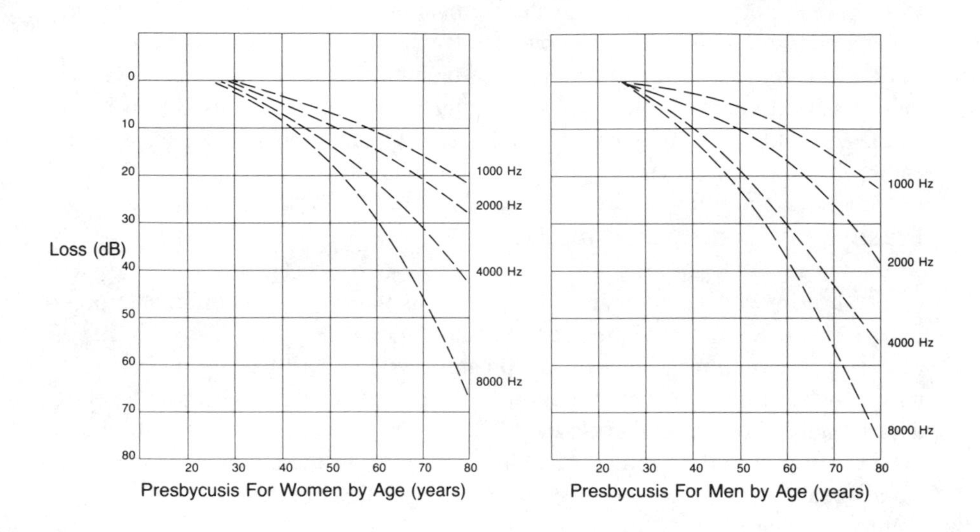

Figure 8. Average presbycusis threshold shifts.

issued because head or ear sizes may fall outside the range that the muff band or cup openings can accommodate.

Earmuffs are good for intermittent exposures due to the ease with which they can be employed. With regard to the wearing of earmuffs, supervision is easy; however, for long-term wear, some employees report that some earmuffs feel tight, hot, and uncomfortable. Earmuff performance is degraded by anything that compromises the cushion to head seal. This includes long hair, sideburns, eyewear, masks, face shields, and helmets. Foam pads that fit over eyeglass temples are available to correct the problem; however, pad effectiveness is often debated. When the use of headgear is required, hard hats with attached earmuffs provide convenience. The selection of foam- or liquid-filled cushions offers similar performance and small and medium volume earmuffs show differences between comfort, acceptance, and price. In summary, good, realistic attenuation is gained by earmuffs over a wide range of noise spectra if they are properly worn, fitted, and maintained.

Earplugs

Earplugs tend to be more comfortable than earmuffs for situations in which protection must be used for extended periods, especially in warm and humid environments. They can be worn easily and effectively with other safety equipment and eyeglasses, and are convenient to wear; however, they are less visible than earmuffs, and therefore their use can be difficult to monitor.

Earplugs come in a variety of sizes, shapes, and materials, but regardless of type, care must be taken in inserting and sometimes preparing them for use. Some plug designs require more skill and attention during application than the wearing of earmuffs. In some instances, a small percentage of users may never wear earplugs correctly, even with the formable "one size fits all" type.

Earplugs may work out of the ear over time and lose a good seal or unfortunately, for the comfort of the employee, may be worn loosely. Foam earplugs provide significant noise attenuation if worn properly. The noise reduction ratings (NRR) for foam plugs are usually higher than other plugs and protection devices, however, this is dependent on the rating laboratory, and some laboratory test results may give undue credit to some devices. The NRR, adopted by EPA in 1979 from earlier NIOSH work in 1975, represents the attenuation that will be obtained by 98% of the users in typical industrial noise environments, assuming they wear the device in the "same manner as the test subjects," and assuming they are accurately represented by the test subjects. These, of course, are significant assumptions and therefore a wide range of laboratory attenuation values can be found for formable plugs, especially for frequencies below 2000 Hz. Unfortunately, most hearing protection devices do not attenuate noise levels below 1000 Hz very effectively. If the OSHA regulation were based on the total energy concept rather than the discounting of noise energy below 1000 Hz and the employment of the dBA measurement scale for regulations, many hearing protection devices would be calculated to have significantly lower noise attenuations. This is a significant point of consideration for management if company employees are exposed to industrial noise spectra with peak levels below 1000 Hz. The selection of hearing protection devices with higher attenuation at lower frequencies would in these instances provide greater total energy protection. Modifications to the Figure 4 calculation of the hearing protection factor could accordingly be made if the A-weighted adjustment was not performed in steps 2 and 3 of the sample calculation. Then a total energy (and not an A-weighted or protected octave band energy level) could be determined, resulting in a calculated total energy reduction.

Due to the considerable variety of earplugs, they can be separated into three types: premolded, formable, and custom molded.

Premolded Earplugs

Premolded earplugs are manufactured from flexible materials such as vinyls, cured silicones, and other elastomeric formulations. Generally, the silicone formulations offer the best durability and resistance to shrinkage and hardening. Most models are available with attached cords to help prevent loss, to improve storage, and to reduce contamination by permitting their being worn like a necklace when not in use. It is difficult to determine employee sizes for premolded earplugs since individual ear canal dimensions vary greatly. Some premolded plugs are prone to shrinkage, cracking, and hardening, and safe use periods vary. As with other protection devices, premolded plugs are subject to employee modifications to improve comfort (at the expense of lower noise attenuation). As a general rule, the greater variety of sizes of a particular premolded plug, the greater the likelihood that different plugs will be required for each ear, and therefore the employment of unmatched sizes.

Formable Earplugs

Earplugs of this variety may be manufactured from cotton and wax, spun fiberglass, silicone putty, and slow-recovery foams. Life expectancies vary from single-use products such as some of the fiberglass down products, to multiple-use products such as the foam plugs. The primary advantage of formable earplugs is comfort; some of the products in this category are considered among the most comfortable, user-acceptable devices sold today. Additionally, formable plugs are generally sold in only one size that usually fits most, but not all, ear canals. This simplifies dispensing, recordkeeping, and inventory problems, but when such products are used, special attention must be given to wearers with extra small or extra large ear canals to ensure that the plugs fit correctly. Since these plugs usually require manipulation by the user prior to insertion, this type of hearing protection may not be the best hearing protection device for some "toxic substance" work environments.

Although cotton alone is a very poor hearing protector due to its low density and high porosity, it can be combined with wax for protection. With slow-recovery foam plugs, better fits and attenuation are found with deeper insertions; however, individual comfort can vary with depth of insertion, the occlusion effect on the individual, and therefore the effectiveness of the device. Since depth of insertion of foam earplugs can be comfortably varied, best fit does not necessarily signify maximum canal occlusion and attenuation. Deeper earplug insertion usually lowers the occlusion effect on the sensitive individual than using, for example, larger volume earmuffs.

Custom-Molded Earplugs

Custom-molded earplugs are most often manufactured from two-part curable silicone moldable material. The silicones are either cured by a catalyst (at the time the impression is taken by the fitter) or returned to the manufacturer. Most ear molds fill a portion of the ear canal that forms the acoustical seal that blocks incoming sounds. Molds have little chance of working loose over wearing time. Considerable skill and time is required to take individual employee impressions. Although a custom ear mold might appear to assure maximum protection, there are disadvantages. The molded earplugs are customized, however, for the individual employee, and this may provide incentive, general comfort, and minimal misinsertion problems. Initial costs for these earplugs are significantly higher; however, over a long period of time the customized plug may be cost-effective.

Semiaural Devices

Semiaural devices, which consist of pods or flexible tips attached to a lightweight headband, provide a compromise between earmuffs and earplugs. They can be worn in close quarters, easily removed and replaced, and conveniently carried when not in use. One size fits the majority of users. Their fit is not compromised by safety glasses or hard hats. The tips can be made from vinyl, silicone, or composites such as foam encased in a silicone bladder, and may cap, or in some cases enter, the ear canal. The tips normally have a bullet, mushroom, or conical shape.

Semiaural devices are principally intended for intermittent use conditions where they must be repeatedly removed and replaced. During longer use periods, the force of the caps pressing against the canal entrance may be uncomfortable, but they can offer adequate protection for those who do employ this device for long periods. Semiaural devices tend to create the most noticeable occlusion effect and consequently distort the wearers' perception of their own speech more than other types of hearing protection. This may be objectionable to some users. Semiaural devices have a wide range of noise attenuation with the greatest attenuation received by the plugs that partially enter the ear canal.

Double Hearing Protection

For very high level noise exposures, especially when 8-hr TWAs are greater than 100 dBA, the attenuation of a single device may be inadequate. For such exposures, double hearing protection, i.e., earmuffs plus earplugs, may be warranted. It is well recognized that double hearing protection does not yield overall

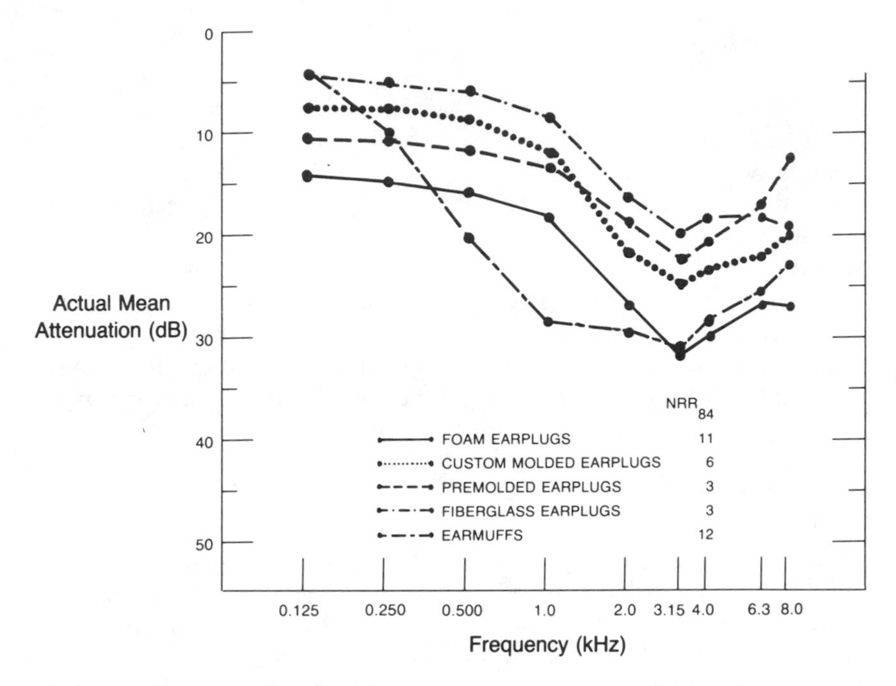

Figure 9. Actual attenuation of five types of hearing protection. (Adapted from Noise and Hearing Conservation Manual, 4th ed., Akron, OH: American Industrial Hygiene Association, 1986.)

attenuation equal to the sum of the individual attenuation of each device. The gain in attenuation (or NRR) for the double protection combinations is 7 to 17 dB when compared with the plugs alone, 3 to 14 dB when compared with the muffs alone, and 3 to 10 dB when compared with the better of the two individual devices.

It is invariably found that the actual performance of hearing protection devices is significantly lower than laboratory measurements with average differences estimated to be as great as 20 dB. The actual attenuation and NRR ratings for 84% of the work force for five types of hearing protection is shown in Figure 9.[4] It should be realized that standard deviations vary with each type of protection. As stated in Vol. 46, No. 11 of the *Federal Register*,[2] employers are required to evaluate hearing protection and attenuation for specific noise environments by methods described in Appendix G of the *Federal Register* and reprinted in the American Industrial Hygiene Association (AIHA) Manual Appendix.[4]

Evaluation of a large number of existing hearing conservation programs suggests that many protectors can provide at least 10 dB of noise reduction for 84% of the work force. A 10-dB attenuation is often all that is needed to reduce noise exposures to an acceptable level. It cannot be overemphasized that a strong attempt should be made to match the protection with the employee's industrial noise spectra, the environment, and the individual. It should always be recognized that the "best hearing protection device" is the one that is employed and employed correctly by the worker.

In compliance with OSHA regulation, an attempt should always be made to correct the noise problems by engineering methods, however, during the interim period, the employment of effective hearing protection devices is the recommended method to protect the worker from occupational noise and hearing loss.

AUDIOMETRIC EQUIPMENT

The purpose of pure-tone air conduction audiometry is to determine hearing sensitivity at different frequencies for which various methods and techniques are available. Regardless of the method, the end result of audiometric testing is to obtain the most reliable hearing responses at the minimum test frequencies of 500, 1000, 2000, 3000, 4000, and 6000 Hz for each ear. From these data it can be determined if a significant hearing loss exists, or if as previously discussed, an STS exists. One of the most difficult decisions to be made in setting up an audiometric testing program is the selection of an audiometer. Three basic types of testing equipment are manual, self-recording ("automatic," Bekesy), and computer-controlled ("microprocessor") audiometers.

Manual Audiometry

The standard audiometric procedure for clinical purposes is manual audiometry. The basic controls of this instrument permit the manual selection of (1) the frequency to be tested, (2) the intensity, (3) right or left earphone, (4) continuous or pulsed tone, followed by (5) presentation of the test signal to the listener. Other features may be added to the manual audiometer, such as provisions for measuring bone conduction, masking, and speech functions. These other features that are employed in diagnostic testing and for the records are not intended for use by the industrial technician.

The manual audiometer is the least expensive equipment for the testing of employee hearing. An advantage of manual audiometry is that the one-on-one or the evaluator and employee testing situation is less impersonal than in automatic audiometry, and is therefore conducive to the establishment of a good rapport and the acquisition of more valid test results. Manual audiometry does, however, have some drawbacks: technicians must manipulate several control switches, and there is the possibility of testing error. Technicians using the manual audiometer may also develop their own self-tailored technique, which results in a lack of standardization from tester to tester and plant to plant, and may create audiometric testing problems when the overall plant or site is evaluated.

Self-Recording Audiometry

The self-recording, Bekesy, or automatic audiometer was originally introduced by Bekesy in 1947. This type of audiometer offers another method of measuring the employee's hearing STS. This method is a tracking procedure in which the employee is allowed to track his/her own thresholds or baseline audiogram by pressing and releasing a response switch. The audiometer automatically increases the intensity of the test signal until the listener presses the button, at which point the intensity of the signal begins to decrease. The listener is required to continually press the button as long as the test signal is heard. When the signal is no longer heard, the response button is released which again automatically increases the signal intensity. There are also automatic functions to this instrument that eliminate some aspects of human error prevalent in manual audiometry.

With the self-recorder, testing problems can be identified by observing the employee's responses. The employee response should be within 10 dB of the previous test, similar to the manual audiometry retest procedure. Many common test problems can be identified on the self-recording audiometer: tinnitus, possible malingering, coordination problems, fatigue, and instruction problems. The employee who cannot, after sufficient reinstruction and assistance, complete the self-recording audiogram successfully should subsequently be retested on a manual audiometer. The grading of self-recording audiograms introduces significant problems and the same potential for human error as recording of manual audiometric tests. The advantage is that it eliminates the retesting and evaluation of a large number of employees of self-recording audiometry who have obviously normal hearing.

Computer-Controlled Audiometry

The availability of the computer microprocessor chip has resulted in the development of a manual audiometer that is controlled by this chip. The chip is programmed to present a routine manual test, record employee responses, and based on a programmed software algorithm, determine the employee's hearing threshold level and shift. One significant advantage of the microprocessor is that its electronic construction ensures greater stability than the more mechanical self-recording and manual instruments. Variation among microprocessors raises some question of potential problems in testing validity. As long as the employee responds properly, the microprocessor is an outstanding instrument in comparison to manual audiometry. Like other instruments, problems exist, however, with the emergence of more sophisticated computer controlled systems, they will be significantly reduced.

WORKER'S HEARING LOSS COMPENSATION

A review of the provisions and statutes in the U.S. reveals many variations, discrepancies, and inadequacies in worker's compensation schedules and the various procedures for determining and evaluating hearing loss claims which is discussed in Chapter 7. Many methods have been employed over the years to estimate hearing impairment but none have been found to be completely acceptable for a variety of reasons. Most of the states, therefore, have various schedules and formulas for determining awards. Each state's jurisdiction is independent; however, hearing loss disability is generally based on the impairment of the ability to understand speech. Although there are several guides for the evaluation of hearing impairment, the NIOSH definition is based on the average loss over three frequencies, 1000, 2000, and 3000 Hz, with the beginning of impairment or handicap set at a 25-dB average level. The American Academy of Otolaryngology (AAO) is based on the average loss over four frequencies by including 500 Hz. Noise exposure usually causes more damage at 3000 Hz than at lower frequencies, therefore the inclusion of 500 Hz by the AAO decreases the number of calculated impairments. The latest survey of state worker's compensation legislation shows the varying degrees of act administration. The hearing loss state statutes are shown in Table 5.[4] Occupational hearing loss awards may be covered by specific or general provisions of the worker's compensation laws, or may be the result of common law action.

HEARING CONSERVATION PROGRAM IMPLEMENTATION

Although audiometric testing presents a difficult challenge in terms of establishing consistent and valid test results, almost anyone can legally perform industrial audiometric testing without formal training or certification. The OSHA regulation presently allows an audiologist, otolaryngologist, or physician to administer audiograms and provide general program supervision. This includes training, supervision, and direction of the audiometric technician. The responsibility ultimately lies with the professional in charge. The quality of tests and the program results will reflect the technique, procedures, and quality controls specified by the assigned professional.

The occupational health nurse plays a key role in an industrial hearing conservation program, because the success of the program greatly depends on his/her competence, skills, and enthusiasm. In addition to audiometric testing, fitting of hearing protection, and keeping records, the nurse in a smaller company often has significant responsibility for the implementation, coordination, and continuing administration of the hearing conservation program. Physician staffing at smaller companies or plants is frequently limited to a few hours per week, and the smaller company physician is mainly concerned with trauma treatment and pre-employment examinations. Physicians usually encourage a hearing conservation program and appreciate the direct involvement of an adequately trained occupational health nurse. In many large companies or plants with full-time medical directors and multiple medical personnel, hearing conservation programs are usually implemented and administered by the medical director with management approval and support. In these companies, the nurse is responsible for the audiometric testing, fitting hearing protection, and recordkeeping, and the assurance that the employee

Table 5. Hearing Loss Statutes in the United States

JURISDICTION	Is occupational hearing loss compensable?	Is minimum noise exposure required for filing?	Schedule in weeks for one ear.	Schedule in weeks for both ears.	Maximum compensation (one ear).	Maximum compensation (both ears).	Hearing impairment formula.	Waiting period	Is deduction made for presbycusis?	Is award made for tinnitus?
Alabama	Yes	No	53	163	$ 11,660	$ 35,860	ME	No	No	
Alaska	Yes	No	52	200	$ 9,800	$ 37,800	ME	No	P	Yes
Arizona	Yes	No	86	260	$ 14,575	$ 43,725	AAO-79	No	No	No
Arkansas	Yes	No	40	150	$ 6,160	$ 23,100	ME	No	No	P
California	Yes	No	50	311	$ 7,035	$ 43,540	AAO-79	No	P	Yes
Colorado	Yes	No	35	139	$ 2,940	$ 11,676	ME	No	No	No
Connecticut	Yes	No	52	156	$ 19,812	$ 59,436	ME	No	No	P
Delaware	Yes	Yes	75	175	$ 17,372	$ 40,535	ME	No	No	P
District of Columbia	Yes	No	52	200	$ 21,490	$ 82,652	NIOSH	6 mo.	P	P
Florida	Yes	No					ME	No	No	
Georgia	Yes	Yes	prop.	150	$ 8,100	$ 20,250	AAOO-59	6 mo.	No	No
Hawaii	Yes	Yes	52	200	$ 15,132	$ 58,200	AAOO-59	No	No	Yes
Idaho	Yes	No		175		$ 27,816	ME	No	P	P
Illinois	Yes	Yes	100	200	$ 14,681	$ 58,722	See Comments	No	No	P
Indiana	No									
Iowa	Yes	Yes	50	175	$ 26,650	$ 93,275	AAO-79	6 mo.	No	Yes
Kansas	Yes	No	30	110	$ 6,810	$ 24,970	AAO-79	No	No	No
Kentucky	Yes	No	78	156			ME	No	P	P
Louisiana	No									
Maine	Yes	No	50	200	$ 22,396	$ 89,584	AAOO-59	30 days	Yes	Yes
Maryland	Yes	Yes	125	250	$ 13,625	$ 72,594	AAOO-59	90 days	Yes	Yes
Massachusetts	Yes	No	150	400	$ 7,425	$ 19,800	ME	No	P	P
Michigan	Yes	No					ME	No	No	No
Minnesota	Yes	No	85	170			AAO-79	3 mo.	No	No
Mississippi	Yes	No	40	150	$ 5,040	$ 18,900	ME	No	P	P
Missouri	Yes	No	40	148	$ 6,300	$ 24,056	AAOO-59	6 mo.	Yes	No

Provision for hearing aid.	Credit for improvement with hearing aid.	COMMENTS
	No	
Yes	No	
Yes	No	
Yes	P	
Yes	Yes	High level of claims activity. Scheduled awards shown are adjusted not only for disability, but also for occupation and age.
Yes	No	High level of claims activity.
Yes	No	High level of claims activity.
Yes	No	Compensation depends upon average weekly wages.
Yes	No	
		Award is calculated on a wage loss basis.
Yes	No	Penalty for willful failure to wear HPDs.
Yes	No	High level of claims activity.
No	No	
Yes	No	High level of claims activity. Uses formula including 1000, 2000, and 3000 Hz average with a low fence of 30 dB. Both ears weighted equally.
		Noise-induced hearing loss not specifically mentioned in the statutes.
Yes	No	Penalty for willful failure to wear HPDs.
Yes	No	Requires verification by speech audiometry.
Yes	Yes	Award based upon 66-2/3% of wages. Recommends AAOO-59 formula, but use is optional.
		Statute revised 1984. Compensable only for traumatic injury due to single isolated incident.
Yes	No	Has a deduction for prebycusis, ½ dB for each year over 40. High level of claims activity.
	No	
Yes	No	Must be total loss in one or both ears to be compensable.
Yes		Compensated only for inability to perform usual and customary work.
Yes	No	No scheduled awards. Compensation may vary depending on whether employee is able to continue working.
Yes	Yes	
No	No	Has deduction for presbycusis, ½ dB for each year over 40.

Table 5 (continued). Hearing Loss Statutes in the United States

JURISDICTION	Is occupational hearing loss compensable?	Is minimum noise exposure required for filing?	Schedule in weeks for one ear.	Schedule in weeks for both ears.	Maximum compensation (one ear).	Maximum compensation (both ears).	Hearing impairment formula.	Waiting period	Is deduction made for presbycusis?	Is award made for tinnitus?
Montana	Yes	No	40	200	$ 5,720	$ 28,600	AAOO-59	6 mo.	Yes	No
Nebraska	Yes	No	50	100	$ 10,000		ME	No	No	
Nevada	Yes	No					ME	No	Yes	No
New Hampshire	Yes	No	52	214	$ 13,320	$ 54,612	ME	No	No	Yes
New Jersey	Yes	Yes	prop.	200	$ 4,320	$ 25,000	See Comments	4 wks	No	Yes
New Mexico	No	No	40	160	$ 11,945	$ 44,795	ME	No	No	Yes
New York	Yes	Yes	60	150	$ 8,100	$ 20,250	AAO-79	3 mo.	No	No
North Carolina	Yes	No	prop.	150	$ 19,600	$ 42,000	AAOO-59	6 mo.	Yes	No
North Dakota	Yes	Yes	50	200	$ 3,000	$ 12,000	AAO-79	No	Yes	No
Ohio	Yes	No	25	125	$ 4,425	$ 22,125	ME	No	No	No
Oklahoma	Yes	No	100	300	$ 16,300	$ 48,900	ME	No	No	Yes
Oregon	Yes	Yes	60	192	$ 6,000	$ 19,200	See Comments	No	Yes	P
Pennsylvania	Yes	No	60	260	$ 20,820	$ 90,220	ME	10 wks	No	No
Rhode Island	Yes	No	17	100	$ 5,400	$ 18,000	AAOO-59	6 mo.	Yes	No
South Carolina	Yes	No	80	165	$ 22,962	$ 47,358	See Comments	No	No	Yes
South Dakota	Yes	Yes	prop.	150		$ 38,100	AAO-79	6 mo.	Yes	No
Tennessee	Yes	No	75	150	$ 14,175	$ 28,350	ME	No	No	P
Texas	Yes	No		150		$ 30,450	AAO-79	No	No	Yes
Utah	Yes	Yes	prop.	100		$ 21,500	AAO-79	6 mo.	Yes	No
Vermont	Yes	No	52	215	$ 14,456	$ 59,770	ME	No	No	No
Virginia	No									
Washington	Yes	No			$ 4,800	$ 28,800	AAO-79	No	No	No
West Virginia	Yes	No	100	260	$ 21,420	$ 55,692	AAO-79	No	No	P
Wisconsin	Yes	Yes	36	216	$ 5,940	$ 35,640	AAO-79	14 days	No	Yes
Wyoming	Yes	No	40	80	$ 9,260	$ 18,520	ME	No	No	No
U.S. Department of Labor	Yes	Yes	52	200			AAO-79	No	No	No
Longshoremen	Yes	No	52	200	$ 30,952	$119,048	AAO-79	No	No	P

*Data compiled by E.H. Berger and Loretah D. Rowland from appendix by M.S. Fox in AAO-HNS (1982), Page (1985), U.S. should only be used as a guide. Consult local WC administrators for confirmation of current statutes.

Abbreviations used in this table are: **AAOO-59**: avg. of 500, 1000, 2000 Hz >25dB; **AAO-79**: avg. of 500, 1000, 2000, 3000 Hz physician; **prop.**: proportionate to compensation for 100% bilateral hearing loss; **P**: possible, depending upon the medical

Provision for hearing aid.	Credit for improvement with hearing aid.	COMMENTS
Yes	No	Deduction for presbycusis, ½ dB for each year over 40.
		Permanent total loss of hearing compensated as permanent total disability.
Yes	No	Degree of disability determined using AAOO-59 as applied to AMA guidelines for whole man impairment.
Yes	No	
Yes	No	High level of claims activity. Uses 1000, 2000, and 3000 Hz average with a 30 dB low fence. Penalty for willful failure to wear HPDs.
		Compensation only for injuries resulting from trauma.
No	No	High level of claims activity. Compensation for tinnitus if accompanied by hearing loss.
No		Penalty for willful failure to wear HPDs.
Yes	Yes	
Yes		Must be permanent total hearing loss in one or both ears.
Yes	No	
Yes	Yes	High level of claims activity. Formula uses average of all frequencies 500-6000 Hz.
No		Award only for total permanent hearing impairment.
No		
Yes	No	Uses 500, 1000, and 3000 Hz average with a 25 dB low fence.
Yes	No	Statute revised 1986. Penalty for willfull failure to wear HPDs.
Yes	No	Medical opinion based on current AMA guidelines.
Yes	No	
Yes	No	
P	No	
		Re 1985 Virginia Supreme Court decision work-related hearing loss is not compensable since it is "an ordinary disease of life." As of 1986 new legislation is pending.
Yes	No	High level of claims activity.
Yes	No	State's commission anticipates 1500-2000 new claims with an average award of $17,000 in 1985.
P	No	High level of claims activity.
Yes	No	
Yes	No	
Yes	No	

Chamber of Commerce (1985), WC regulations (when available), and phone calls to WC offices around the country. This table

>25 dB; **NIOSH**: avg. 1000, 2000, 3000 >25 dB; **ME**: medical evidence, hearing loss formula up to the discretion of the consulting evidence.

has not been exposed to noises that would produce a temporary threshold shift before the hearing test. Unless a plan has been implemented involving audiometric testing on a continuing, predetermined schedule, the nurse often finds it necessary to do a large number of audiograms in a short period, perhaps only a few days. This, of course, can create problems for the medical team and company production efforts.

From the beginning management must be aware of the problems that can arise in this and many other areas due to an inadequately trained staff. For example, audiometric tests completed by an individual with less than acceptable skills may serve as a basis for later compensation claims and OSHA citations. Good, reliable recordkeeping, no matter how devoted the individuals are, is one of the most important aspects of any hearing conservation program. This includes an otologic history, which lists present and previous noise exposure, job classification, and audiological record. Continual wearing of hearing protection by employees must be stressed and is considerably dependent on the counseling skills of the physician and nurse; however, enforcement is the responsibility of the industrial hygienist, management, or safety personnel. The rules must be tough, and in some companies continued employment is contingent on an employee's compliance with this type of regulation. If the company has a safety director or industrial hygienist, his/her role will probably be to direct engineering noise controls, noise measurements, and monitor worker noise exposures. The nurse should have a record of noise and exposure measurements in order to determine the degree of hazard and type of hearing protection needed. All must be familiar with all regulations and their complications.

Although motivation is a team effort, the physician, nurse, and industrial hygienist very often have the best opportunity to encourage the employee to wear hearing protection. This, of course, does not eliminate the need for ongoing group education. These sessions should follow certain principles: "Keep It Simple, Short, Meaningful, and Motivating." Worker education and training are discussed in other chapters of this book.

SUMMARY OF KEY POINTS

1. Up to 27.5 million adults have some degree of hearing impairment.
2. Compensation cost for occupational hearing losses today can reach $20 billion.
3. Occupational noise exposure has been found to have an association with hypertension and elevated blood pressure.
4. A hearing conservation program must be initiated if employees have average noise exposure levels of 85 dBA.
5. A conductive hearing loss is related to flat or ascending audiometric slopes.
6. A sensori-neural hearing loss is usually related to a higher frequency hearing loss or descending audiometric slopes.
7. A noise that induces temporary hearing losses is probably capable of causing permanent hearing losses.
8. Most significant noise-induced hearing losses initially occur at 3000 to 4000 Hz.

9. Hearing losses due to aging are similar to those caused by excessive noise.
10. In the selection of hearing protection, a wide variety of factors must be considered.
11. Most hearing protection devices do not attenuate noise levels below 1000 Hz very effectively.
12. If the hearing protection required reaches a significant level, octave band noise measurements of the work area should be taken to determine effective protection.
13. Realistically, most hearing protection devices perform significantly lower than laboratory measurements.
14. Earmuffs and foam earplugs appear to have the best realistic average attenuation.
15. Human error and problems exist with all types of audiometry.
16. A 10-dB average hearing threshold shift at 2000, 3000, and 4000 Hz is considered significant.
17. Employers shall make hearing protectors available to all employees exposed to an 8-hr TWA of 85 dBA or greater at no cost to employees.
18. Most states have schedules and formulas for determining employee worker's compensation hearing loss awards.
19. Company worker noise exposure enforcement levels may be different than governmental regulations.
20. The occupational health nurse plays a key role in most industrial hearing conservation programs.
21. Although eliminating significant noise sources should always be paramount, it should always be recognized that the best hearing protection is the one that is employed correctly.

REFERENCES

1. Hearing Status in the United States and the Auditory and Nonauditory Correlates to Occupational Noise Exposure, EPA Study Report 1982.
2. Occupational Safety and Health Administration, Occupational Noise Exposure and Hearing Conservation Amendment, Fed. Regist. 46(11): 4078-4181 (1981); 48(46): 9738-9783 (1983).
3. Criteria for a Recommended Standard Occupational Exposure to Noise, HSM 73-11001, National Institute for Occupational Safety and Health, U.S. Government Printing Office (1972).
4. Noise and Hearing Conservation Manual, 4th ed. (Akron, OH: American Industrial Hygiene Association 1986).

SELECTED BIBLIOGRAPHY

1. McGuire, J. L. "The Effects and Dynamics of Worker Exposure to Industrial Noise," PhD Thesis, University of Michigan, Ann Arbor, MI (1981).
2. D. Henderson, R. Hamernik, D. Dosangh, and J. Mills, Eds. *Effects of Noise on Hearing* (New York: Raven Press, 1976).

3. Olishifski, J. B., and Harford, E. R. *Industrial Noise and Hearing Conservation* (Chicago: National Safety Council, 1975).
4. DeWeese, D. D., and Saunders, W. H. *Textbook of Otolaryngology* (St. Louis: C V. Mosby, 1977).
5. Yost, W. A., and Nielsen, D. W. *Fundamentals of Hearing* (New York: Holt, Rinehart, and Winston, 1977).
6. The Industrial Environment—Its Evaluation and Control, U.S. Department of Health Education and Welfare (1973), chaps. 23 and 24.

CHAPTER 12

Trip, Slip, and Fall Prevention

William Marletta

INTRODUCTION

Trip, slip, and fall accidents comprise 16.4% of all work-related injuries. Falls are the second largest source of accidental death in the U.S.[1] Falls do not occur only from high places. Statistics taken from New York State accident reports show that 13.3% were same-level falls, while only 7.8% were from elevated places.[2] According to a report provided to the Consumer Product Safety Commission (CPSC) and the National Bureau of Standards (NBS), there were 31 million minor stair fall accidents in 1975 alone.[3] It is clear that a fall prevention effort is needed to reduce this type of accident in the workplace.

The prevention of trip, slip, and fall accidents begins with an understanding of their causes in order to provide insight as to the methods of control. As all such accidents involve the pedestrian interacting with a surface, there is both the human factor and the surface interaction to consider. One of the major causes is related to the expectation of the potential accident victim.

When walking conditions are contrary to those anticipated, the probability of an accident is significantly increased. Sudden changes in surface coefficient, surface design, or the introduction of a foreign substance create the potential for a trip, slip, and fall accident. A significant causal relationship is established between the "unexpectedness" of the event and the accident. This basic theory of "expectation" cannot be overemphasized with respect to trip, slip, and fall occurrence.[4] The following analysis of factors involved illustrates contradiction of expectation.

WALKING

A basic understanding of the mechanics involved with the walking process can help to explain vulnerability to accident. When walking, the natural motion transfers body weight from one foot to another. At the middle of the walking cycle the body weight is somewhat evenly distributed between both feet before a shifting of the body weight occurs. To create forward movement, the leg behind provides a force by pushing off the ball of the foot and transfers weight toward the big toe as the rear leg rises and swings forward. The weight balance shifts between the two legs. The rear propelling leg then becomes the forward leg, moving forward and landing on the heel of the foot (see Figure 1). The heel lands on the forward floor surface at an angle as an increasing proportion of the body weight is shifted. During this process, the entire body, including the leading foot moves forward.[6] This is a simplified description of what is actually a very complex musculoskeletal and neurological control process. The walking process, which we take for granted, is the result of thousands of calculations made by the brain responding (within milliseconds) to information provided to it, thus coordinating the neuromuscular control necessary to accomplish the walking cycle.[6] A change in conditions contrary to expectation will result in the gait that our brain intended for one set of conditions now being used for another set of conditions.

SLIPS

When a person walking anticipates that a surface is slippery, they will try to walk more carefully. On ice or snow, the pedestrian accommodates by adjustment of his/her gait. For example, the lead foot does not come as far forward and lands almost straight down. This reduces the contact angle between the heel and the floor and reduces the horizontal component of force, making it possible to walk on a surface with a lower coefficient of friction (COF).[5]

Friction is commonly defined as the relative resistance to movement between two surfaces in contact.[7] Static friction is the resistance at the instant relative motion begins between the two sliding surfaces in contact.[8] Dynamic friction is the resistance to movement, once sliding has begun, between the two sliding surfaces in contact. Although there has been considerable controversy, most researchers agree that it is the static friction that is largely involved with the initiation of common slip occurrences.[6,9,10,12] The COF is the ratio of horizontal force required to move one level surface over the other to the total vertical force perpendicularly applied to the two surfaces. The COF serves as an indicator of the relative friction between two surfaces in contact (COF = horizontal force/vertical force).[9] A higher COF provides greater slip resistance and less hazardous walking conditions. The walking process accommodates the anticipated COF conditions.

The heel of the forward foot lands on the floor at an angle. The angle is largely

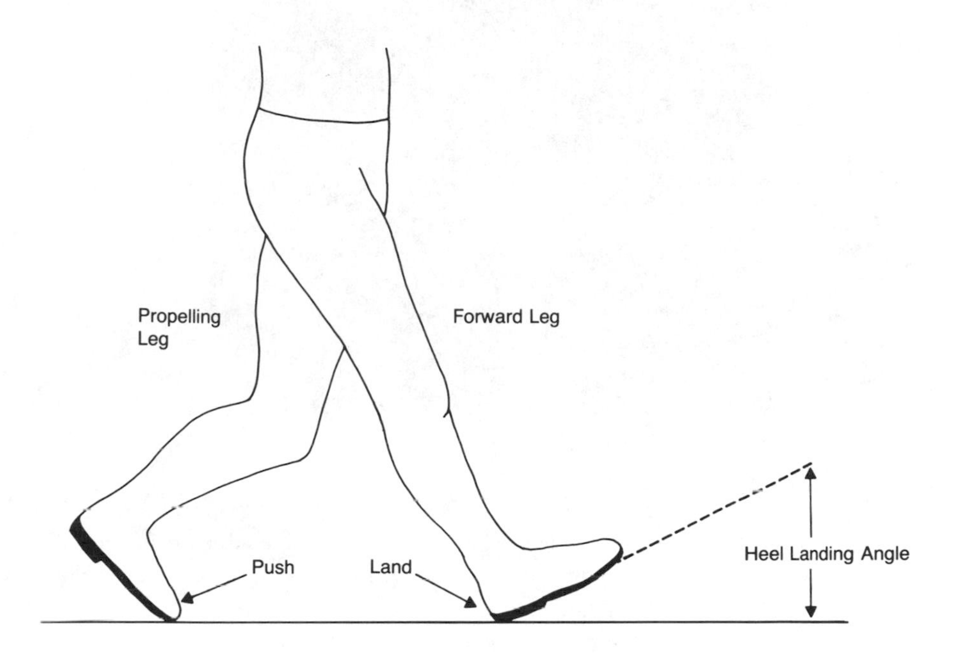

Figure 1. Heel strike during walking.

dependent on factors including the length of the legs, the length of the stride,[5] and (to a lesser extent) the particular personal characteristics of the walker. It is easy to see that as the angle decreases the vertical component (downward) friction force is increased and the horizontal component (forward) is reduced. Thus, one can adjust to walking on a surface with a lower COF.

People walking quickly take longer strides. Longer strides increase the angle between the floor gripping surface and the heel's contact angle, the reason longer strides need a higher COF for safe walking. Therefore, what was a safe surface for one type of walking may not be safe for another. One can walk on ice without slipping by walking slowly and taking shorter strides, thus reducing the heel contact angle and the horizontal force component.[5]

Contradiction of expectation of the COF often produces drastic fall results. Thus, the unexpected change in COF between surface types can be a hazard in itself. For example, a restaurant floor could be covered one half by carpeting and one half with linoleum tile. Although both floor surfaces normally can be considered safe walking surfaces, there is a significant difference in their COFs. The pedestrian traveling from the carpeted area to the linoleum will automatically com-

Figure 2. Change in walking surface friction.

pensate for the change in COFs as the brain makes complex adjustments needed to maintain proper balance. These subtle stride changes can be seen by observing slow motion videotapes of pedestrians.

The real problem is evident in situations in which the pedestrian does not expect the change to the lesser COF (the brain therefore cannot adjust the stride). Frequently, the cause of slip accidents is "unexpectedness." This is exhibited by the unusually high incidence of slip and fall occurrences associated with stepping from carpets or mats onto lower COF floor surfaces (see Figure 2). Floor surfaces with uneven wax application, floor surfaces where water is present and unobserved, fine sand particles on a smooth but properly waxed floor, fine sand particles on athletic polyurethane floors, nonuniform riser and tread designs, and stairsteps with forward sloping treads are all examples of hazardous conditions which can cause slips due to "disappointment" of expectation. These transitions become hazardous due to the fact that the pedestrian is not expecting the change and is unable to adjust his gait to the new surface condition. Perception of changes in walking surfaces conditions is often masked by factors such as inadequate lighting, momentary distractions (e.g., argument), intentional diversions (e.g., advertising), and others.[4]

TRIPS

This contradiction of expectation explains trips as well as slips. Safe practice requires that potential hazards be highlighted to improve perception or removed to avoid a pedestrian encounter with an unexpected event. Commonly encountered trip hazards (Figure 3) include foreign objects in the path of travel, such as a loose wire crossing a hallway, a curled edge of carpet, walkway depressions,

Figure 3. Dangerous sidewalk condition.

raised sidewalk projections, a pothole in a darkened parking lot, a single-step riser, etc. Such trip hazards should be eliminated to provide an unobstructed surface or sufficiently highlighted to attract the pedestrian's attention.

Trip Prevention

Vertical Transitions

Trips, like slips, are likely to occur with change in expectancy. In most occurrences, if the trip hazard were perceived by the pedestrian, the expectation would not be contradicted or the hazard would have been totally avoided.

Abrupt vertical transitions of relatively minor dimensions are capable of producing serious trips and falls. Slow-motion analysis of the path of a pedestrian foot in transit reveals that the heel of the foot skims across the floor surface as close as 1/4 in.;[13] therefore, if the walking surface provides an irregular protrusion of more than 1/4 in., which is not perceived by the pedestrian, a trip hazard exists. With respect to floor transitions, it is generally accepted safe practice that changes in level up to 1/4 in. may be allowed without edge treatment. Studies of the human gait have shown that a change in level of 1/4 in. is not normally capable of inducing a trip.[13] Changes in levels of more than 1/2 in. are significant trip hazards[6] and a transition should be made (Figure 4) with the use of a ramp that follows the requirements of the applicable building construction code; however, changes in level between 1/4 in. to 1/2 in. should be beveled with a slope no greater than 1:2 (Figure 5).[14,15]

Standards for ground and floor surfaces are incorporated in the *Federal Register* (49 CFR 31528) for Uniform Federal Accessibility Standards utilized by the General Services Administration, the Department of Defense, the Department of Housing and Urban Development, and the U.S. Postal Service.[15] In addition, ANSI (American National Standards Institute) Standard A117.1-1980 (revised from 1961), "Specifications for Making Buildings and Facilities Accessible to and Usable by Physically Handicapped People," intends for these standards of tolerance to specifically facilitate persons who have difficulties or physical limitations that hinder their use of the environment.[14]

Trip Hazards

The trip and fall theory is basic: any time man alters the floor surface conditions from those expected by the pedestrian, a hazardous condition is created. Trip occurrences often result from unevenly transitioned floor surfaces (holes, chips, depressions), cracked sidewalk conditions, inadequately marked speed bumps, improperly attached metal carpet trim, improperly placed or designed parking lot wheel stops, improperly laid or torn carpet surfaces, and improperly placed electrical extension cords. Office environments are notorious for the number of trip obstacles responsible for accidents.

Telephone cords at the sides of desks should be neatly tied up and away. Special ties are available for this purpose. Electrical floor receptacles represent difficult to see obstacles and should be properly located away from corridors, aisles, and traveled walkways. Floor receptacles that become exposed due to the relocation of desks or office equipment should be capped flush with the floor or marked with cones or similar warning signs. Coffee carts offer potential for spilled liquids that should be promptly cleaned up. Stairs must be kept clear of debris, papers kept clear from walkways, file drawers should be kept closed, extension cords should be carefully placed (taped or marked if required), to mention just a few of the most commonly overlooked hazards. Many times these hazardous conditions are all too casually observed.

One-Step Riser Design

The single-step riser has long been recognized as a potential hazard to pedestrians because of its difficulty of recognition, partially due to its brief span of transition. Again, the primary accident cause is usually related to a change in expectation. When there are no visual cues provided, such as a handrail, the single step is especially likely to be undetected by pedestrian users. Also, if there is a similarity of surface colors and lack of contrast between the surface levels, the

Figure 4. Carpet to tile transition.

Figure 5. Dangerous raised sidewalk condition.

ability of pedestrians to quickly perceive the transition is diminished. Some building codes prohibit the use of single-step risers in the path of an exit, except at grade level door exits. Although the use of a one-step riser may be anticipated at grade level exterior door exits, it should be avoided wherever possible. Safe practice is to consider the use of a ramp wherever one step risers are contemplated.[4]

The potential hazard associated with one-step riser design can be reduced with the use of visual cues such as direct lighting sources, handrails, or warning signs. Visual perception of the step can also be enhanced with proper hazard marking as identified in the ANSI Standard, "Safety Color Code for Marking Physical Hazards" (Z53.1-1979).[16] Contrasting colors and patterns should be integrated in decorating finishes to maximize the contrast between floor levels.

MEASUREMENT OF SLIP RESISTANCE

The degree of slip resistance of a floor surface, shoe surface, or coating application often requires measurement. The measured COF provides a numerical value from 0 to 1 which can be used as a scale reference. The larger the value, the higher the degree of slip resistance. Higher values indicate less slippery surfaces. Two different types of friction, static and dynamic, are commonly tested. Most slips involve the static measurement of resistance at the instant relative motion begins between the two sliding surfaces in contact. It is the static measurement that measures the friction force the moment the heel begins to slide. Although early research focused on dynamic testing, many recent studies have concluded that static measurement is more pertinent to most slip occurrences.[6,9,10,12]

Devices for Testing Slip Resistance

Numerous devices have been developed for the measurement of slip resistance and fall into three basic categories: pendulum devices, articulated strut devices and drag testers (sled).[17]

Pendulum testers measure the frictional properties of the swinging shoe as it contacts the surface. The pendulum method was originally developed by Dr. Percy Sigler in 1945 while employed by the National Bureau of Standards. The Sigler Pendulum Tester (Figure 6) was used to measure dynamic friction in the 1940s and 1950s.[6] Its use has been largely superseded by static testing devices.

The second category of testing devices is known as the articulated strut. The two machines most widely acknowledged within the articulated strut category are the James Machine Tester and the Brungraber Tester. Articulated strut devices operate by applying a known vertical force through a strut which has mounted a plate with the sole specimen (i.e., shoe sole) to be tested. The plate with the sole specimen is pressed against a horizontal table fixed with test flooring. The table moves horizontally at a determined speed, causing the strut to move from the perpendicular position until a slip is recorded. The James Machine (Figure 7) is a large test instrument, developed in 1945 by Dr. Sidney James while working at Underwriters Laboratories (UL), which measures the static COF of surfaces.[7] It is not suitable for field use due to its size; however, it is a widely recognized testing device with standards developed for use and evaluation of the test data. The James Machine is referenced in the ASTM standard D2047, "Static COF of Polish-Coated Floor Surfaces as Measured by the James Machine."[9] The Brungraber Tester is also an articulated strut device with an advantage of being portable.[12]

The third category, drag testers, utilizes a known weight provided with a shoe

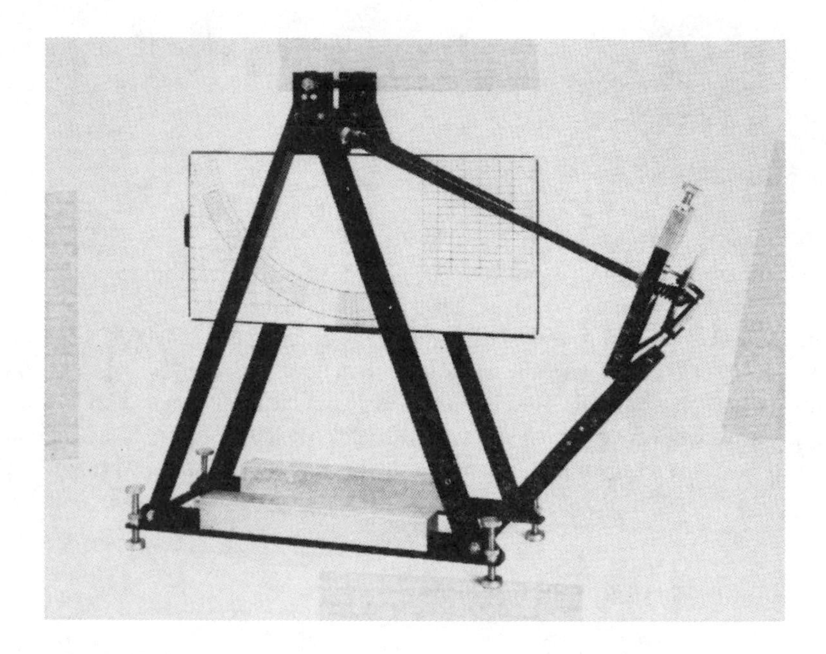

Figure 6. Sigler Pendulum Tester.

sole material sample attached to the bottom. The material is pulled across a surface while measuring the force required to create a slip. These devices can be used to test static or dynamic friction. The Horizontal Pull Slipmeter (HPS) is a drag device designed for use in the laboratory and field (Figure 8). Samples are mounted (usually leather) directly in the base of a slip index gauge and are used as frictional material for evaluating various floor conditions. The slip meter is pulled across the floor surface at a constant weight, speed, and direction. The applied force is increased to the point that the device records a slippage. A reading is taken directly from the slip index gauge. Differences in measurements due to operator technique are reduced by the use of a motor driven unit that controls the speed and direction at which the force to overcome friction is applied.[18]

The HPS is specified by the American Society for Testing and Materials (ASTM) for use in ANSI/ASTM Standard Test Method F609-79 to determine static slip resistance of footwear, heel, sole, and related materials on smooth walkways.[8] ASTM Committee D-21 (Polishes) has recommended a test method for measuring the "Slip Resistance of Polish Coated Floors" with the HPS. The Occupational Health and Safety Administration (OSHA) has evaluated the HPS for use in field testing safety conditions in the workplace. Military Specification MIL-D-23003A for "Deck Covering Compound, Non-Slip, Rollable" refers to the use of a modified HPS test.

The Technical Products tester (Figure 9) is a less expensive drag device, that

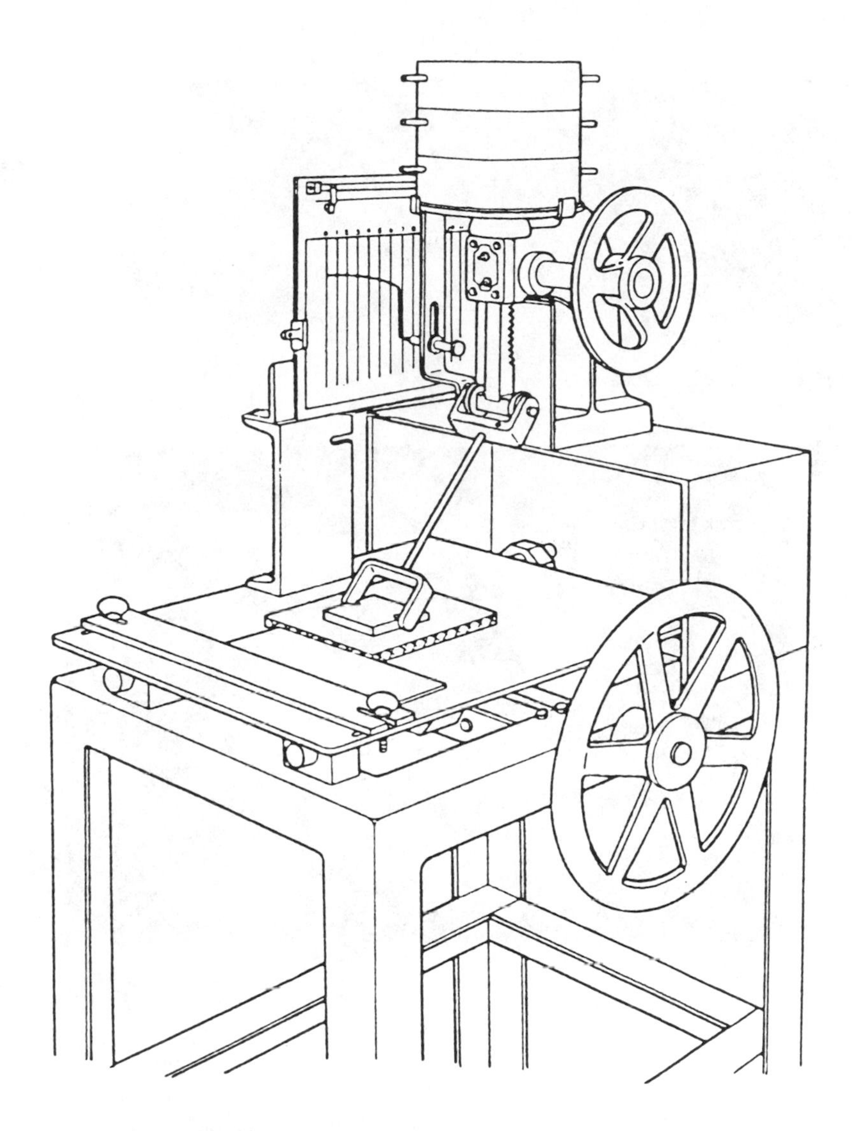

Figure 7. James Machine.

has contact pads on the bottom of the tester but that is manually pulled across the surface. It does not offer the uniform velocity, smoothness of operation, and assured pull parallel to the direction of surface as would be provided by the automated pull box devices. Many other similar types of drag testers are also in use.

Different slip testing machines will produce different results.[6] Slip testers that do not conform to ASTM standards will produce results that cannot be compared to those standards. Differences in slip testing equipment as well as procedure will

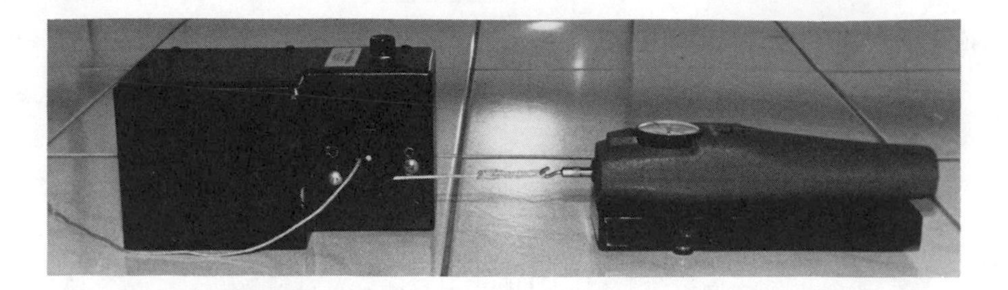

Figure 8. HPS slipmeter.

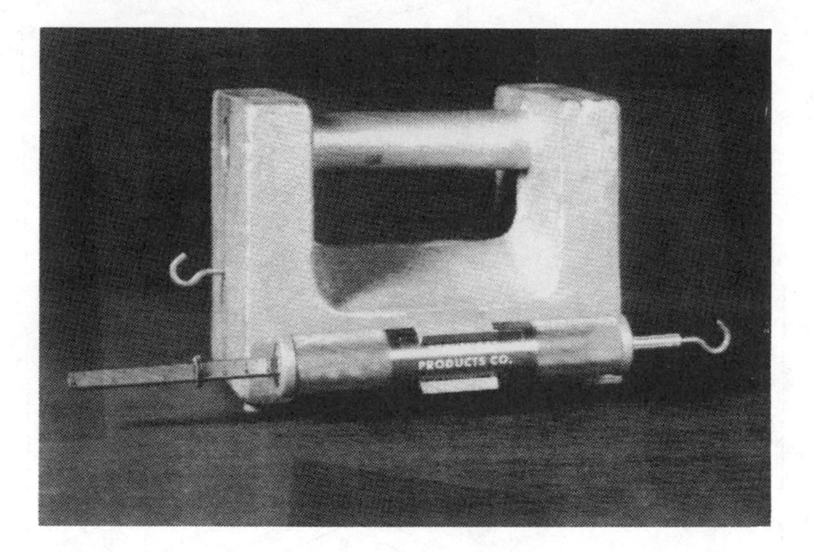

Figure 9. Technical Products Model 80 Floor Friction Tester.

affect the validity of the results. Differences in temperature and humidity may be found to affect slip resistance as well.[18] Although some slip devices have an internal calibration, comparison to a known source should be performed before and after each test. Unless the reliability of the slip testing equipment is known, a valid comparison to a standard cannot be determined.

INTERPRETATION OF SLIP TEST RESULTS

The greater the COF or antislip index, the less slippery the surface. A highly glossed surface is not necessarily a slippery surface. There is no easy way of telling (i.e., merely by observing the surface reflection) whether the floor surface is slippery. Many safe floor waxes and finishes are commonly available that

Table 1. COF Interpretations and Examples

Slip Resistance (COF)	Interpretation	Example
0.70 and above	Extremely safe	Rough brushed concrete
0.60 to 0.69	Very safe	Vinyl tile, quarry tile
0.50 to 0.59	Safe	Dry terrazzo, dry marble
0.40 to 0.49	Dangerous	Wet terrazzo, wet marble
0.35 to 0.39	Very dangerous	Wet terrazzo, wet marble
0.00 to 0.34	Unusually dangerous	Ice, soapy water

Note: These examples are shown as typical of the COF category slip resistance with leather illustrated for relative comparison; however, it is recognized that significant variances will exist due to condition, wear, differences in materials, etc. (Table COF cateogory and interpretation adapted from the source.[6])

provide brilliant and desirable floor shines. It is generally accepted that a COF greater than 0.5 (or antislip index 5.0) will produce a relatively safe surface (Table 1). This concept is recognized in standards published by the ASTM[9] and UL.

Slip Mechanism

The COF is largely affected by the interaction between a heel material and a floor surface. However, slipperiness can be affected by factors other than the reaction of these materials. For example, foreign materials on either of the mating surfaces can adversely affect the degree of slip resistance. Sand and gravel on a smooth floor surface can act like ball bearings in their ability to reduce friction. Temperature and humidity can also produce a significant effect. Therefore, "unless all factors are taken into proper consideration, acceptable results from the measurements of slip resistance will not necessarily assure resistance to slipping"[8] under field conditions.

Surface Materials

Terrazzo is a mosaic floor material composed of marble chips, granite, onyx or silica. The floor material is bonded with Portland cement, poured in place, cured, and polished when dry. The installed terrazzo is frequently sealed. Although most terrazzos are safe for normal pedestrian use when dry, *terrazzo is slippery when wet*. Terrazzo is identified as a "high risk" material for stair tread surfaces;[3] however, it can be manufactured with slip-resistant additives to produce a floor that is safe when wet. Terrazzo floors can be made safe through the use of slip-resistant mats. Grooving, slip-resistant strip applications, and grit additives are alternative methods. Terrazzo floors at building entrances require water-absorbing mats (Figure 10) to absorb moisture tracked on the bottoms of pedestrian shoes.[20,21]

Marble as a flooring material will vary according to the type, condition, and degree of wear. Marble, as well as many other stone surfaces, is basically non-porous and resists water absorption. Smooth marble floors are slippery when wet.

Figure 10. Use of mats to cover terrazzo entrance to building.

Rough finish concrete surfaces provide good resistance to slipping when dry and wet; however, smooth finished concrete surfaces often are slippery when wet and should be avoided in outdoor walkways.

Ceramic tile floors come in many different types with varied surfaces and slip-resistant properties. The properties of each vary considerably, although clean, virgin tile usually will produce slip resistance levels greater than 0.6. Glazed smooth surface tiles have low COFs and can be slippery when wet. Some ceramic tiles are manufactured with particles imbedded in the surface for increased safety. Other ceramic tiles alter the smooth surface by introducing a tractional pattern to produce a safer surface. Manufacturers' recommendations should always be reviewed regarding areas for safe use; however, the manufacturer often overstates the recommended safe use and fails to indicate that certain tiles can be slippery with water or other substances. Grease on or absorbed into ceramic tile (i.e. restaurant applications) can produce very slippery conditions when wet or mixed with commonly used cleaning chemicals.

Smooth ceramic tiles can be treated with acid etching techniques to improve the slip resistance of tile already installed in the field. The use of rubber-backed mats, slip-resistant strip applications, grooving or other controls is necessary in areas where wet conditions are anticipated.

Figure 11. Cafeteria ice dispenser over a tile floor.

Water and Foreign Debris

Safe practice requires that building entrances having smooth, nonporous surfaces such as terrazzo, most varieties of marble, ceramic tile, and many other floor materials, be covered by absorbent mats on wet days when water can be expected to be tracked inside. Similarly, usage of these floor materials in cafeterias where spillage of food, ice, and liquids (Figure 11) can be anticipated is risky.[20,21] These areas require frequent inspection, prompt maintenance, and the strategic placement of mats or other slip-resistant floor controls. One of the most commonly overlooked slip hazards exists in supermarkets where produce aisles are rarely

Figure 12. Grape hazard in grocery store produce aisle reduced with use of mats.

covered with sufficient slip-resistant mats (Figure 12). Safe practice requires that produce aisles have mats, in addition to frequent inspection and prompt maintenance of the area, because the use of mats represents a minimal cost investment that is more than justified by the reduction of risk (Figure 13).

FLOOR AND DECK PAINTS

Increased attention should be given by manufacturers of floor and deck paints to provide safer floor surfaces. Sometimes a floor product, which has been well designed from the standpoint of durability, is not a safe walking surface. Frequently, the coatings tested have a slip resistance below the 0.5 COF value when dry! Many paint and application products contain no warnings that they can be slippery when used outdoors. Painting of floor surfaces often reduces the COF of the surface. Repeated paint applications on brushed concrete, for example, will have the effect of filling in the microscopic pores that are vital to the tractional properties of the surface.

Consumers Union has published paint evaluations of the slip resistance of different porch and deck products including latex and oil-based formulations. The report confirms other testing which supports that latex products are less slippery than the oil-based paint products tested. More importantly, even though some

Figure 13. Multiple mats at entrance for slip control.

Figure 14. Sand additive makes paint slip resistant.

products tested in the report were much better than others, *none of the products were considered acceptable.*[17,22]

Floor paints can be inexpensively made nonskid with the addition of sand or other grit particles that can be mixed with the paint. Skid-Tex (Gamma Laboratories, San Fernando, CA) is one such product commonly available for control of slips and falls. Most experienced professional painters will use a slip additive for floor paint applications in recognition of the potential hazard (Figure 14).

SNOW AND ICE

As snow and ice adhere to walking surfaces (drastically reducing the COF of the surface), pedestrians are faced with special walking problems. The traction obtained is no longer the result solely of the contact between the shoe and the floor. The ice between the two surfaces is in direct contact with the shoe sole. This will produce COF values that are unusually dangerous. Although traction considerations in shoe or floor design are important factors, immediate attention should be given to the proper removal and maintenance of public walking areas. Proper maintenance techniques include plowing, shoveling, deicing, salting, and sanding, as needed. Preplanning ensures the best areas for snow pile deposits. If possible, snow storage piles should be located downslope, with sufficient capacity for drainage, to avoid problems with melting and refreezing in pedestrian paths.

Salt application is commonly used as an ice remover although it has some distinct limitations. Salt mixed with ice creates a solution that has a lower freezing point than water. Therefore, the salt will cause snow and ice to melt when applied

Table 2. NEISS Survey of Stair Accidents

Incident Type	Incidents per Year
Flight uses	1,953,000,000,000
Noticeable missteps	264,000,000
Minor accidents	31,000,000
Disabling accidents	2,660,000
Hospital treatments	540,000
Related deaths	3,800

Note: The estimated incident occurrence on stairs (1975) is the result of data compiled by the Consumer Product Safety Commission from NEISS; from A Survey of Stair Use and Quality conducted by Carson Consultants, Inc., of Milwaukee, WI; and from videotapes of stair use studied by the NBS. The NBS reports that NEISS compiled information of reported injuries treated in 119 hospital emergency rooms across the U.S.[3]

to roads and sidewalks. When temperatures fall below 25°F the salt is not as effective. The ice formed tends to prevent the salt from going readily into solution.[23] Also, the use of salt applied on the surface of the ice will not provide needed traction. Salt can harm vegetation and can cause corrosion as well. Sand provides excellent traction on ice and snow surfaces, does little harm to vegetation, and is relatively inexpensive. Alas, it will not melt snow. Most municipalities use a blend of salt and sand. In addition, chemical deicing products are commonly available for use on sidewalks.

STAIRS

Since 1973, the CPSC has gathered product-related accident and injury data through the National Electronic Injury Surveillance System (NEISS). According to the NBS report prepared for the CPSC (NBS Building Science Series 120 *Guidelines for Stair Safety*, 1979), since 1974 "stairs, ramps, and landings are among the two most hazardous consumer products in the United States." The statistics published in that report (see Table 2) are astounding: 31 million minor accidents and 3800 deaths reported.[3]

The NBS *Safety on Stairs* Series 108, 1978, reports that "stairs are hazardous," and in a study of 253 residential respondents to a survey (n = 52), a total of 87 repairs were needed to make their stairways safe. The 68 full replacement repairs (not including minor repairs) most often noted the need for structural changes such as handrails and tread materials.[27] The high accident rates identified with stair use have directed establishment of strict tolerances for step dimensions and criteria for safe design.

Stairs are comprised of a series of steps or flights of steps connected by level landings (Figures 15 and 16). Guardrails are provided as a protective barrier against falls from open sides. In addition, handrails are installed as a means of

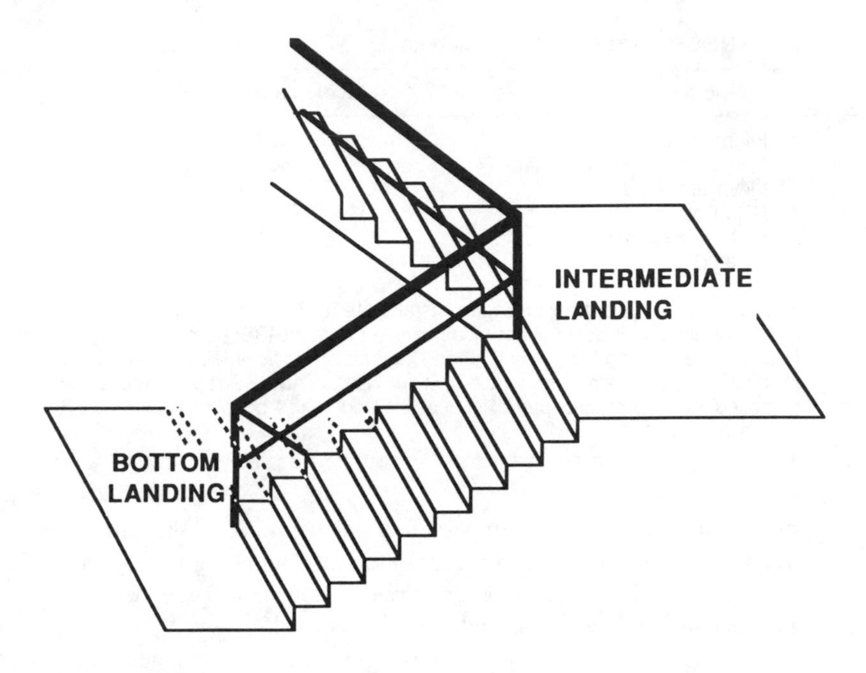

Figure 15. Top view diagram of typical stair with landings.

support, providing a safe handhold. Handrails are usually mounted on brackets to a wall or partition, paralleling the stair slope, and are designed for the grasp of the stair user.[24] A baluster is a vertical member that supports a handrail or guardrail (a railing with its supporting balusters is referred to as a balustrade).

The step is the basic unit of a stair consisting of a tread and a riser. The tread is the walking surface of the step defined as the distance from the front of the step to the riser wall. This should not be confused with the run which is the horizontal distance between risers (the tread distance minus the nosing). The rise is the vertical distance between treads and the riser is simply that portion of the step that is vertical between two treads. Many stairs have a nosing or projection of the front portion of tread over the riser wall (see Figure 17).

Step Geometry

Step geometry refers to the relationship between the rise and the run which comprises the basic slope of the stair. Early studies (reported circa 1672 by Sir Francois Blondel, director of the Royal Academy of Architecture, Paris) proposed that the rise and run of a stair should be related to the human stride.[25] Blondel devised a formula still recognized today, although slightly modified, as a basic method in determining step design:

$$2 \times \text{riser} + \text{run} = 24\text{-}25 \text{ in.}$$

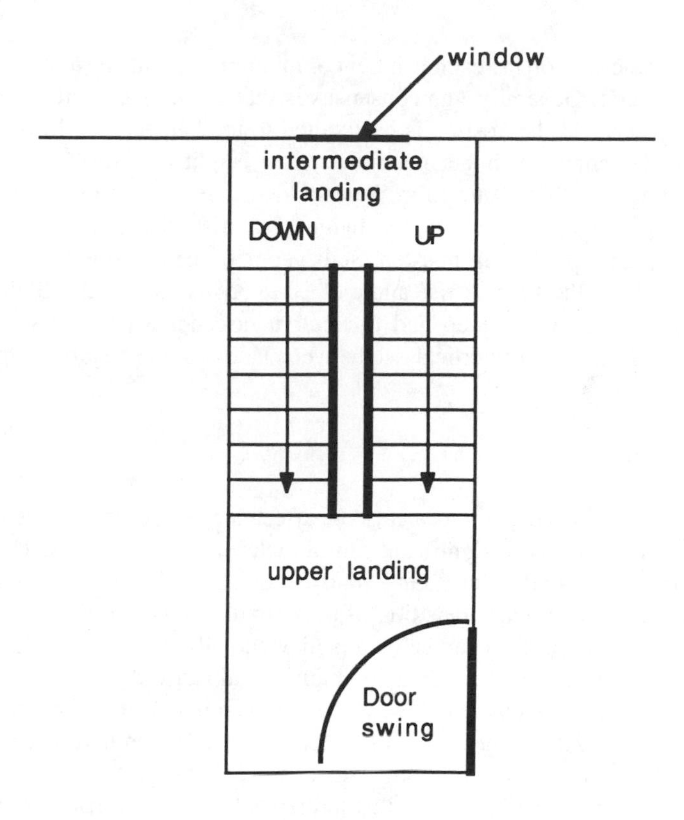

Figure 16. Typical stair design (side view shown).

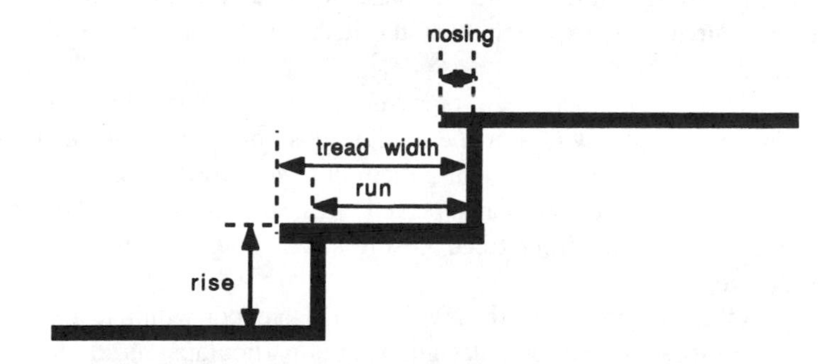

Figure 17. Rise, run, tread width, and nosing.

This equation is true even though anthropometric changes in height have occurred in the adult population and today's concept of the inch is slightly greater.[25]

Safe step geometry refers to the combination of rise and run utilized in design. The Blondel formula is suitable for a limited range of dimensions and requires that minimum and/or maximum rise and run criteria be established. Codes and standards vary somewhat on the establishment of minimum and maximum riser height and tread criteria. Generally, the consensus is that a 7-in. riser with a 101/2-in. tread, irrespective of the nosing, is best suited to the human gait. Most national building codes stipulate that a maximum riser height of 71/2 to 8 in. and a minimum tread depth of 9 to 10 in. be provided; however, many codes do not specifically establish a minimum riser height. Low riser design is hazardous and is known to cause pedestrian misstep. Stair geometry can be observed by measuring the slope of the stair. A rise angle of 30 to 35° is preferred.[28] Stairways of 40° are considered very steep and difficult to ascend. Most codes require a maximum of 10 to 12 ft of vertical rise between landings and headroom to exceed 61/2 ft.

Step Design

The pedestrian's expectation can also be affected by improper step design. The altering of step design is a significant causal factor in stair accidents. The human gait will best accommodate a 61/2 to 7 in. rise; however, whatever the riser height, the natural gait will attempt transition. If the riser design is too large or too small the natural walking process can be disrupted. Generally, in consideration of stair descent, Rosen states, "if the riser is too high, it will cause the foot coming off it to land further out on the step below. If the riser is too low, it will cause the foot coming off it to land further back on the tread surface."[6] The high riser situation causes the ball of the foot to land further forward where there is inadequate support. In contrast, the low riser situation tends to involve difficulties with the opposite foot. The heel becomes caught clearing the tread surface due to its already improper positioning toward the back of the tread.[6] Both situations are different than our mental expectation of the riser height and can result in trip occurrence.

A similar mechanism for accident occurrence exists with tread design. Minor variations in tread depths will alter the landing position of the ball of the foot. Tread depths that are too short will cause the foot to slip off the nosing. Altering a tread design so that it is longer than expected "will cause the heel of the opposite foot to get caught on the front tread nosing" as it unexpectedly attempts to complete the step.[6]

Changes in the riser height or tread width can also occur within a stair system and are referred to as nonuniform design. This is particularly hazardous as the pedestrian expects the next riser or tread to be the same as the one before it. This hazard is recognized by most building codes which require that variance between risers or treads be less than 3/8 in. (Some codes require uniformity to within 1/8 in.) Nonuniformity produces changes contrary to our mental expectation of the riser height or tread width. This change, when not perceived, prevents necessary accommodation in the gait, which often results in a trip and/or a fall.[4]

Stairs with two or more risers usually require a handrail set 30 to 34 in. from the nosing;[24,28,] however, as already mentioned, the use of a handrail is desirable even for one step risers as it will help serve as a visual cue. A handrail is important as it will serve to help prevent the initial occurrence of a slip/misstep as well as provide a means of recovery for the pedestrian once a fall has been initiated.

A handrail should be capable of supporting a 200-lb load as per ANSI A12.1-1973 "Safety Requirements for Floor, Openings, Railings and Toeboards." Adequately supported handrails are most important to the elderly who will sometimes use the handrail to pull on in assisting themselves up the staircase. Most codes require handrails for both sides of the stairs. Stairs <44 in. in width may be required only one handrail. Stairwidths >88 in. may be required to have an additional intermediate railing.[24] It is equally important, especially to the elderly, to design the handrail without requiring the user to break the handhold. Clearance between the handrail and all objects should be at least 11/2 in.[24,28] The use of ornamental, oversized handrail construction does not permit powerful grasping of the handrail and should be avoided. Many codes do not adequately address this issue in detail; however, a round (or similarly appropriate shape) handrail with a diameter of 11/2 to 2 in. is well suited for safe use.[28] In addition, safe practice and many codes require that the handrail system be extended 12 in. past the top and bottom risers. The extension of the handrail system is an important feature as it helps stabilize the pedestrian on the landing area and provides a means of recovery in an area of the stair where a majority of accidents occur (top and bottom three steps).

Inherent to winders and spiral stairs is the intended design characteristic of making one side of the tread wider than the other. Most codes prohibit their use in required exit stairs although some make exception for ornamental use or use where the exit stair is not required by the code. Unfortunately, most winding stairs are central and represent the main exit. Spiral and winder stairs often lack proper slope design, lack adequate tread width dimensions (on the narrow side of the tread), lack uniformity, have outer handrails out of reach at the winder, and are difficult for most pedestrians to maneuver. Spirals and winders are often permitted by codes in single family dwellings in consideration of the familiarity of the user with the environment and the reduced volume of traffic anticipated. Regardless, they are dangerous to pedestrian users and should be avoided in stair design.[25]

A common accident causal factor, frequently overlooked, is the improper installation or maintenance of stairway carpet.[3] If stairway carpeting is loose or improperly attached, excess carpeting may exist on the step edge (Figures 18 and 19). Improper carpet installation may be caused by excessive padding, inadequately stretched carpet, or improper attachment of the carpeting to the stair. Selection of high pile carpet also should be avoided in commercial stair applications.[3,14,15] Excess carpet causes the tread depth support to appear longer than it actually is and when the foot lands on the tread edge it is insufficiently supported. This causes the foot to slide off the edge. If the pedestrian's expectation of the presence of a supporting tread fails to be maintained, the condition is ripe for a slip occurrence on the stair tread surface.

Many floor surface materials that are considered safe for other walking applications provide inadequate traction and are unsafe for use on stairs or steps. The ball of the foot contacts the nosing at an angle which reduces the surface area of

Figure 18. Excess stair carpet over nosing.

the sole contact and which provides a friction force downward as well as horizontal component forward in the walking direction. The mechanics of descending a stair system includes the pedestrian's weight pressing the ball of the foot against the tread nosing while the opposite foot has not yet reached the tread below. It is at this point that the pedestrian's stability is most vulnerable (with only one point of contact at the stair). Therefore, the COF needed for safe pedestrian walking is greater on a stair tread surface than that of a flat walking surface. Most codes recognize the need for slip-resistant tread materials and require that treads and landings be slip-resistant. The COF of treads can easily and inexpensively be

Figure 19. Loose carpet attachment.

improved by many techniques, including tread replacement with slip-resistant treads, tread resurfacing (i.e., vinyl- or rubber-tractioned top), addition of nonslip strips, application of slip-resistant epoxy, or grooving.

Stairs or steps should be designed to transition from level landings at the top and bottom.[28] A stair should never lead to a ramp. Doors should never open onto stairs. Level platforms should always be provided for each side of door openings. Most codes require a minimum 5 ft top landing, 5 ft for intermediate landings,[15] and 6 ft for bottom landings.[35] Platforms or landings more than 4 ft high should have a standard guardrail.[24]

Stairs and exits should be well illuminated to allow the pedestrian the opportunity to adequately perceive the step edges as well as identify any possible hazard that can effect his/her ability to safely transition the stair. Many codes require a minimum illumination of 1 footcandle (fc) at the tread level. Some codes require 5 fc illumination at required exits. These are certainly minimal criteria for the illumination of stairs; however, many stairs fail to provide even these basic requirements. The Society of Illuminating Engineers (IES) recommends light intensity levels of 10 to 20 fc in stairways.[26] It is recommended that an alternate source of illumination be provided in stairways to help reduce objectionable shadows that

Table 3. High Risk Stair Accident Factors[3]

Variable	High Risk Factors
Steps	Top and bottom 3 steps represent 70% of stair accidents
Width	60, 61 and 66 in.
Riser height	Lower than 61/4 in.
Tread depth	12 in. or shorter
Nosing	No nosing high risk
Stair surface	Terrazzo

can be cast with some single illumination sources. The quality and character of the light source selected (i.e., direct or indirect) is also an important consideration as direct light can provide strong shadows as opposed to more diffuse illumination sources.

The NBS (Guidelines for Stair Safety NBS Series 120) has made numerous recommendations based on its evaluation of studies on the subject of stair falls. Some of the high risk factors identified are listed in Table 3. The visibility of step edges is often adversely affected by the use of distracting patterns (Figure 20). It is important to delineate the edges of each tread to allow stair users to view and perceive the tread edges. Avoid use of treads that are surfaced with floral patterns, pictorial, and busy randomized patterns that distract from the perception of the edge of the treads.[3] If necessary, mark the edge of the step with a safety stripe in a contrasting color (safety yellow).

RAMPS

A ramp is defined as a walkway surface with a slope of more than 1:20. Safe practice requires that a ramp be designed, if feasible, wherever a one-step riser is contemplated. Ramps are usually a good substitute for unsatisfactory step geometry as well. The increased use of ramps in building design has enabled access of handicapped persons to facilities. Although local codes vary, the American National Standard "Specifications for Making Buildings and Facilities Accessible to and Usable by Physically Handicapped People," ANSI A117.1-1980, and the Uniform Federal Accessibility Standards are widely used reference sources.[14,15]

The slope of the ramp is a critical aspect of safe design. The slope ratio is usually expressed in terms of feet of vertical rise (H) in relation to the feet of horizontal travel (L). In Table 4 and Figure 21, the slope ratio is expressed as H:L. To convert the slope angle in degrees to a slope ratio, utilize the formula: tangent (ramp slope angle) = (H/L). Many codes and safe practices require that pedestrian ramps be designed with a maximum slope ratio of 1:8 (approximately 7°). Designs exceeding 1:8 are steep and do not provide the user the opportunity to stop and rest. Handicapped ramp designs should not exceed a maximum slope ratio of 1:12 (approximately 5°).[14,15,35] Commonly accepted ramp slope criteria are summarized in Table 5.

Figure 20. Busy pattern on stair treads making difficult proper step perception.

Table 4. Commonly Used Ramp Slope Conversions (Approximate)

Slope Ratio	Degrees	Notes
1:4	14.0	
1:6	9.5	
1:8	7.1	Maximum pedestrian ramp slope
1:10	5.7	
1:12	4.8	Maximum handicap, new construction
1:15	3.8	
1:20	2.9	Walkway

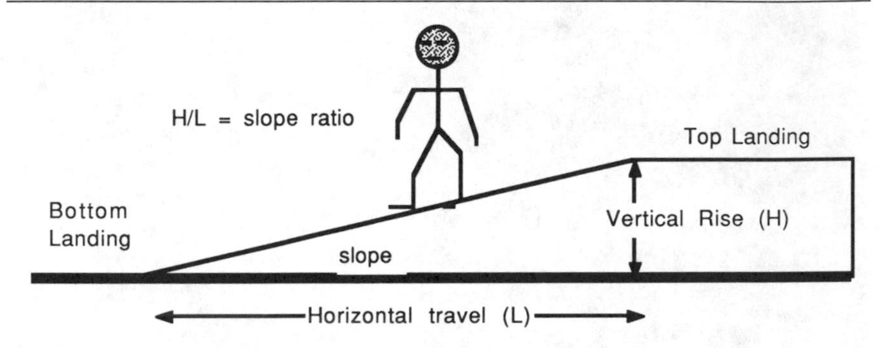

Figure 21. Ramp slope ratio.

Ramps should always have a level top and bottom landing. Doors should never swing directly onto a ramp but rather onto a level landing. Level platforms should be provided on each side of door openings. Ramps that are too short to detect visually can be hazardous to pedestrians. Ramps should transition a minimum of 3ft in the direction of travel. Minimum ramp widths of 44 in. are generally required. Platforms more than 4 ft high should have a standard guardrail.[24] Generally, landing depths should be designed with 5 ft minimum at the top landing, 5 ft minimum at the intermediate landing,[15] and 6 ft minimum at the bottom landing.[35] Requirements for handrails vary; however, many codes require a minimum of one handrail on a ramp greater than a 1:15 slope and some require handrails on both sides. Most codes require handrails to be 30 to 34 in. above the ramp surface. Handrails should extend 1 ft beyond the top and bottom of the ramp surface.

Most nationally recognized building codes require a nonslip surface on ramp and landing surfaces. A nonslip surface is generally required to have at least a 0.5 COF under dry conditions. Nonslip surfaces can be easily obtained by using a brushed concrete surface, frictional strip applications, crosscutting into the surface, or by applying other treatments or additives to the surface designed to increase surface friction.

Table 5. Commonly Accepted Ramp Slope Criteria

Slope Ratio	
1:8	Maximum pedestrian slope and existing handicapped
1:12	Maximum handicapped slope, new construction
1:20	Walkway

GUARDRAILS AND HANDRAILS

The guardrail is a protective barrier railing that is provided around an open-sided landing or stair to prevent persons from falling. A guardrail can commonly consist of a top railing with vertical support railings between the top and bottom or can be designed with a horizontal intermediate railing between the top railing and bottom; however, the purpose of the guardrail is to guard via provision of a sufficient safe enclosure.[24]

Local building codes vary with respect to specific design requirements; however, a nationally recognized standard has been established by ANSI (A12.1-1973) entitled, "Safety Requirements for Floor, Openings, Railings and Toeboards." The ANSI standard and safe practice requires that a guardrailing have a top rail, intermediate rail, or equivalent protection, and shall have a minimum vertical height of 36 to 42 in. A stair railing should be of similar construction, however, the vertical height should not be more than 42 in. or less than 30 in. measured directly above the tread at the stair nosing.[24]

Safe practice requires that guardrails be on all open sided flooring, wall openings, ramps, platforms, or runways to prevent falls. A handrail is a single support member mounted on brackets on a wall or partition, installed as a means of support to furnish ramp or stair users with a handhold in the event of a slip or trip. Handrails and stair railings should be provided with a clearance of not less than 11/2 in. between the railing and any other support. They should also be rounded or otherwise designed to furnish the user with an easy-to-grasp handhold; capable of sustaining a load of a least 200 lb in any direction[24] and continuously attached so that it offers no obstruction that would cause a user to break his/her handhold from along the rail. This is especially important to elderly users who may have difficulty in accessing the stair and who rely heavily on the handrail for support. Note that the presence of a handrail, usually set 30 to 36 in. from the nosing (depending on the source referenced) provides a rail for pedestrian hand support and stability, but which can be considered inadequate by itself as a safe measure to enclose and guard the stair.

Some codes specifically require that stairs have walls, grilles, or guards at the sides in addition to handrails. Often, codes do not specify an intermediate rail on the guardrail system because of the ease with which persons (especially children) fall through. Standards refer to specifications regarding the spacing of supports in view of the fact that it is easy for a child or adult to slip through excessively spaced supports.

LADDERS

ANSI has developed nationally recognized standards regarding ladders, which OSHA has also adopted. In addition, the National Safety Council (NSC), American Ladder Institute (ALI), and UL also publish information regarding ladder design, use, or maintenance.[29] The American Society of Safety Engineers (ASSE) presently serves as secretariat for two of the ANSI Standards discussed here: Portable Wood Ladders, ANSI A14.1, and Portable Metal Ladders, ANSI A14.2.

The information provided in this discussion of ladders is intended to highlight some of the more important and basic practices for safe ladder use and design. Reference is made to some of the more commonly used ANSI standards, although other ANSI standards are written for specific ladder applications (i.e., marine ladders, fire ladders, etc.). This brief overview is not intended to replace the need to reference applicable standards or reference sources available. Specific ladder design applications should refer directly to the appropriate standards. For our purposes, we consider ladders as divided into four main categories, as listed in Table 6,[28] although it should be recognized that other ANSI standards (e.g., ANSI 14.5 -1982 Safety Requirements for Portable Reinforced Plastic Ladders) are available for specific applications.

Job-made ladders are temporary (temporary, because they may be used only until a permanent ladder can be manufactured and installed) custom made ladders that are provided during construction or demolition operations.[32] Portable ladders can be easily moved or carried whereas fixed ladders are rather permanently attached. The side members of a ladder are called rails and are joined by steps or rungs.

Fixed Ladders

Some of the more important requirements for fixed ladders are that they be able to withstand a single concentrated load of 200 lb. Rung minimum diameters of 3/4 in. metal or 11/8 in. wood are required. Side rails should be at least 16 inches apart and rungs spaced no more than 12 in. apart. Clearance of the ladder on the climbing side should provide a minimum of 21/2 ft, although 24 in. is acceptable when unavoidable obstructions are encountered (where deflector plates are provided). Clearance of 7 in. should be provided from the back of the ladder to the nearest permanent object[29] (Figure 22).

Fixed ladders should have a preferred pitch 75 to 90° from the horizontal and should have siderails that extend 31/2 ft above the landing platform. Fixed ladders should maintain a clear width of 15 in. on each side of the center line of the ladder (unless within a cage). Fixed ladders should be designed to minimize accumula-

Table 6. **ANSI Ladder Standards**

Type of Ladder	ANSI Standard
Portable reinforced plastic	ANSI A14.5-1982[36]
Job made	ANSI A14.4-1984[32]
Fixed	ANSI A14.3-1984[29]
Portable wood	ANSI A14.1-1982[30]
Portable metal	ANSI A14.2-1982[31]

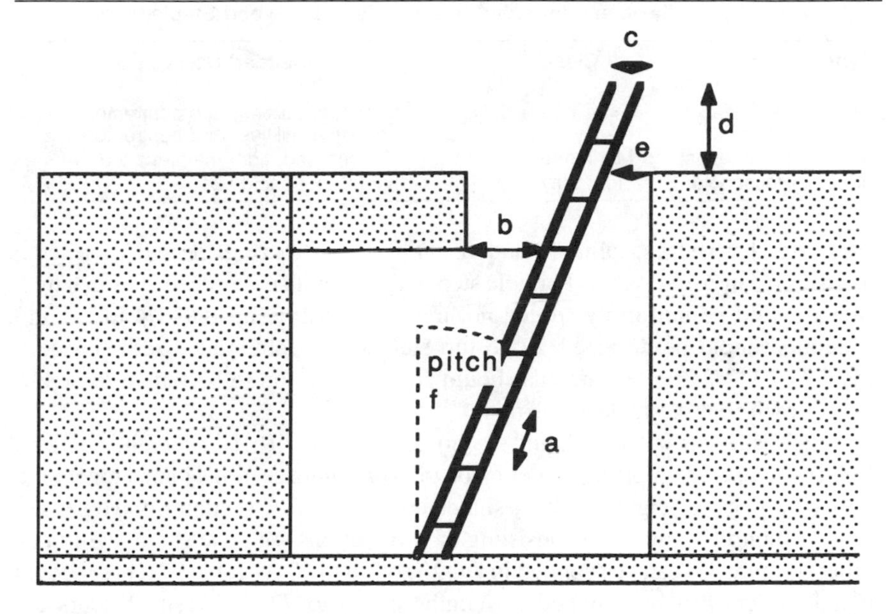

Figure 22. Ladder spacing requirements. (a) Rung spacing, (b) clearance on the climbing side, (c) rail spacing, (d) side rails extension above landing, (e) clearance from back of ladder, and (f) pitch (shown here from vertical).

tion of moisture which will reduce the potential for rust, rot, or corrosion. They should be painted or treated. Fixed ladders which are 20 ft or more in height should be provided a cage. Cages should begin 7 to 8 ft from the bottom landing and continue past the top landing a minimum of 31/2 ft.[29]

Portable Wood Ladders

Industry recognizes four general classes of portable wood ladders, summarized in Table 7. Stepladders are self supporting portable ladders, nonadjustable in length, with a hinged back and flat steps (Table 8).[30] The top cap is the uppermost horizontal piece of a portable stepladder. A sign or label containing either general information regarding the ladder or a hazard warning should be attached to the ladder.[30]

Portable wood ladders are required to be designed from seasoned wood parts,

Table 7. Type, Class, Use, and Range of Portable Wood Stepladders[30]

Type	Use	Class	Max. Weight (lb)	Range (ft)
IA	Industrial	Extra heavy duty	300	3 to 20
1	Industrial	Heavy duty	250	3 to 20
II	Commercial	Medium duty	225	3 to 12
III	Household	Light duty	200	3 to 6

Table 8. Type, Class, and Intended Users of Portable Wood Stepladders[30]

Type	Use	Class	Intended Users
IA	Industrial	Extra heavy duty	Industry, utilities, and contractors
I	Industrial	Heavy duty	Industry, utilities, and contractors
II	Commercial	Medium duty	Offices and light maintenance
III	Household	Light duty	Light household use

free from sharp edges, splinters, and visual irregularities (e.g., decay, knots, etc.), as specified in ANSI A14.1. Portable stepladders should not exceed 20 ft in length. Steps should be uniformly spaced not more than 12 in. apart (+1/4 in., -1 in.). Siderails should be designed at the top step with a minimum width (inside to inside) of 111/2 in. The siderails should spread from top to bottom at least 11/4 in. for each foot of stepladder length.[30]

Single straight ladders should be no longer than 30 ft in length. Sectional ladders (nonself-supporting ladder made in even length sections joined together to obtain the desired height) can be designed up to 31 ft. Two section wood extension ladders (nonself-supporting consisting of two sections with one sliding within the other) (Figure 23) can be supplied in lengths of up to 60 ft (Table 9). Three section metal ladders can be supplied in lengths of up to 72 ft. Trestle ladders and extension trestle ladders are designed to support planks or scaffold boards but have rungs that are not sufficient to safely support a user's feet. Painter's ladders are special purpose ladders designed without the top cap. Painter's ladders are required to conform to type II stepladder requirements and should not be made longer than 12 ft. Step stool size ladders, less than 32 in., have special requirements for safe design referenced in ANSI A14.1-1982.[30]

Straight ladders should be placed so that the horizontal distance from the base to the vertical plane of support is approximately 1/4 the ladder length between supports. It is an important safe practice that straight ladders extend a minimum of 3 ft past the landing.[28] Ladders should never be used in lieu of a properly designed horizontal structure such as a scaffold or a platform.

Portable Metal Ladders

Portable single metal ladders (Table 10), like wood single ladders, should not be longer than 30 ft. Single metal ladders should have a minimum of 12 in. between railings. Each rail of a metal stepladder, single, or extension ladder (Table

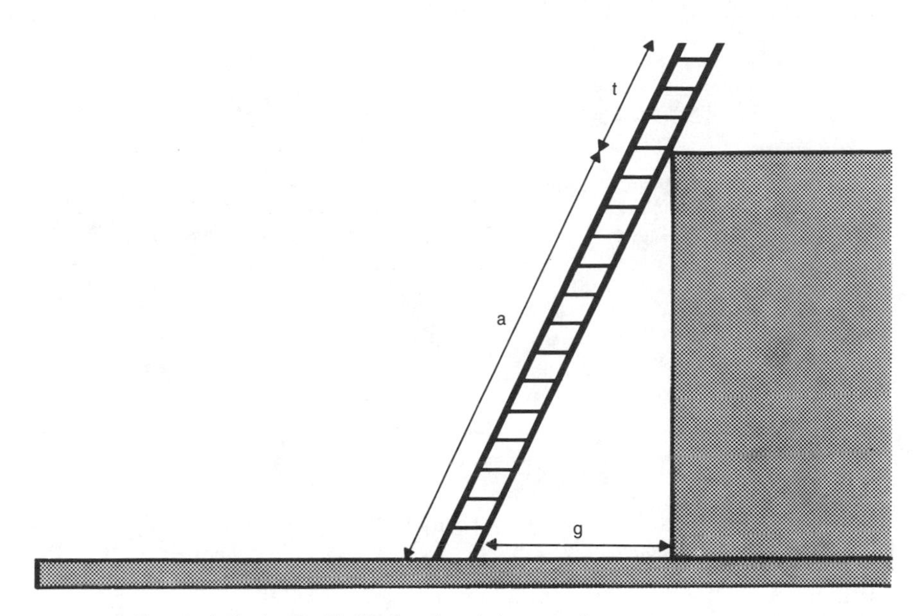

Notes: a is the length of ladder from base to top support.
g should be a/4 from the vertical plane of the top support to base of the ladder.
t should extend a minimum of 3 ft. beyond the top support.

Figure 23. Proper support position for extension ladder. (Adapted from Accident Prevention Manual for Industrial Operations, 7th ed., Chicago: National Safety Council, 1978.)

Table 9. Maximum Single and Two Section Ladder Lengths[30]

Type	Class	Max. Single Length (ft)	Max. Double Length (ft)
IA	Extra heavy duty	30	60
I	Heavy duty	30	60
II	Medium duty	20	40
III	Light duty	14	28

11) should be provided a means of slip resistance secured to the base of the rail. Metal stepladders are required to have tread widths with a minimum of 3 in., and metal stepladders longer than 20 ft should not be supplied. The spacing of rungs or steps should have 12 in. centers (±1/8 in.). Rungs, steps, and platforms should be "corrugated, serrated, knurled, dimpled, or coated with a skid-resistant surface."[31]

Both wood and metal portable ladders require appropriate "DANGER" and "CAUTION" warning labels alerting consumers of the specific hazards associated with their use. The ANSI Standard is specific as to the provision of numerous

Table 10. Metal Single Ladder Lengths[31]

Type	Class	Max. Load (lb)	Max. Height (ft)
IA	Industrial extra heavy duty	300	30
I	Industrial heavy duty	250	30
II	Commercial medium duty	225	24
III	Household light duty	200	16

Table 11. Metal Extension Ladder Lengths[31]

Type	Class	Two section (ft)	Three section (ft)
IA	Extra heavy duty	60	72
I	Heavy duty	60	72
II	Medium duty	48	60
III	Light duty	32	NA

Table 12. Recommended Safe Use of Ladders [29-33]

Always face the ladder when ascending or descending
Use both hands when possible
Carry hand tools in a pouch holster - do not carry tools by hand
Hoist heavy objects as opposed to carrying
Report defects or deterioration promptly
Tag damaged or defective ladders "CONDEMNED - DO NOT USE"
Remove damaged or defective ladders from use immediately after tagging
Keep shoes and footwear clean
Use only proper footwear
Only one person on the ladder at a time
Maintain safe distance from electrical current
Ladders should not be used in a horizontal position as a scaffold or platform
Never place a ladder in front of a door
Never place a ladder against a window pane
Never slide down a ladder
Do not climb a ladder higher than third rung from the top
Never use a defective ladder
Always place a ladder on solid and level footing
Secure by tieing or other fastening ladder when accessing high locations
Inspect ladders carefully before and after use
Keep ladders free of grease, oil, and slippery debris

warnings such as "Danger — Do Not Sit or Stand on Top Cap," "Danger — Do Not Stand Above This Step," "Danger — Watch for Wires," "Danger — Place Ladder on Level Surface," and others. The standard also recommends the addition of labeling to identify the ladder size, type, duty rating, manufacturer, month and year of manufacture, model, and indication of ANSI standard compliance.[31]

Many accidental falls from ladders are caused by unsafe use of the equipment. It is always important for users to follow recommended safe practice as the equipment design is only a part of the accident avoidance problem. Table 12 provides a list of some important practices to remember regarding the safe use of ladders.

SUMMATION

This chapter has discussed the importance of considering "expectation" in all trip, slip, and fall prevention. Many times potential hazards in the field are overlooked or are unobserved. Frequently, the trip, slip, and fall hazard is a result of poor design. This review has covered several of the major contributing structures such as floor surfaces, stairs, ramps, and ladders, as well as the effects of foreign substances.

The intent of this chapter was to make safety practitioners aware of the basic causes and methods of control of common trip, slip, and fall occurrences. The information presented here on the subject needs to be taught as basic principles in the field of safety and accident prevention. It is hoped that this will serve as a foundation that other persons, including safety practitioners, may utilize in slip, trip, and fall prevention.

REFERENCES

1. *Accident Facts,* National Safety Council, Chicago (1987).
2. "Compensated Cases, Closed, 1979," Research and Statistics Bulletin #40, New York State Workers Compensation Board, Albany (1980).
3. Archea, J., B. L. Collins, and F. I. Stahl. "Guidelines for Stair Safety", NBS Building Science Series 120, National Bureau of Standards (1979).
4. Marletta, W. "The Effect of Expectation on Slip, Trip and Fall Accidents," *ASSE J.* (1989).
5. Sherman, R. M. "Professional Safety," *ASSE J.* 31(2):40 (1986).
6. Rosen, S. I., *The Slip and Fall Handbook* (Columbia, MD: Hanrow Press, 1983).
7. Pooley, R. W. "Measurement of Frictional Properties of Footwear and Heel Materials," in *Walkway Surfaces: Measurement of Slip Resistance,* ASTM STP 649, Anderson/Senne, Eds. (Philadelphia: The American Society of Testing and Materials, 1978).
8. "Static Slip Resistance of Footwear Sole, Heel, or Related Materials by Horizontal Pull Slipmeter (HPS)," ASTM Standard F609-79 (Philadelphia: American Society of Testing and Materials, 1988).
9. "Static COF of Polish-Coated Floor Surfaces as Measured by the James Machine," ASTM Standard D2047-82 (Philadelphia: American Society of Testing and Materials, 1988).
10. Andres, R. O., Kreutzberg, K., and Trier, E. M. "An Ergonomic Analysis of Dynamic Coefficient of Friction Measurement Techniques," Technical Report, University of Michigan (1984).
11. Strandberg, L. "On Accident Analysis and Slip Resistance Measurement," *Ergonomics* 26(1): 11-32 (1983).
12. Chaffin, D., and R. O.Andres. "Evaluation of Three Surface Friction Devices for Field Use," Technical Report, University of Michigan (1982).
13. Turnbow, C. E., *Slip and Fall Practice* (Santa Anna, CA: James Publishing Group, 1988).

14. "Specifications for Making Buildings and Facilities Accessible to and Usable by Physically Handicapped People," ANSI A117.1-1980, American National Standards Institute (1980).
15. "Uniform Federal Accessibility Standards," 49 CFR 31528.
16. "Safety Color Code for Marking Physical Hazards," ANSI Standard Z53.1-1979, The American National Standards Institute (1979).
17. Guevin, P. R. "Review of Skid and Slip Resistance Standards Relatable to Coatings," *J. Coatings Technol.* 50:33 (1978).
18. Irvine, C. H. "A New Slipmeter for Evaluating Walkway Slipperiness," *Materials Res. Standards* 7(12):535-542 (1967).
19. Irvine, C. H. "Evaluation of Some Factors Affecting Measurements of Slip Resistance of Shoe Sole Materials on Floor Surfaces," *Materials Res. Standards* 4(2):133 (1976)
20. "Floor Mats and Runners," Data Sheet I-595 (Chicago: National Safety Council, 1986).
21. "Falls on Floors," Data Sheet I-495 (Chicago: National Safety Council, 1986).
22. *Consumer Reports* 42(7):417-419, (1976).
23. Feldman, D. *Why Do Clocks Run Clockwise?* (New York: Harper & Row, 1988), pp. 12-13.
24. "Safety Requirements for Floor, Openings, Railings and Toeboards," ANSI A12.1-1973, National Safety American National Standards Institute (1973)
25. Sinnott, R., *Safety and Security* (London: Van Nostrand Reinhold, 1985).
26. Kaufman, J. E., Ed. *IES Lighting Handbook - Reference Volume,* (New York: Illuminating Engineering Society, 1984).
27. "Safety on Stairs," Carson, D. H., J. C. Archea, S. T. Margulis, and F. E. Carson, Eds., National Bureau of Standards, (1978).
28. *Accident Prevention Manual for Industrial Operations,* 7th ed.(Chicago: National Safety Council, 1978).
29. "Standard for Fixed Ladders," ANSI A14.3-1984, American National Standards Institute (1984).
30. "Standard for Portable Wood Ladders," ANSI A14.1-1982, American National Standards Institute (1982).
31. "Standard for Portable Metal Ladders," ANSI A14.2-1982, American National Standards Institute (1982).
32. "Standard for Job Made Ladders," ANSI A14.4-1984, American National Standards Institute (1982).
33. "Portable Ladders," NSC Data Sheet 665 (Chicago: National Safety Council, 1977).
34. Pater, R. "Fallsafe: Reducing Injuries from Slips and Falls," *Professional Safety* 30(10):15 (1985).
35. Building Officials & Code Administrators International, Inc., *The BOCA Basic National Building Code/1984,* 9th ed., Country Club Hills, IL (1983).
36. "Standards for Portable Reinforced Plastic Ladders," ANSI A14.5-1982, American National Standards Institute (1982).

CHAPTER 13

Fire Protection

Joe Levesque

INTRODUCTION

Fire is part of everyday life. It heats homes, cooks food, and provides energy for industry; however, that flow of energy can be misdirected and can lead to much damage. The historical record reflects this point (Table 1). Equally a part of our lives are the methods by which we protect ourselves from fire, ranging from the use of noncombustible materials, control of smoke movement through buildings, fire resistance ratings of walls, and capacity of exits, to electrical capacities of wiring and design of suppression systems.

BUILDING CODES AND DEFENSE IN DEPTH

The ways in which we protect ourselves from fire have been developed by reactions to fire disasters, usually in retrospect. All too often a second disaster occurs that motivates people into meaningful action. Regrettably, most of these reactions occur in the form of building codes. Codes motivate people to take action, as opposed to adopting common sense and responsibility for their own safety. To compound the problem, codes are not often retroactive when it comes to building design. This leaves many structures "grandfathered" and below current standards. Codes do have their place in providing guidance for those seeking a

Table 1. Examples of Fires with Life Loss

Examples of High Loss of Life Fires

Date	Location	No. of Deaths
February 11, 1987	Vacant warehouse, GA	3
November 8, 1982	Jail fire, Biloxi, MS	29
November 21, 1980	MGM Grand Hotel, Las Vegas, NV	85
July 24, 1980	Metal Products Mfg., New York City	11
November 5, 1978	Department store, Des Moines, IA	10
December 22, 1977	Grain elevator explosion, Westwego, TX	36
May 27, 1977	Beverly Hills Supper Club, So. Gate, KY	165
December 16, 1972	Steel plant, Weirton, WV	21
September 24, 1972	Ice cream parlor, Sacramento, CA	22
December 1, 1958	Our Lady of Angels, Chicago	95
July 28, 1945	Empire State Building, New York City	14
November 28, 1942	Coconut Grove Nightclub, Boston	492

Examples of High Dollar Loss Fires

Date	Location	Loss ($ millions)
November 14, 1987	Chemical plant, TX	$154
June 6, 1982	Retail distribution center, Falls Township, PA	$100
November 25, 1982	Bank building, Minneapolis	$91
February 27, 1975	Telephone Exchange, New York City	$60
May 11, 1969	Nuclear weapons plant, Golden, CO	$60
October 8, 1871	The Great Chicago Fire (17,430 buildings, 250 deaths)	$168

level of performance; however, code requirements tend to be used as design guides and not as the minimum level standards.

The Code of Hammurabi (2250 B.C.) contains the first building code in recorded history, and was developed to address the building problems in Egypt. Today, however, there are professional organizations that develop standards, recommended practices, and suggested codes. Consumers must be aware of the type of organization presenting the material: manufacturer-sponsored organizations, consensus groups (comprised of a balance of manufacturers, insurance companies, fire protection professionals), and insurance company standards, to name a few. These codes are adopted in whole by local authorities or are edited in part. It is always advisable to consult with the local governing authorities as to their applicability in individual cases as they may impose additional requirements and/or interpretations on the code. One prominent agency is the Occupational Health and Safety Administration (OSHA). Through the enactment of legislation by Congress, OSHA[1] establishes minimum levels of safety which include fire protection. In addition, an insurance carrier may impose requirements above and beyond codes to reduce risks and lower insurance premiums.

Fire protection is defined as the conservation of life, property, continuity of operations, and the protection of the environment. Since the goals are so diverse and the importance of each is so high, it requires "defense in depth," which, in the vernacular, means, "if one thing does not work I want another to be there as a barrier. If that does not work, I want another..." The number of levels is dependent upon the importance to the operation. A practical example is a computer room. No smoking is permitted in the computer room, thereby reducing the chance of fire. If fire does occur, the example room has self-extinguishing waste pails. If the fire starts outside the pail, e.g., in an equipment rack, the operators of the computer room have been trained to call the fire department, shut down power, and use a fire extinguisher. If the operators are not in the room and there are no security patrols, the smoke detection system calls the fire department. If the equipment rack design included neither less combustible cables nor less combustible printed circuit boards, the fire may continue to spread before the arrival of the firefighters. By the time fire spreads to another rack, the overhead automatic sprinkler system has either contained or extinguished it. The fire preplan for the facility guides the fire department to the sprinkler control valve (shut only after the blaze is confirmed to be out) and to the electrical shutoffs. Despite several barrier failures, the "defense in depth" approach of this example has limited the fire loss to one rack instead of an entire room.

The key factor in the strategy of "defense in depth," or systems approach, is professional, educated judgment. The "code compliance attitude" is left behind. By making educated judgments the facility's management can evaluate the return on their investment in fire protection. Compliance with codes provides protection against fines, and to a lesser degree against legal liability. Compliance with insurance company requirements can also result in lower premiums.

SCOPE OF THE FIRE LOSS PROBLEM

Historically, the fire protection field has dealt with property loss, loss of life, and interruption to operations. The first two are apparent to most, and the third is subtle, but deadly. Regretably, there are instances when a fire is small in area and only damages one piece of equipment, but halts the entire operation. Imagine a fire destroying the electrical transformer to a facility. It takes 3 months at the very least to install a replacement. During this time business is affected, commitments for product go unfulfilled, and the staff remains on salary along with all the normal overhead. This condition is known as "interruption to operations" (also known as "business interruption" or "programmatic interruption"). In hard economic times a company may examine the costs of reopening a damaged facility. It may be economically sensible to relocate or abandon the operation. This is an intangible factor that must be thought about prior to the occurrence of fire.

Environmental awareness is the current vogue, which has found a place in the fire protection field. Consider the case of the plant that the fire department allowed

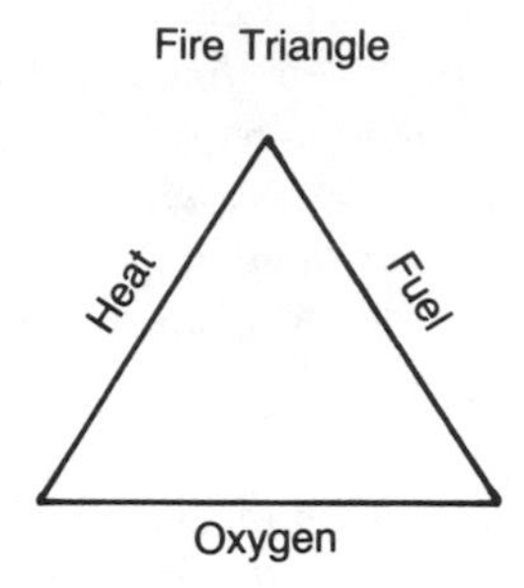

Figure 1. Triangle of fire.

to burn. The extensive chemical runoff from the hose streams presented a tremendous risk to the surrounding community because the plant was located above the sole water supply for the town. The fire chief weighed the risks and benefits and decided to protect the town by letting the plant burn. This is an extreme example, but also consider a situation in which a fire involves an oil-filled electrical disconnect switch. The oil is a fire-retardant type called PCB, popular in the 1960s and 1970s, but now considered detrimental to the environment. The oil is released during the fire and seeps into the ground. While the cost of the switch itself is several thousand dollars, the removal of the dirt and restoration is in the several tens of thousands of dollars.

Fire prevention demands forethought. The impact is not always obvious and safety professionals must be constantly aware of the hazards, the impact, and the risks.

BASIC FIRE THEORY AND FIRE PROTECTION CONCEPTS

One definition of fire is the rapid oxidation of materials (i.e., fuel) that liberates heat and light. Fire is an endothermic reaction, i.e., it requires a positive energy flow to maintain its ongoing reaction. In explaining this reaction, fire can be seen as a triangle (Figure 1) whose sides consist of fuel, heat, and oxygen. The lack of any one side prevents the fire from occurring or continuing. Another common explanation is the tetrahedron of fire (Figure 2) in which the requirements for chain reactions have been added. This evolved from the need to explain the highly effective special extinguishing agents such as halon or dry chemical (discussed later). For simplicity the remainder of this chapter uses the triangle concept.

In designing for fire prevention, the objective is to insert elements that pose a barrier to any three of the sides of the triangle. In practical applications it is difficult to limit oxygen as a factor; unless there is an opportunity to store metal bars under 12 ft of water, the occasion is rare. Practical examples of this are the use of noncombustible construction to eliminate fuel prior to a fire, having construction fire rated (a method of providing a thermal barrier) to reduce heat

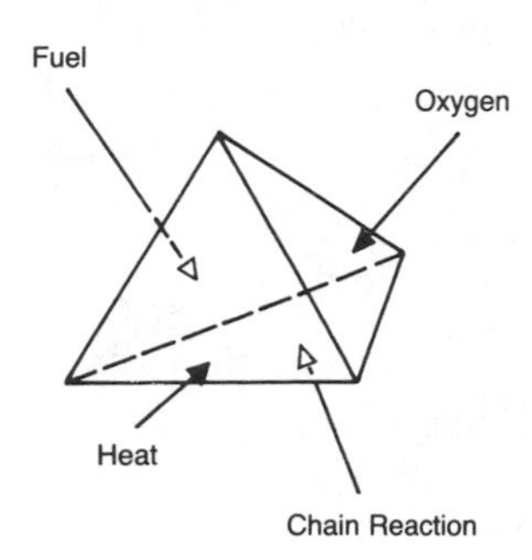

Figure 2. Tetrahedron of fire.

transfer to adjacent areas during a fire, and using water spray (hose streams or automatic sprinklers) to absorb heat and exclude oxygen as steam develops.

Another concept is the differentiation between static and dynamic protection. Static protection is a mechanism that does not require change or action as the result of a fire. A good example is a fire wall. It exists prior to the fire and does not change as it is exposed to heat and smoke. This type of mechanism tends to be highly reliable based on the lack of moving parts and its inherent simplicity. Another example is the use or substitution of less flammable materials in construction and equipment. Because of its reliability, static protection should be thought of as the basic, inner core of defense. Dynamic protection is a mechanism that reacts to a fire to fulfill its function. The majority of fire protection systems are dynamic: fire alarms, sprinklers, fire departments, self closing lids on flammable liquids tanks, etc.

The discussion has thus far focused on the equipment used in fire protection. There is also an operational side to the protection equation. The broad category can be called prevention. Prevention is not allowing the fire to occur. By administrative control and/or training, the prevention program ensures that there are insufficient fuels to burn, the ignition sources have not been introduced, and the workplace fire risk is being managed. The prevention program is no less important since it is oriented to the "human element" and is not as easily defined as a fire extinguisher. Without the trained employee, or the right maintenance, that fire extinguisher may as well be a doorstop.

FIRE PREVENTION PROGRAMS

Table 2 contains a generic listing of the elements of a good fire protection program. In the Appendix to this chapter, checklists are also provided to aid in developing an individual program to meet the needs of a particular business.

Table 2. Generic Elements of a Fire Protection Program

Safety policy (organization goals and responsibility)
Controlling hazardous operations/materials
Controlling cutting/welding
Controlling flammable liquids
Maintaining the level of fire protection (expert maintenance)
Controlling the electrical hazard
Housekeeping policies designed to limit:
Combustibles (prompt removal and disposal)
Smoking (designated area)
Quantities of flammable and easily ignited wastes
Limitation on storage heights
Emergency organization
Quality assurance
Self-inspections program (see checklist)

Organization/Responsibility

Organization is important to understanding individual responsibility in any program. This understanding must range from management (whether delegating authority or providing funds) to the first-line supervisor directly in charge of the worker. Each facility or operation must have its structure customized. Large facilities may employ persons full time for training, fire inspection, and fire system maintenance, while one person in a smaller operation may shoulder all of these responsibilities. A well-thought out program will also divide duties between the oversight and the operations groups. One who is responsible for performing the task should not be responsible for a safety review of that task. Table 3 lists generic safety functions and a brief description of the responsibilities of persons involved.

Controlling Hazardous Operations/Materials

How does one know if one has a hazardous operation or material? This requires a detailed self-examination. Materials can be assessed using the Material Safety Data Sheets (MSDS) outlined in other chapters. The MSDS will identify the chemicals as flammable or requiring special treatment. Do not assume that an operation using a "safe" material is also safe. For example, hydraulic fluid has a flashpoint well above the flammable range at room temperature, but when passed through a pinhole leak at 5000 psi, hydraulic fluid is easily ignited and can produce a 75-ft long blow torch.

The evaluation of operations is more intensive. First, one must investigate the existence of a code or a recommended practice. If there is an associated industry-recognized hazard, there will be a written guidance document. Table 4 lists several National Fire Prevention Association (NFPA) standards that apply to several kinds of facilities. The best approach is to check the list. An example of an unexpected vulnerability is water cooling towers. They are always filled with running water

Table 3. Generic Safety Functions and Fire Protection Responsibilities

Upper management: responsible for ensuring the implementation of the safety program. They must make policy decisions based on information supplied by staff on what level of performance is acceptable and what risks are acceptable. They must demonstrate support through policy, financial, and supervisory input.

Line supervisors (direct employee supervisor): responsible for the daily implementation of the safety program as defined by management and bringing to the attention of management information regarding deviations from accepted norms to ensure management is aware of all risks.

Facility and/or operational safety: a second level of management between first line supervisors and upper management that aids in identifying the risks and hazards associated with the facility and operations to upper management. They have the day-to-day oversight responsibility of the first line supervisor of ensuring the proper implementation of policy. They also track trends and report to upper management needed changes and direction of resources. When actions at the first line supervisors' level do not agree with policy, the facility/operations safety segment is responsible for corrective action, either by redirecting first line management to implement existing policy or by altering the policy in accord with what occurred. Any change in policy must be approved by upper management in order that unnecessary risks are not being assumed.

Safety office: an independent oversight group that conducts inspections and reviews to ensure the consistent implementation of the established program. This group reports to upper management and is also available as a resource when developing and/or changing the program.

Maintenance: responsible for prioritizing identified safety deficiencies and correcting them. Also responsible for maintaining safety systems (i.e., fire alarms, fire suppressions, fire walls, fire doors).

Training: responsible for ensuring that the procedures and levels of performance established by management are passed to the appropriate persons and that they are trained as needed.

Emergency organization: team of knowledgeable persons trained in the emergency functions required (see section on emergency organization).

to dissipate unwanted heat for building or operations, but there are always dry spots. Cooling towers in full operation have burned to the ground.

Once the hazard has been defined, generic principles can be applied. Segregation or separation of individual hazards from each other is a common principle. In Saudi Arabia, the most practical application of this principle is distance between refineries; however, in Lower Manhattan (New York), fire walls and fire floors provide the necessary segregation.

The degree of protection must correlate with the hazard. The level of protection or resistance to fire passage in walls and building construction is known as fire rating. Fire-rated construction is most commonly expressed in "hours" based on a standard furnace test. The most commonly accepted test is presented in NFPA 251[2] (also known as ASTM E119, see Figure 3). The wall or roof/floor assembly

Table 4. NFPA Codes Commonly Applicable

NFPA Code	Title
10	Portable Fire Extinguishers
12	Carbon Dioxide Extinguishing Systems
12A	Halon 1301 Fire Extinguishing Systems
13	Installation of Sprinkler Systems
14	Installation of Standpipe and Hose Systems
17	Installation of Dry Chemical Extinguishing Systems
30	Flammable and Combustible Liquids Code
31	Installation of Oil Burning Equipment
33	Spray Applications Using Flammable and Combustible Liquids
43A	Storage of Liquid and Solid Oxidizing Materials
43C	Storage of Gaseous Oxidizing Materials
51B	Cutting and Welding Processes
54	National Fuel Gas Code
58	Storage and Handling Liquefied Petroleum Gas
69	Explosion Prevention Systems
70	National Electrical Code
70B	Electrical Equipment Maintenance
72	Series Fire Alarm Systems Codes
75	Protection of Electronic Computer/Data Processing Equipment
77	Static Electricity
78	Lightning Protection
79	Electrical Standard for Industrial Machinery
85	Series Furnace Explosion Prevention Codes
90A	Installation of Air Conditioning and Ventilation Systems
101	Safety to Life from Fire in Buildings and Structures
110	Emergency and Standby Power Systems
214	Water Cooling Towers
231	General Storage
231C	Rack Storage of Materials
232	Protection of Records
497M	Classification of Gases, Vapors, and Dusts for Electrical Equipment in Hazardous (Classified) Locations
601/602	Guard Service/Operations in Fire Loss Prevention

is constructed as part of one side of a furnace (top for floor/ceilings, side for walls). The assembly is exposed to a predetermined time vs temperature curve (which has no real relation to a potential fire). There are five basic criteria that must be maintained: ability to support load, maintaining low temperature on unexposed side, resistance to heat and flame passage, maintaining low surface temperature on steel members, and maintaining integrity of wall to hose streams. Assemblies are typically rated in 15, 30, and 45 min, and 1, 2, 3, and 4 hr units. One-hour assemblies are used for low level segregation within similar occupancies (e.g., separating a mechanical equipment room from an office area). Two-hour walls are used for higher degrees of separation (e.g., enclosing a flammable liquids room from a production area). Three- and four-hour walls are used primarily for maximum loss protection, when the building is expected to burn to the ground and it is to be stopped without question (e.g., a 4-hr fire wall divides a warehouse worth $50 million in half, limiting the maximum loss to $25 million). Always consult the

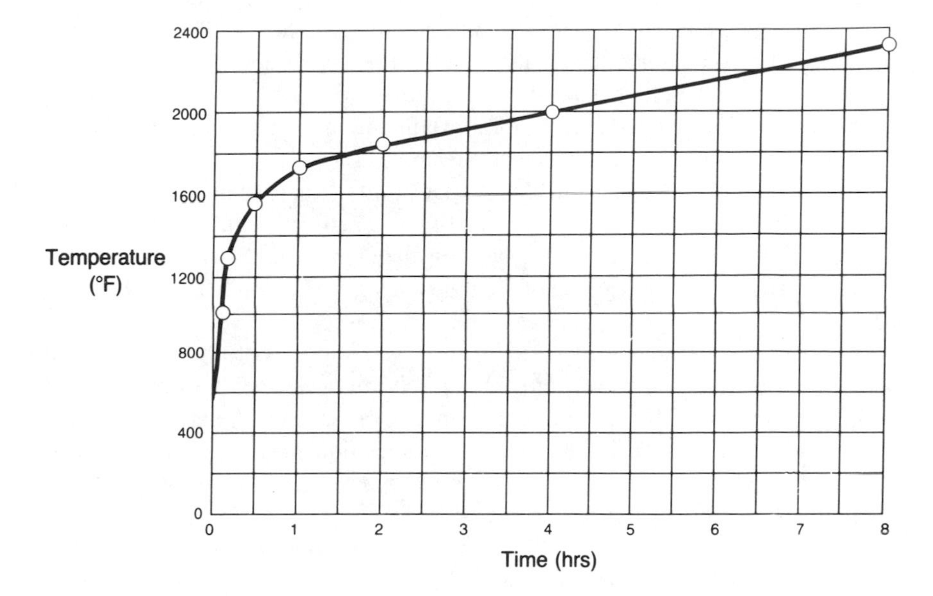

Figure 3. ASTM E-119, the standard time-temperature curve.

governing authority for the proper rating required for the individual application.

Certain generic construction assemblies have accepted fire ratings. A good source is the *Fire Resistance Ratings* booklet published by the American Insurance Services Group.[3] Specific approval of assemblies can be found in the UL *Fire Resistance Directory*[4] or *Factory Mutual Approval Guide.*

If a wall is built, electrical lines, steam and sprinkler pipes, and other utilities must be encased within. In a fire wall, these wall penetrations are required to have equivalent fire ratings, rules, and guides to ensure they have reasonable integrity. Penetrations can also be doors and air handling ducts. Because the areas of these penetrations are normally a small portion of the total wall area, their ratings are not as severe. They are typically rated as "20 min," "1/2 hr," "1 hr," "11/2 hr," and "3 hr." The 20 min, 1/2 hr, and 1 hr door ratings are normally for 1 hr walls. 11/2 hr door ratings are for 2 hr walls. One 3 hr door is for a 3-hr wall and two 3 hr doors (one each side) normally protect a 4-hr wall. NFPA 80, "Standard for Fire Doors and Windows", is a good reference for installation guidelines.

Controlling Cutting/Welding

Cutting and welding operations are the most identifiable ignition source for industrial fires. By instituting a simple permit system that would require a cutting/welding operator to obtain a safety review, the risks can be reduced tenfold. The nature of the operation involves open flames and hot/molten beads of metal traveling 35 ft from the area. Not all sparks die harmlessly. These small beads can nestle between combustible materials and smolder for hours. Then, when the operation is completed and the area vacated, a fire may erupt.

Protecting against these hazards is the jurisdiction of management. There should be no exemptions from permits. Maintaining safe operations calls for constant management interest and supervision.

The welder must understand his/her responsibility: where to get a permit (with the associated review), the importance of a fire watch, the importance of maintaining equipment in good condition, respect for the combustible gas cylinders, and the need to ensure the segregation of combustibles from the torch and molten slag produced.

The safety officer should be responsible for issuing a permit which should specify the required conditions for carrying out the operation. This implies that operation should cease if the conditions cannot be maintained. A detailed program can be found in the National Fire Protection Association's recommended standard 51B, "Cutting and Welding Operations." A checklist outlining the basic contents of a permit has been provided in Table 5.

Thoroughness counts when dealing with cutting/welding reviews. An operation planned for the second floor of a metal parts warehouse may not pose an obvious hazard, but when the first floor is a spray paint operation, any opening, even a 1/4-in. crack, can allow sufficient material to pass through, resulting in a fire. Similarly, a welding operation on a metal wall may not appear hazardous; however, the extreme heat and the high thermal conductivity can cause ignition of combustible materials on the opposite side (this includes expanded plastic insulation).

Controlling Flammable Liquids (FL)

A flammable is a liquid that liberates sufficient vapor so that it ignites at room temperature (100°F or less) and is one of the most common and most powerful hazards. A common term for the point at which a liquid can be ignited is "flash point," i.e., combustible liquids liberate sufficient vapor for ignition at 100 to 200°F (or, 100 to 200°F flash point). For protection of operations, any liquid heated to within 50°F of its flash point should be considered flammable. Recalling the concept of the fire triangle, the only event needed for the ignition of a fire (using a flammable liquid in air) is the introduction of an ignition source. In addition, the fact that it is a liquid allows the material to spread when spilled. This increases the surface area capable of burning and the heat release rate. Once ignited, the volatile flammable liquid will absorb heat quickly and produce additional vapors which in turn burn. A flammable liquid pool fire will consume fuel and burn at a rate of 1 to 2 in. of depth per hour. The heat release and the energy release rates are thus easily calculable.

To demonstrate the hazard of flammables, consider a 20 × 50 ft room. A 1-pt bottle of liquid is toppled and spills, covering about 6 ft^2 and ignites. The fire will burn without producing enough heat or smoke to injure an occupant on the other

Table 5. Cutting/Welding Checklist

Date Issued:__________Date Expires:__________
Issued to:____________________
For Location:__________________
Permission is Granted For:______________________________

THIS PERMIT IS ONLY VALID FOR THE OPERATION AND LOCATION ABOVE

Special Conditions:____
Fire Watch Necessary____________(Yes/No)
Fire watch to remain ___________ hours after completion of work
(Normally 1—2 hr)
Issued by:______________________(Safety Officer Signature)
Read/Understood by:___________________ (C/W Sup'v. Signature)

Safety Officer Checklist:

Occupancy

Is protection fully in service (sprinkler, detection)?
Are there hazardous operations that should be shut down and/or protected (flammable dip tanks, spray booths)?

C/W Operations

Is the C/W equipment in good condition?
Is the welder qualified to operate the equipment?

Protection of Combustibles

Have combustibles been removed for a 35-ft. radius?
If removal is not practical, have combustibles been covered with sand, water, or welding blanket?
Are all openings in walls and floor within 35 ft sealed?

Fire Watch

Is a properly equipped, trained, and dedicated fire watch posted?
Does the fire watch know to summon help first, then use the extinguisher or hose?
Does the fire watch know how long to remain after operations stop?

Work on Ceilings or Walls

Is construction noncombustible on both sides?
Have combustibles been moved away from the opposite side?

Work on Enclosed Equipment

Have containers been purged of flammable vapors?
Has equipment been cleaned of combustible materials?
Is there a confined spaces entry condition?

side of the room. On the other hand, a 1-gal bottle containing a flammable is spilled and covers 50 ft^2. The heat release is sufficient to make the once-safe room untenable within 2 min.

Table 6. Maximum Flammable Liquid Quantities

Type of Occupancy	FP <140°F	FP <200°F
Low hazard (office, residential or assembly, etc.)		
Not in FL cabinets or rooms	10 gal	60 gal
Total quantities	25 gal	60 gal
Moderate hazard (shops, nonchemical labs, manufacturing, etc., where FL use is incidental to operation)		
Not in FL cabinets or rooms	25 gal	100 gal
Total quantities	75 gal	100 gal
High hazard (chemical labs, operations, etc., where FL use is principal to operation)		
Unsprinklered areas	2 gal/100 ft^2	4 gal/100 ft^2
Total quantities	75 gal	100 gal
Sprinklered areas	2 gal/100 ft^2	4 gal/100 ft^2
Total quantities	150 gal	200 gal

Notes: Quantities for the first and second categories are based on NFPA 30 and OSHA 29 CFR 1910.106. The third category is based on the least hazardous chemical classification in NFPA 45. Limitations apply to a fire area defined by 2-hr fire barriers. Flammable liquids room and cabinets are defined in NFPA 30. The quantities for workplaces in categories one and two are not to exceed a 1-day supply based on OSHA 29CFR1910.106(e)(2)(ii)(b).

The above example illustrates a very important principle: *reduce the risk as much as possible*. OSHA requires the limitation of a maximum of a 1-day supply for any facility not normally using flammable liquids. By reducing storage from 1 gal to 1 pt, for example, the impact on the occupant dramatically decreases. The effort is one of policy and awareness. Evaluation and determination of the level of need is the best defense.

There are two main approaches in protecting against the flammable liquids hazard. First, protect against ignition. This is done by isolating fuel from ignition. Several typical quantity limits are listed in Table 6. An important example is electrical equipment since it is a common source of ignition. When electrical equipment is in an area that can have flammable vapors; refer to NFPA 497[7] for general electrical classifications, NFPA 33[8] for spray paint operation, and NFPA 30[9] for general arrangements of storage and use. Class I Division I equipment is intended for areas in which vapors are always present (in tanks, by vents, by spray paint operations). Class I Division II is a class of equipment to be used around areas in which sufficient vapors to ignite are always present (in tanks, by vents, by spray paint operations). The difference, besides increased cost, is that Division II does not protect against the simultaneous fault in the electrical equipment and the flammable liquids operation. A Division I light fixture will have low surface temperature in order to avoid igniting a liquid as will a Division II fixture. The difference is that the Division I barrier will not introduce ignition if there is a reasonable mechanical failure, such as breaking the lens on a lamp. A Division II fixture does not guarantee that protection.

Because flammable liquids burn so easily, there must be an assumption that despite all reasonable precautions, there will be a fire. Thus, the second level of protection is suppression. Suppression takes many forms: the use of thermal-operated (i.e., fuseable link), self-closing lids on tanks and self-closing valves which are suitable for small, localized quantities of flammable materials. Installation of sprinkler systems can extinguish combustible liquids and control damage for flammables. Carbon dioxide systems are often an effective element when dealing with particular hazards such as printing presses, flammable storage rooms, and dip tanks.

Maintaining The Level of Fire Protection

A fire protection system (i.e., sprinklers, carbon dioxide, dry chemical, water supplies to hydrants, alarm systems) represents a large financial investment and can be installed once the significant hazards have been identified. Because one improperly closed valve could negate any benefit, it is as important to have fire protection supervision as it is to physically install protection.

The initial portion of the supervision program should be the surveillance of the system. All critical valves and operable devices should be locked into proper position. Identification of the device and warning tags to caution users about unauthorized operations should be attached to critical components. By providing a sign at a sprinkler control valve stating "Keep Open — Control Entire Building Sprinklers — Contact XXXX Prior To Operating" can communicate the importance of the system to all workers. Since signs do not always work, a further level of protection is warranted. A mechanical barrier is provided by locking down valves and critical components. The keys to the locks should be in the sole possession of authorized personnel (i.e., Emergency Organization, Maintenance Staff).

When fire protection equipment is taken out of service or impaired (the reliability or proper functioning reduced below the normal operating level, e.g., removing backup batteries, replacing smoke detectors with heat detectors while a smoke producing operation is underway), there should be a formalized procedure of review, tracking, and confirmation of restoration. Most insurance companies have specific programs for their clients which consist of a pre-review of the extent and need for a shutdown. The review gives consideration to removing as little protection as possible, ensuring all preparatory work has been done, alternative protection has been provided when possible (e.g., connecting a length of hose to the 2-in. drain or fire department connection on a sprinkler system to replace the main water line during construction), and the cessation of any extremely hazardous operations.

Once a shutdown is about to occur, the tracking portion of the system requires notifying a third party (a safety officer or insurance carrier) and notifying the Emergency Organization so that in the event of a fire the system can be restored or the fire preplan can be adjusted to compensate for the deficiency. To compensate for the loss of service, a special watchman tour should be made of the areas without protection. The fire protection system that is affected should be tagged

"Out of Service." This is especially important for operations that will not be restored by the same workpersons and/or will last for more than 1 day. Additional preplanned actions occur such as working continuously until the system is restored and providing partial restoration. (If sprinkler piping blanks are used to cap or isolate piping, they should be closely tracked by numbering each and having red tags attached.)

Once the system is restored, all tags should be collected and counted to verify that none were overlooked, the third party should be recontacted and informed of restoration, the safety officer should verify restoration, locks should be replaced, and necessary confirmation tests should be conducted. Confirmation tests vary depending on the type of fire protection system involved. Sprinkler systems should have flows conducted through the 2-in. drains and the pressure results compared with previous results. Alarm systems should be tested using routine testing procedures, including sending test alarms to central alarm stations or fire departments.

Fire protection systems are safety systems. They require the highest degree of assurance that when they are impaired these impairments are minimized and restoration is prompt. An impaired system is one barrier less in the line of defense. Most unrestored impairments occurred for legitimate reasons and without malice; for instance, no one realized the importance of the system and the required follow through. A formalized system of reviews will address this concern.

The second portion of the supervision program is periodic inspection and maintenance. A competent inspector who knows what the purpose of the system is and how it operates must be selected. If the expertise is not available in-house, engage a competent outside contractor. Document the inspections of sprinkler valves, special suppression systems, fire alarm systems, and fire extinguishers. Table 7 provides a sample of inspection and testing frequencies for several common systems (check with the local governing authority and insurance carrier for their requirements).

CONTROLLING THE ELECTRICAL HAZARD

Electricity is a basic component of every structure and every piece of equipment. When the flow of energy is not as desired, an unwanted fire may ignite. Controlling the energy flow is the prime element in this aspect of protection. Mechanical barriers are provided that protect against electrical shorts (i.e., insulated conductors, grounded equipment cases) to the systems. Secondary barriers prevent overcurrent conditions (circuit breakers, fuses, limited power supplies). Arrangements of electrical systems are complex. Details are provided in the local electrical codes that specify type of wires, size, thermal charateristics of insulation, degree of mechanical protection, and overcurrent protection. Qualified personnel are required to install and maintain the system. Approved equipment ("approved" as defined by the local governing body, most often Underwriters Laboratory[10-12]) must be used in an approved application.

Table 7. Sample Fire Protection Inspection and Test Frequencies

	Sprinkler systems
Valves	Inspect weekly Operate yearly to lubricate Test tamper switches twice yearly
Waterflow	Test alarms monthly
Dry pipe	Trip test annually Inspect internally every 5 years
	Fire detection/manual alarms
Heat detectors	Restorable type, test every 6 months Nonrestorable type, sample 2% every 5 years
Smoke detectors	Test every 6 months
Manual boxes	Test every 6 months
Fire alarm panel	Full functional check annually
Batteries	Check liquid level on lead acid weekly Check charge of each cell quarterly Discharge test annually
Annunciation	Quarterly

HOUSEKEEPING

In a fire protection program, housekeeping is needed to minimize or eliminate the amount of fuel by frequently removing combustibles and properly disposing of the materials.

The housekeeping program requires direct involvement of the supervisors. Good daily work habits are required to ensure the removal of oily rags, wastepaper, and packing cartons. Special containers are required for oily rags (these must be metal and self-closing) and nightly removal of contents is required. Self-extinguishing waste cans are recommended for sensitive electronic areas (computer rooms) or areas in which normal contents are easily ignited. Periodic removal of paint deposits and cleaning of oil-laden vapor removal equipment, etc., are essential parts of the routine.

SAFE STORAGE AREAS

Limits on storage heights must be defined and observed. The higher the storage, the more severe the potential fire. Likewise, the closer the aisle spacing, the more severe the fire (smaller fire breaks). Sprinkler systems should be designed to

Table 8. Emergency Organization Assignments

Emergency organization supervisor: responsible for staffing, ensuring training, and drills of the organization. Provides overall direction during the emergency.

Alarm notification: once a fire is discovered, the alarm notifier telephones the fire department. Do not rely on fire alarm transmissions. They can fail and there is no room for such a risk.

Sprinkler valve verification: verifies the sprinkler system valve is open and/or special suppression systems are in service and have operated. The person assigned this function should be knowledgeable in the location and operation of the fire protection systems for the facility (perhaps the valve inspector or maintenance personnel).

Fire department escort: meets the fire department and directs them to the incident scene. This may be the person responsible for coordinating pre-fire visits by the fire department.

Plumber: a person capable of prompt restoration of the sprinkler systems. Has a supply of spare sprinkler heads, ladders, wrenches, etc. Also familiar with building and process services and capable of effecting a safe shutdown.

Electrician: person familiar with the fire alarm system and capable of prompt restoration. Has spare fire alarm devices and necessary equipment. Also familiar with process and building service and capable of effecting safe shutdown.

Salvage: team capable of cleaning up, preserving, and reclaiming materials. They have been provided with tarps, mops, etc..

Note: During idle hours, watchman services may have to provide some or all of the above functions.

protect the most severe storage arrangement possible in a building, although this is sometimes impossible due to the limitations of water supply system strengths. The potential for sprinkler systems becoming the limiting factor should be identified. Any area with a restriction should be designed for its maximum safe height. Idle pallets are a very common hazard because their physical construction resembles logs in a fireplace. Due to their intense burning potential they should not be stored over 5 ft high unless the sprinkler system is specially designed for this hazard. Another common hazard is storage in the aisles, providing fuel for jumping a "fire break." Aisle storage that provides combustible continuity from one storage section to another should be avoided.

Storage areas used for wastes must also be evaluated. Their proximity to buildings and their separation should be based on their combustibility and fire resistance of the exterior walls.

Control of smoking is important. Safe areas, where smoking is permitted, should be designated and posted. Likewise, high hazard, flammable storage areas or sensitive computer areas should be posted with "No Smoking" signs. Proper ash receptacles should be provided. Sand ashtrays or trays with positive "V" grasps are preferred. All smoking areas should be separated from combustible storage and flammable operations by substantial partition walls.

EMERGENCY ORGANIZATION

Establishing a pre-fire plan is a simple and very effective component of any fire protection program. There are seven assignments that should be customized for each facility. Having these assignments designated, with alternates (Table 8), ensures proper employee reaction during a fire.

QUALITY ASSURANCE IN FIRE PROTECTION

Quality assurance is "getting what you want and what you paid for," and has long been an element in fire protection. The proper design and testing of equipment to ensure its safe application has occurred through nationally recognized testing laboratories, the most recognizable of which is Underwriters Laboratory and the UL label. "UL" indicates approval by placing a label or approval mark on equipment, such as an extension cord. Another approval indication is "listing" (such as fire wall assemblies) where the details of construction are documented, but the unit does not bear a mark or label. Whether the device is a fire alarm box or a circuit breaker, it must be used under circumstances specified in its listing or approval category (i.e., its intended use). Listings and approvals place restrictions on the use. Common restrictions for circuit breakers restrict their use to certain manufacturer's panels. For example, even though the circuit breaker may fit into another manufacturer's panel, it may not have been manufactured with the tab to prevent overload of the panel box with too large a current capacity. Always obtain approved materials and find out the approval restrictions (manufacturer literature often is required to indicate listing restrictions).

Another aspect of quality control is control of change. When operations undergo change, an opportunity for error is introduced. When a change occurs, knowledgeable people should be included in the process to ensure the change does not affect safety. The level of the review depends on the importance of the system being changed (the replacement of a pencil sharpener probably does not require review, however, the replacement of a variable motor driver set on the main production system in the plant does).

One effective method of instituting control is with the flow of money. Change does not occur without cost. Reviewing purchase orders, and being connected to the project approval stream, can aid in targeting safety concerns. For example, an order for ceiling tiles may lead to the discovery that fire-rated ceiling tiles have been removed from an area exposing sensitive process cabling to potential fire damage. Another effective method is being in the plan review chain. An organization with a formal quality assurance program or structure plan review system can provide key information concerning major changes.

FIRE PROTECTION/DETECTION SYSTEMS

Fixed fire protection systems are often the last line of defense that is within human control. The point at which this system is needed is when the fire occurs and the system must then work (similar to a parachute). The requirements and

Table 9. Fire Detector Characteristics

	Heat Detectors
Fixed temperature	Low false alarm rate, low cost, not sensitive to slow smoldering fires
Rate of rise	Quick detection of flash fires, high false alarm rate, low cost
Fixed temperature/rate of rise	Early detection on most fires, high false alarm rate
Rate compensated	High cost, early detection
	Smoke Detectors
Ionization	Good detection of clean burning fires
Photoelectric	Good detection of smoldering/smoky fires
Projected beam	Large area protection
	Flame Detectors
Flame "signature"	High cost, independent of ceiling heights

justifications for installation are complex and vary with codes and insurance carriers; however, this is beyond the scope of this chapter.

DETECTION SYSTEMS

Fire detection systems are alarm systems (Table 9). They react to fire conditions but do not put out fires. They sense the by-products of fire, such as heat, smoke, or light generated by it. These by-products must reach the detectors in a timely fashion in order to activate the system. Ceiling heights, air movement, and area covered per detector affect their response times. The listing for detectors are for response times equal to sprinkler heads. When ceilings are high, air movement is increased or early detection is desired (recommended), and detector spacing must be reduced.

Smoke detectors provide early warning by detecting particles produced during burning. Ionization detectors have a small radioactive element in a chamber that ionizes the air. A small current is passed through the chamber and is monitored for change. Particles entering the chamber absorb the ions and the current through the chamber decreases. Small particles below that detectable by the human eye will trigger the alarm.

Photoelectric detectors use a beam of light within a chamber. The light is either deflected onto or away from a sensor. Change past a certain threshold produces an alarm. Because this mechanism is optical, it is more sensitive to large particles visible to the eye. Projected beam units operate on similar principles, but the area protected is the "chamber."

Flame detectors are optical units, similar to cameras. They detect a certain "flame signature" (a predefined mix of light spectrum, intensity and flicker). They overlook the protected area, particularly large areas in which fast-developing fires are expected (e.g., airport hangers, gasoline terminals).

SPRINKLER PROTECTION

With a 95% plus reliability factor, sprinkler protection is the most effective suppression system. In a life cycle analysis, sprinklers are cheaper due to low maintenance compared to detection.

The first of the two common types of sprinkler systems is the wet pipe system. Wet pipe sprinklers have a constant level of water in the pipe. A fire melts a thermal link (fuseable link). Only the heads in close proximity to the fire will activate. They discharge 15 to 30 gpm (gallons per minute) per head which is much lower than the 150 to 500 gpm sprayed from fire department hose streams. Water flow is detected by alarm devices in the piping network.

A dry pipe sprinkler system is used in areas in which freezing is a concern. The pipe network is filled with air. The air holds back the water at the dry pipe valve via a differential valve using a mechanical advantage. Fusing of a sprinkler head releases the air and trips the valve. Due to the more complicated valve arrangement, the system is slightly less reliable and requires a higher level of maintenance. The dissipation of the air also delays the discharge of water, therefore, several additional sprinkler heads are required to operate. Alarms are tripped by a device located at the sprinkler valve station.

SPECIAL SUPPRESSION SYSTEMS

Unlike the comparatively unlimited water supplies of sprinkler systems, special suppression systems have limited quantities of agent. They are most often used to protect localized hazards, such as flammable liquid dip tanks, cooking grills, computer rooms, and printing presses. The agent is discharged upon activation of a detection system. This can be something as simple as a fuseable link holding back a counterweight, or as complex as an elaborate smoke detection system requiring two detectors to alarm as a prerequisite to discharging the agent. In addition, the agent normally has to be confined to allow for complete extinguishing (often called "agent soak time"). Because of the complexity of the conditions required for extinguishing, special suppression systems require a higher level of maintenance and are less reliable than sprinklers.

Several common agents are dry chemical, carbon dioxide, and halon. Dry chemical is a powder used to control two-dimensional fires. Often used on flammable liquids, it is easily deflected from the target by winds. Clean-up efforts are labor intensive. Carbon dioxide is a gas system that is most often used in contained areas. Clean-up is not labor intensive. In order to be effective it must reach concentrations from 30 to 50%. It is hazardous to any occupants in the area of discharge due to the exclusion of oxygen. Systems require 30 sec or more delays from alarm to discharge to allow occupants to escape. During discharge large volumes of air must be displaced, the temperature drops, and vision is obscured. Halon is a gas system. Halon 1301 is most often used in 5 to 8% concentrations. It acts by inhibiting chain reactions and has no immediate or prolonged effects to personnel in the 8% or less range. It is easily cleaned up and has long been the solution for computer room protection (in addition to sprinklers and smoke detection); however, it is five times heavier than air and quickly leaks out of cracks and openings. There is currently concern about its effect on the ozone layer and global warming. At the time of this writing, the Environmental Protection Agency (EPA) has not prohibited its use; however, it appears that there will be at least a high tariff on new systems to protect against their casual installation.

APPENDIX

Self-Inspection Checklist

✓ Housekeeping proper?
  Aisles defined and clear?
  Storage heights acceptable?
  Excess combustibles removed?
  Limits on smoking present?
✓ Heating and fuel fired equipment adequately maintained?
✓ Electrical wiring permanent and well maintained?
✓ Electrical equipment, fixtures, appliances in good condition and maintained?
✓ Flammables limited as much as practical?
  Large amounts in flammable liquid storage cabinets or rooms?
  Safety cans being used when practical?
  Dispensing operations have dead man valves, drum vent bungs, grounds, bonding cables, adequate mechanical ventilation?
✓ Hazardous materials/operation are properly arranged and operated as per developed procedures?
  Have changes in hazardous material/operations been reviewed?
  Are adequate emergency procedures present?
  Are all shutoffs (gas, electrical, process flow) labeled and unobstructed?
✓ Is fire protection equipment (extinguishers, sprinklers, alarms) adequately maintained and in good repair?
✓ Are protection control valves locked open and inspected weekly?
✓ Are fire walls, doors in good repair?
✓ Are fire doors operable?
✓ Are exit signs clearly visible from all locations?

- ✓ Are emergency lights (if required) in good operating condition?
- ✓ Are all exit doors clear of obstructions and unlocked during occupancy?
- ✓ Are all exit paths clear of obstructions and debris, and the walking surfaces in good repair?
- ✓ Have exterior fire exposures changed?

Valve Supervision Program

Facility Location:________________________________

Date:__________ Inspected by:____________

Inspected item	Item location	Normal Condition (Open/Shut)	Found

Retain the above form for 1 year.

Questions for Evaluating the Sprinkler Supervision Program

Are valves inspected weekly?
Are all valves listed?
Any forms missing?
Is there the appearance of "use" during the inspection?
Are valves locked in the open position?
Do they have signs stating what they control?
Are all valves accessible for emergency operations?
Is the valve inspector properly instructed?
Is there an alternate inspector?
Do the inspector and plant personnel know the impairment procedures?
Does the plant's Emergency Organization know the location of all valves?

Suggested Library from the National Fire Protection Association Battery March Park, Quincy, MA 02269 (800-334-3555)

Fire Protection Handbook (general reference)*
National Electrical Code Handbook
Electrical Installations in Hazardous Locations

---

* If you buy only one book, this should be it.

Life Safety Code Handbook
Automatic Sprinkler System Handbook
Flammable and Combustible Liquids Code Handbook
Liquified Petroleum Gases Handbook
National Fuel Gas Code Handbook
Fire Protection Guide on Hazardous Materials
Fire Alarm Signalling System Handbook
Conducting Fire Inspections, A Guide for Field Use
NFPA Inspection Manual

Suggested Sources of Information

American Concrete Institute, P.O. Box 19150, 22400 West 7 Mile Rd., Detroit, MI 48219 (313-532-2600). Group specifically related to the design, construction, and maintenance of concrete products such as those used in fire barriers.

American Gas Association, 1515 Wilson Blvd, Arlington, VA 22209, (703-841-8400). Group specifically assembled to promote the safe use of gas systems, transportation, and distribution.

American Insurance Association, 85 John Street, New York, NY 10083 (212-669-0400). Group assembled to coordinate safety standards in building designs.

American Petroleum Institute, 2101 L Street, NW, Washington, D.C. 20037 (202-682-8000). Industry group assembled to publish safety standards related to the petrochemical industry.

American National Standards Institute, 1430 Broadway, New York, NY 10018 (212-354-3300).

American Society for Testing and Materials, 1916 Race St., Philadelphia 19103 (215-299-5400). Group assembled to develop and publish safety standards covering a broad range of topics.

Compressed Gas Association, 1235 Jefferson Davis Highway, Arlington, VA 22202 (703-979-0900). Industrial organization assembled to promote the safe handling and use of compressed gases.

Factory Mutual Engineering Association, 1151 Boston-Providence Turnpike, P.O. 9102, Norwood, MA 02062 (617-762-4300).

National Electrical Manufacturers Association, 2101 L Street, NW, Washington, D.C. 20037 (202-457-8400).

National Fire Protection Association, Battery March Park, Quincy, MA 02269 (800-344-3555).

National Fire Sprinkler Association, P.O. Box 1000, Paterson, NJ 12563 (914-878-4200).

Occupational Safety and Health Administration, CFR 1910, available through the Superintendent of Documents, U.S. Government Printing Office, Washington, D.C. 20402.

Underwriters Laboratory, Inc., 333 Pfingsten Road, Northbrook, IL 60062 (708-272-8800).

With the exception of two states, each state has a state fire marshal's office. This is a good start to finding out what the local governing body requires.

REFERENCES

1. For details, see CFR 1910 "Occupational Safety and Health Standards," Subparts E,H,L,M,S. Available from the Superintendent of Documents, U.S. Government Printing Office, Washington, D.C., 20402. The text of this chapter does not address the details of OSHA Subparts. It attempts to guide the reader in developing an effective program, however, strict compliance with OSHA does not guarantee this.
2. NFPA 251, Standard Methods of Tests of Building Construction and Materials, National Fire Prevention Association.
3. Fire Resistance Ratings, Engineering and Safety Services, American Insurance Group, Inc., 85 John Street, New York, 10038.
4. Underwriter's Laboratories, Inc., Fire Resistance Directory, 333 Pfingsten Road, Northbrook, IL, 60062.
5. Factory Mutual Engineering Association, Approval Guide, 1158 Providence Turnpike, Norwood, MA, 02062.
6. An alternate reference is ANSI 249.1, "Safety in Welding and Cutting."
7. NFPA 497, "Recommended Practices for Classification of Class I Hazardous Locations for Electrical Installations in Chemical Plants."
8. NFPA 33, "Safety Applications Using Flammable and Combustible Materials."
9. NFPA 30, "Flammable and Combustible Liquids Code."
10. Underwriter's Laboratory, Electrical Appliance and Utilization Equipment Directory.
11. Underwriter's Laboratory, Inc., Electrical Construction Materials Directory.
12. Underwriter's Laboratory, Inc., Hazardous Locations Equipment Directory.

PART IV

Personal Protection

While workplace or engineering controls, discussed previously in this book, are the preferred ways to control a worker's exposure to hazards, personal protective equipment (PPE) may sometimes also be used to limit exposure. The two chapters in this section present two different uses of PPE, namely, protection from chemical exposures and protection from physical injuries:

Personal Protective Equipment for Preventing Chemical Exposures reviews the use of PPE for limiting chemical exposures, including the limitations regarding when chemical exposure PPE (such as respirators) can and cannot be used.

Personal Protective Equipment for Preventing Physical Injury similarly reviews the use of other PPE, namely, steel-toed shoes and hard hats, etc., to prevent physical injury.

CHAPTER 14

Personal Protective Equipment for Preventing Chemical Exposure

Andrea Bailey

INTRODUCTION

Personal protective equipment (PPE) can be used as a final line of defense for protecting the employee against potentially harmful conditions in the work environment. Federal regulations and good industry practice recommend that engineering controls and work practice changes be considered primarily as methods for protecting workers against hazardous materials or conditions. Engineering controls include enclosure of the hazardous material/process or improved ventilation. Work practice modifications (administrative controls) may include rotation of workers through high exposure jobs (so the exposure is split up or "shared" among more two or more workers), changing the way they perform their jobs so that contact with the hazardous agent is minimized, or possibly allowing for remote control of the hazardous operation. In instances where engineering or work practice controls cannot effectively reduce the potential for exposure to harmful materials or conditions, where exposure conditions are temporary, where the cost of such controls may be too high to justify their use, or where a second line of defense or "redundant control" is desired, PPE can be utilized. For example, PPE is frequently used for very short-term operations where engineering controls may be a very expensive alternative.

PPE includes respirators, dermal protection (full body suits, sleeves, gloves, and boots), eye/face protection, and hearing protection (discussed more thoroughly in Chapter 11). It should be kept in mind that while the use of PPE may seem a convenient alternative to employers, it can be burdensome to employees. Certain types of PPE, such as respirators or protective clothing, can become hot, heavy, or uncomfortable and workers may avoid their use. Worker acceptance is an important factor for the successful use of PPE. Protective devices must be worn correctly and consistently and must be properly maintained in order to achieve the optimum level of protection. If employees find the necessary protective equipment uncomfortable and do not understand the reasons for its use, they are likely to misuse the device and consequently not receive the necessary protection. Therefore, training is an important part of a successful PPE program. This chapter describes the different types of PPE, explains important elements of a PPE program, and identifies situations where PPE is likely to be used.

TYPES OF PERSONAL PROTECTIVE EQUIPMENT

Respiratory Protection

The major route of exposure to a harmful airborne material is inhalation. The respiratory system may be physically damaged or irritated by the agent, or it may allow the material to become absorbed in the lung and transported throughout the body. A respirator is a device used to prevent harmful materials from entering the respiratory system. In environments where there is the potential for inhalation of toxic or irritating dusts, aerosols, mists, fumes, vapors, or gases, a respirator can be selected to provide a clean source of air to the wearer. To achieve this, the respirator can filter hazardous agents from contaminated workroom air (air-purifying respirator), or it can provide clean air from a source separate from the work environment (air-supplied respirator). A number of respirator types are available to provide protection in a variety of situations. Table 1 outlines the different types of respirators and facepieces.

Air-purifying respirators purify the air through a system of cartridges or canisters. Because these respirators purify the existing air rather than supply air, there are limitations associated with their use. Table 2 summarizes conditions under which air-purifying respirators should not be used for protection. Most air-purifying respirator facepieces operate under negative pressure relative to the environment. Air is drawn into the facepiece via the negative pressure created when the user inhales. This requires slightly more effort than would be necessary if a respirator were not worn. For this reason, some medical conditions (generally heart or lung problems) may restrict the individual from using this type of respirator. A respiratory protection program should include a medical evaluation of personnel to screen for those conditions that are not compatible with the use of this type of respirator. Another disadvantage of the negative pressure respirator is that any leaks or imperfect seals in the system will allow contaminated air to be drawn in during inhalation.

Powered air-purifying respirators operate in a manner similar to negative

Table 1. Respirator Types

1. **Types of respirators**
   - A. Air-purifying (can be negative or positive pressure)
     1. Particulate cartridge — mechanically filters:
        - a. Dusts
        - b. Mists
        - c. Fogs
        - d. Fumes
        - e. Aerosols
     2. Chemical cartridge — sorbent system removes:
        - a. Gases
        - b. Vapors
   - B. Air-supplied (generally positive pressure) — may be used for particulates, gases, and vapors
     1. SCBA — user carries air supply
     2. Air line — airhose connects to a supply of fresh air
     3. Combination of SCBA and air line
2. **Respirator facepieces**
   - A. Tight-fitting
     1. Full-face
     2. Half-mask
     3. Quarter-mask
   - B. Loose-fitting
     1. Hood
     2. Helmet
     3. Blouse
     4. Full-suit

Table 2. Limitations of Air-Purifying Respirators

Where there may be an insufficient concentration of oxygen, for instance, in a confined space such as a tank
Where the environment is considered immediately dangerous to life or health (IDLH) (i.e., the hazardous agent exists in a concentration that within 30 min may produce irreversible health effects or impair the individual's ability to escape)
Where the nature of the contaminant is unknown
Where the concentration of the contaminant is unknown or where it may exceed the maximum capabilities of the respirator
Where the contaminant cannot be effectively filtered or collected on a sorbent material
Where the contaminant has inadequate "warning properties," and the unit has no "end-of-service-life" indicator. A material with poor warning properties cannot be detected, either by odor or by irritation at concentrations below that which may cause harmful effects. If the odor threshold of an agent is greater than the permissible exposure limit (PEL), the agent is considered to have poor warning properties

pressure masks; however, they supply purified air to the facepiece under positive pressure. A blower unit pushes the contaminated workplace air through the filtering element and delivers the clean air to the facepiece. These respirators may be slightly more comfortable and do not provide the minor resistance to breathing that negative pressure respirators do. Because the facepiece is under *positive* pressure, contaminated air does not leak into the facepiece; hence, greater protection is afforded.

Particulate respirators use mechanical filters to filter out dusts, fogs, mists,

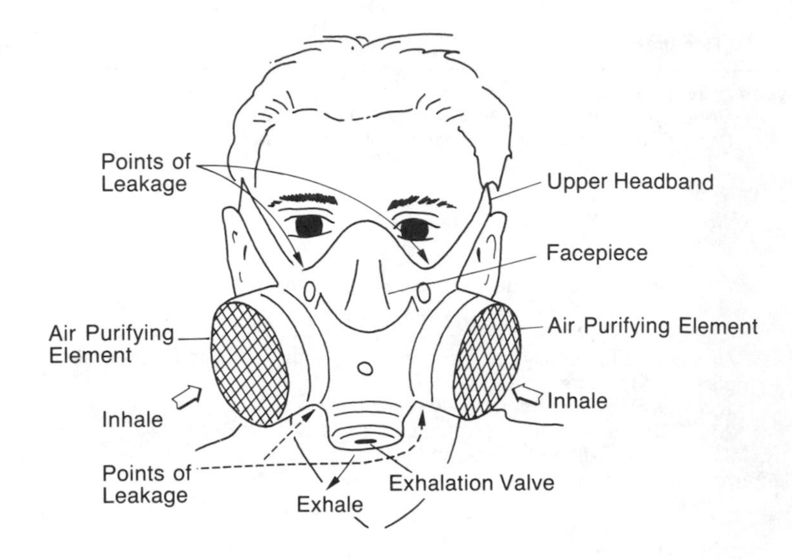

Figure 1. Airflow patterns in a half-mask respirator.

fumes, and aerosols. High-efficiency filters are also available that provide greater filtering capacity than normal particulate filters. Air flows into the mask through filter cartridges attached to the facepiece as the user inhales. Exhaled air is passed out of the respirator through an exhalation valve that closes when exhalation is completed (see Figure 1). These valves should be carefully checked for signs of dirt, since even a single particle on the valve can prevent full closure of the valve following exhalation. A valve left partially open may allow unfiltered air into the mask during inhalation.

Gas or vapor respirators (sometimes referred to as chemical cartridge respirators) remove gases and/or vapors by passing the contaminated air through cartridges or canisters filled with charcoal or other sorbent material that absorbs, adsorbs, or reacts with the contaminant, thus allowing only "clean air" to pass through the respirator. Some sorbents are highly specific for a particular compound and may give no protection against another hazardous agent. Therefore, when using a chemical cartridge respirator, it is essential to select the proper cartridge to provide the necessary protection.

The various types of chemical cartridges can be identified by their color code and the wording on the label. Different categories of cartridges include those for acid gases, chlorine, organic vapors, ammonia gas, radioactive materials, and particulates. Some cartridges also provide protection against more than one of the categories mentioned above, such as an organic vapor/acid gas cartridge. In addition, two cartridges can, in some cases, be "piggy-backed" to provide protection against, for instance, toxic particulates *and* vapors. Table 3 provides a summary of available cartridges and their appropriate uses. An industrial hygienist should be consulted on proper cartridge selection.

Table 3. Color Codes for Common Respirator Canisters and Cartridges

Black	Organic vapors
White	Acid gases
Yellow	Organic vapors and acid gases
Green	Ammonia
Purple	Radionuclides
Gray	Particulates

"Breakthrough" is the term used to describe the situation in which the sorbent system of a chemical cartridge respirator becomes depleted and the gas or vapor is allowed to penetrate the mask. Many respirators, if designed for use under highly dangerous conditions, can be equipped with canisters that change color as the sorbent is spent to warn the user that the canister should be changed. This is referred to as an "end-of-service-life" indicator. Users of respirators that do not offer this feature must rely upon the odor or irritant properties of the gas/vapor penetrating the sorbent system as the only warning that the filtering capacity has been exceeded. It is for this reason that it is important that the hazardous material can be detected at concentrations below those that may be harmful to the employee. Methylene chloride is an example of a material that can be effectively filtered through a charcoal cartridge, but cannot be detected when breakthrough occurs. Chemical cartridge respirators are not recommended for protection against methylene chloride or for other materials with poor warning properties.

Air-supplied respirators are those that supply a clean source of breathable air separate from that in the work environment. The air may come from a cylinder of compressed air that the user is wearing, or it may come from a portable or stationary compressing system. The air is delivered, usually under positive pressure, to the facepiece of the respirator. These devices offer essentially no resistance to breathing and generally provide cooler air than other respirators. Supplied-air respirators include self-contained breathing apparatus (SCBA), air-line respirators, and a combination of the two.

The SCBA provides a "transportable" supply of breathable air and affords protection against toxic gases and oxygen deficiency. The supply of air is carried in a tank on the user's back and supplies the facepiece with a positive pressure source of air. Workers who may need to use an SCBA must be formally trained in donning procedures and the use of the respirator. These devices provide a very high level of protection, but can be cumbersome to put on and wear. SCBAs can weigh over 30 lb and impose an added workload on the wearer due to this weight. Once in place, however, the SCBA user's freedom of travel is not limited by an airhose connected to a stationary source.

Air-line respirators supply air to the worker through airlines or hoses. These lines tie into an air compressor located in a clean environment. Air-line respirators can be used where air-purifying respirators are not sufficient and where the atmosphere is not immediately dangerous to life or health (IDLH).

Combination self-contained and air-line respirators may be used in IDLH

Table 4. Advantages and Disadvantages of Air-Purifying and Supplied-Air Respirators

Respirator Type	Advantages	Disadvantages
Air-purifying	Lightweight	Those under negative pressure cause some breathing resistance
	Does not restrict movement	Limited to certain types of environments (see Table 2)
		Cartridges specific for certain types of agents
	Simple to use	Requires annual fit testing
Air-supplied (SCBA)	Provides highest level of protection; may be used to protect against highly toxic agents and oxygen deficiency	Heavy and cumbersome to wear
		Limited air supply
	Does not provide breathing resistance	May require frequent refresher training in order to use proficiently
Air supplied (air line)	Less heavy and cumbersome than SCBA	Not approved for IDLH conditions
Airline	Unlimited air supply	Airline restricts mobility

atmospheres. This combination consists of an air-line respirator and a small cylinder of compressed air to provide an emergency air supply to allow for escape if the regular air line is cut off.

Advantages and disadvantages of air-purifying and air-supplied respirators are summarized in Table 4.

Respirator Facepieces

The facepiece of the respirator is worn over some portion of the user's face. The facepiece types discussed here cover, at minimum, the nose and mouth. Respirator facepieces may be either tight- or loose-fitting. Tight-fitting facepieces can be full-face, half-mask, or quarter-mask. In order to receive maximum respiratory protection when wearing a tight-fitting facepiece, an effective seal preventing the entry of contaminated air into the mask must exist between the mask and the user's face. Some individuals cannot obtain a good seal with these facepieces, either due to facial configuration or due to beard growth.

Respirator fit testing is performed to evaluate the fit of tight-fitting facepiece respirators. "Qualitative" fit testing can be conducted by passing an irritant smoke near the face of the employee wearing the respirator. If any smoke enters the facepiece, the individual will experience irritation and perhaps cough. Any irritation on the part of the user during this procedure indicates fit test failure and the person should not be assigned that particular respirator. The individual, however, may be tested with various brands of the same type of respirator in an attempt to find a model that will work well. Other "easily detectable" agents such as banana oil or saccharin may be used for qualitative fit testing. "Quantitative" fit testing is a more sophisticated procedure that provides better information. During a quantitative fit test, concentrations of the test atmosphere are measured simultaneously inside and outside the mask to determine the amount of the contaminant leaking into the mask. This procedure gives more accurate information on how well a particular respirator works for a given individual. All fit testing procedures should be conducted by an industrial hygienist or a trained technician.

If a worker is bearded or unshaven, or if respirator fit testing has indicated that a good seal cannot be obtained, the worker cannot use a tight-fitting facepiece. Bearded workers should not even be fit tested with a tight-fitting respirator facepiece, since beard growth changes constantly and a successful fit test one day does not guarantee an effective fit in the future. Employees should also be aware that other types of protective equipment (such as safety glasses or hearing protectors) may interfere with the facepiece/face seal.

In addition to the following descriptions, Figure 2 illustrates the three types of tight-fitting facepieces:

1. **Quarter-mask** air-purifying respirators cover the mouth and nose. The lower edge rests between the chin and the mouth. This mask may be more easily dislodged than the other two and provides a lower level of protection.
2. **Half-mask** air-purifying respirators fit over the nose and mouth and under the chin. This respirator provides better protection than the quarter-mask, but not as great as the full-face respirator. It does not afford any type of eye protection.
3. **Full-face** respirators cover the face from the hairline to the chin and across the face from ear to ear. Full facepiece respirators may be air purifying or air supplying and provide a better protection factor than either the half- or quarter-mask varieties. This type of facepiece also affords eye protection against chemical splashes, irritating or toxic agents, and the impact of flying objects.

Loose-fitting respirators include hoods, helmets, blouses, and full suits. As the name implies, they do not form a tight seal with the face and may be more comfortable (and cooler) to wear for extended periods of time. Clean compressed air is pumped into the enclosure through a tube. It is important that the airflow is great enough that contaminated air from the outside does not flow in through the neck or waist opening. Bearded individuals (who cannot achieve an effective seal from tight-fitting facepieces) can successfully utilize a loose-fitting respirator. The disadvantages of this type of equipment are that storage and cleaning procedures may be less convenient and the equipment may be slightly more cumbersome to wear.

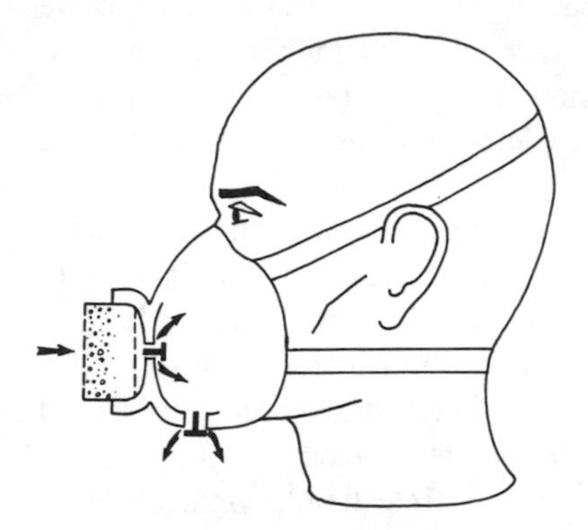

Quarter-mask Facepiece

Half-mask Facepiece

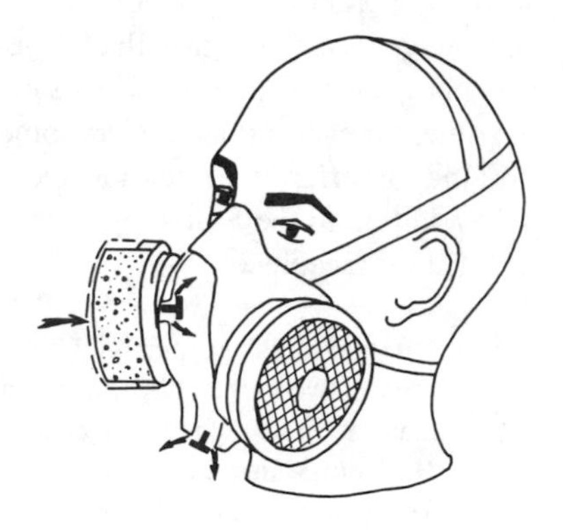

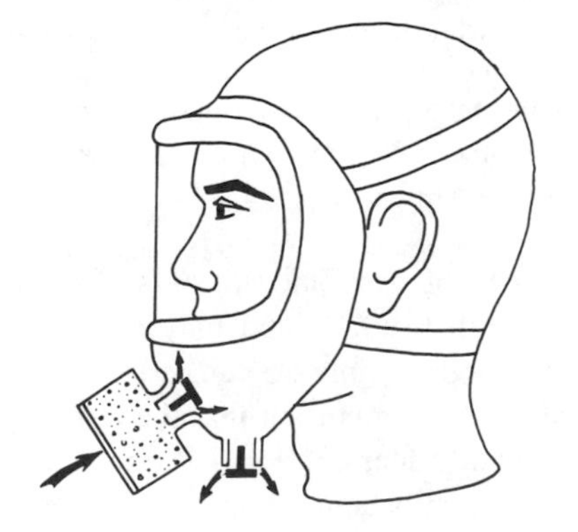

Full Facepiece

Figure 2. Respirator configurations (*NIOSH Guide to Industrial Respiratory Protection*, 1987).

Respirator Protection Factors

Anyone wearing a respirator in a hazardous environment needs to know the amount of protection expected from the equipment. Protection factors provide a measure of the efficiency of each type of respirator and can be determined by quantitative fit testing. The protection factor can be defined as:

$$\text{Protection Factor} = \frac{[\text{Concentration of the contaminant outside of the facepiece}]}{[\text{Concentration of the contaminant inside of the facepiece}]}$$

For example, if a given respirator has a protection factor of 10 and is worn in an environment where the concentration of the contaminant is 200 ppm, the user would expect the concentration of the contaminant inside of the facepiece to be no greater than 20 ppm. The current protection factors established by the National Institute for Occupational Safety and Health (NIOSH) may be found in the 1987 *NIOSH Guide to Industrial Respiratory Protection.* A summary of these protection factors are presented in Table 5.

Assigned protection factors, as used by industrial hygienists and safety professionals for respirator selection, were developed in 1972 by the Los Alamos National Laboratory under contract to NIOSH. These protection factors were based upon laboratory evaluations of respirators and were assumed to provide a reasonable margin of safety. In general, negative pressure respirators provide a lower degree of protection than positive pressure devices. As mentioned previously, when a negative pressure respirator is worn, an imperfect seal can result in the contaminant being drawn into the mask when the user inhales. When positive pressure exists inside the facepiece, the risk of introducing contamination into the mask is greatly reduced.

Respirator Maintenance

A respirator maintenance program is necessary to ensure that respirators continue to function as effectively as when they were first manufactured. Such a program must include the following elements:

1. **Inspection** - OSHA (1910.134) requires that "all respirators be inspected before and after each use" and that those not used routinely, such as emergency escape and rescue devices, "shall be inspected after each use and at least monthly." The need for such an inspection cannot be overstated. Workers wearing damaged/malfunctioning respirators may be in greater danger than if they wore no respirator at all, since they may have a false sense of security and think they are being protected when they are not. A proper inspection program can identify defective respirators before they are used.

Table 5. Protection Factors for Commonly Used Respirators

Respirator	Protection Factor
Air-purifying quarter-mask respirator with particulate filter cartridges	5
Air-purifying half-mask respirator with particulate filter, gas/vapor cartridges, or a combination of both	10
Air-purifying full-face respirator with particulate filter cartridges	10
Supplied-air half-mask respirator operating under negative pressure	10
Powered air-purifying respirator with a loose-fitting facepiece and particulate filter, gas/vapor cartridges, or a combination of both	25
Supplied-air respirator with a loose-fitting facepiece operating under positive pressure	25
Air-purifying full-face respirator with high-efficiency filter cartridges	50
Powered air-purifying respirator with a tight-fitting facepiece and high efficiency particulate, gas/vapor cartridges, or a combination of both	50
Supplied-air respirator with a tight-fitting facepiece operating under negative pressure	50
Supplied-air half-mask respirator operating under positive pressure	1,000
Supplied-air full-face respirator operating under positive pressure	2,000
Self-contained breathing apparatus (SCBA) (full-face and operating under positive pressure)	10,000

2. **Cleaning** - OSHA states "routinely used respirators shall be collected, cleaned and disinfected as frequently as necessary to ensure proper protection is provided" and that emergency use respirators "shall be cleaned and disinfected after each use." Respirator facepieces can generally be washed with warm, soapy water, rinsed thoroughly, and allowed to air dry. Valves should be checked to ensure dirt particles have not lodged beneath, thus allowing unfiltered air to enter the facepiece during inhalation.
3. **Repair** - OSHA states that "replacement or repair shall be done by experienced persons with parts designed for the respirator." It should be noted that in addition

to violating this requirement, substitution of parts from different brands or models of respirator invalidates the approval of the device. Spare respirator parts should be kept on hand, and respirator maintenance personnel should be thoroughly trained in respirator repair.

4. **Storage** - OSHA requires that respirators should be protected in storage against dust, sunlight, heat, extreme cold, excessive moisture, and damaging chemicals. In addition, common sense dictates that respirators should be stored in a manner to prevent contamination of the device. Respirators should be stored in a plastic bag (if possible) inside of a rigid container.

Eye and Face Protection

Inhalation is not the only potential route of entry of a toxic agent into the body. Many solid, liquid, and gaseous agents can be easily absorbed through the eyes or the skin. The moist surface of the eye can dissolve airborne chemicals and deliver them to the rest of the body via the bloodstream. Biological agents, such as viruses, can potentially enter the body through the mucous membranes of the eye. In addition, the eyes are particularly vulnerable to assault by acidic or caustic materials, flying objects, and to some forms of radiation. For this reason, eye protection should always be worn whenever there is a possibility of harmful exposure to a chemical, biological or physical agent. Eye protection can be achieved by using safety glasses, chemical splash goggles, or some types of respirators. The use of faceshields can provide protection to the eyes as well as to the face.

Safety glasses can be worn to protect the eyes from the impact of flying objects, from the effects of most types of nonionizing radiation, and to some extent chemical splashes. When safety glasses are worn without sideshields, liquid splashes or flying particles approaching from the side can potentially reach the eye. If sideshields are attached to the temple spatulas of the frames, greater protection against liquid or solid agents approaching from lateral directions is afforded.

Goggles provide greater protection against chemical splashes than safety glasses with sideshields due to the fact that they fit more closely to the face and form a more effective barrier against foreign materials. Goggles are also impact-resistant. Different types of goggles are available depending upon the type of hazardous condition warranting protection (i.e., flying particulates, liquid splashes, etc.).

Faceshields provide good protection to the eyes, face, and neck against flying particles, sprays of hazardous liquids, splashes of molten metal and hot solutions. They are not recommended for eye protection against the impact of hurtling objects, therefore it is generally necessary to wear safety glasses underneath the faceshield. Faceshields can also be used to provide antiglare protection where required.

While respirators are generally not considered "eye protectors," full-face respirators and air-supplied hoods do provide good protection against splash and against irritant gases and vapors. (In the event that a highly irritating vapor were to be generated, either of these devices would be an acceptable second choice to a SCBA.) While full-face respirators are impact resistant, air-supplied hoods are not, and safety glasses may need to be worn underneath.

Hearing Protection

Hearing protection should be worn in "high noise" areas to protect individuals from "noise-induced hearing loss," a permanent decrease in the ability to hear resulting from excessive exposure to noise. Different hearing protection devices vary in their abilities to reduce noise. Hearing protectors include a variety of earplugs and earmuffs. Additional information on hearing protection can be found in Chapter 11 of this book.

Dermal Protection

Barrier clothing is used to provide dermal (skin) protection against a number of hazardous materials and conditions. Protective clothing may be used to insulate workers against extremely hot or cold conditions or materials. Certain items of clothing may protect the body from cuts, bruises, and abrasions when handling heavy, sharp, or rough materials. Chemical impervious clothing may be needed to protect against chemical agents that can be irritating or corrosive to the skin or that may be absorbed through the skin. Types of barrier clothing range from gloves and boots to smocks, aprons, and full-body suits. In addition, some manufacturers offer selected items of clothing to protect certain portions of the body, such as sleeves to protect the upper and lower arm.

Barrier clothing should be selected according to the particular hazardous agent to be protected against. Gloves, in particular, undergo extensive testing to evaluate degradation and permeability characteristics of the material in the presence of a number of hazardous agents. Degradation testing is conducted to determine how quickly a chemical agent will degrade or destroy the protective fabric. Permeability testing indicates "breakthrough time" or how quickly the chemical will penetrate through the material (this may occur before the material is totally degraded). When selecting a garment, boot, or glove to provide protection against a harmful agent, those materials that may be quickly permeated or degraded by the chemical agent in question should not be used under most circumstances. Again, the expertise of a trained industrial hygienist should be used when selecting protective clothing.

Protective clothing can be made from a variety of materials including cotton, glass fibers, orlon, nylon, and Teflon®.* Coating the material with various plastics,

---

* Registered trademark of E. I. du Pont de Nemours and Company, Inc., Wilmington, DE.

rubber or Neoprene will make the fabric impervious to many chemical agents. Garments can also be fabricated from plastic sheeting using glued or heat-sealed seams. Disposable garments may be recommended when working with highly toxic or hazardous agents to avoid the problem of handling and laundering contaminated clothing. Special laundering procedures should be used for nondisposable items of protective clothing that have been contaminated with a hazardous agent. Contaminated work clothing should never be worn home from work since it could be hazardous to persons handling it there.

Correct use of barrier clothing includes appropriate procedures for removal of the contaminated equipment. Before the contaminated equipment is removed, it should be thoroughly washed off. If respiratory protection is also being used, it should be the last item of equipment removed. Gloves should be removed after boots and other protective garments have been removed and discarded.

Nondisposable items of clothing should be thoroughly decontaminated prior to storing for future use. Equipment should be properly maintained and inspected prior to reuse.

A PERSONAL PROTECTIVE EQUIPMENT PROGRAM

Employers who require the use of PPE for work activities should develop an effective PPE program. Table 6 summarizes important elements of such a program. Employees who are required to use the protective equipment should understand the objectives of the program and should be allowed an opportunity to provide feedback on their experience with the equipment. Worker acceptance is an important part of a successful PPE program. Unless employees are properly educated on the need for PPE and "buy in" to the program, compliance is likely to be a problem and effective use will not be made of the equipment available.

Employee training is an important part of the education process. Training should be completed prior to actual PPE use and should be repeated on a routine (generally annual) basis. As well as educating the employee about the importance of the equipment that will be worn, training allows the user to become familiar with the equipment in a nonhazardous situation. A successful PPE training program might address the following elements:

- A description of the hazard/condition necessitating protective measures
- A description of what has been/can be/cannot be done about the hazard
- An explanation of why a certain type of PPE has been selected
- A discussion of the capabilities and the limitations of the devices selected
- A demonstration of the correct use of the equipment, including donning and doffing (removal) procedures
- An opportunity for employees to practice wearing the equipment
- A discussion of cleaning, inspection, and maintenance procedures

When respiratory protection is necessary, each employee must undergo annual

Table 6. Elements of an Effective PPE Program

Element
Training
Should be conducted by an industrial hygienist or trained technician
Additional training provided to management and line supervisors on correct use of PPE and selection
Overall training on the importance of and need for PPE
Initial training prior to the use of new equipment
Periodic refresher training on equipment used
Fit testing (respiratory protection only)
Should be conducted by an industrial hygienist or trained technician
Initially with each new type/model of respirator
Annually
Medical evaluation
Screen individuals for medical conditions (such as heart/lung problems which may be aggravated by the use of certain types of PPE)
Equipment selection
Initial selection should be made by an industrial hygienist
Supervision may be trained to select appropriate equipment from that previously approved by industrial hygienist depending upon work conditions
Equipment maintenance
Cleaning/decontamination procedures
Inspection
Storage

respirator fit testing. This procedure involves either a qualitative or a quantitative determination of the fit of the chosen respirator to the face of the individual who will be wearing it. Each employee should be fit tested with the same model of respirator that he will be using. As mentioned previously, bearded individuals cannot achieve an effective seal with the mask and should not be assigned to wear a tight-fitting facepiece. Employees who may be required to wear a negative pressure respirator or a SCBA should be medically evaluated to identify any health conditions that may prohibit the use of these types of equipment.

Selection of PPE should be performed by a person knowledgeable of the various protective devices and of the hazards in the work environment. Attention should be paid to the degree and type of protection needed from the equipment. Examples of quantitative assessments of protection offered include the assigned protection factor (in the case of a respirator) or the noise reduction rating (in the case of hearing protectors). Companies that require use of PPE may wish to conduct in-house research on the effectiveness of the equipment for specific work applications. It is recommended that the initial selection of equipment should be made by an industrial hygienist. Supervisors should be trained to select the appropriate item of PPE based upon actual work conditions from a "preapproved" list. For example, the industrial hygienist might select various glove types to be used when handling a number of different solvents in a chemical production facility. The industrial hygienist would provide information to the line supervisor on which glove is best used with each type of solvent. It would be up to management to recommend to the employee the glove that should be used, for example, when working with chlorinated solvents.

SELECTION OF AN EXPOSURE CONTROL METHOD

The Occupational Safety and Health Administration (OSHA) requires that PPE should be used as a "last resort" to protect employees against exposure to harmful conditions. Whenever possible, the hazardous process should be engineered to isolate or remove the hazard from the work area of the employee. If this is not possible, work practices should be changed to either move the employee from the hazardous area or to limit the time spent in the area. PPE is considered by OSHA to be an acceptable alternative only if engineering controls and administrative controls (work practices) cannot feasibly reduce employee exposures to an acceptable level. In many cases, employers may view PPE as an easy and inexpensive alternative to engineering out the hazard. In the long run, however, the cost of the equipment and the time invested in training and maintenance requirements may prove this solution to be less than "cheap." For example, if a one-time expense of engineering a "contained system" that isolates the contaminant from the employee were weighed against the cost of respirators and dermal protection as well as the time necessary for training, fit testing, and equipment maintenance, the former solution may be found to be cheaper and safer. In addition, OSHA is currently defining specific conditions under which PPE will be considered an acceptable alternative to engineering or work practice controls. The outcome may well be that PPE will not be allowed as the only method for exposure control in a great many situations.

PPE does provide a relatively simple, inexpensive, short-term solution for many exposure problems. It can be useful for an operation of very short duration in which it may be impossible to contain the hazardous material and difficult to effectively ventilate. In most cases, it also incorporates an element of flexibility; the employee can safely work in the midst of the hazardous area. In certain operations where exposure conditions simply cannot be engineered out, PPE may be the only solution.

On the other hand, PPE is all too frequently the only solution considered by employers. Unless careful attention is paid to selection and proper use and maintenance of the equipment, workers might not achieve effective protection, while at the same time they may experience a false sense of security. Human error or carelessness should be expected to occur when reliance is placed upon PPE. In addition, employers who require workers to rely solely on PPE for exposure control and implement no engineering controls whatsoever can place employees in an extremely dangerous situation if PPE failure occurs.

HAZARDOUS WASTE OPERATIONS AND EMERGENCY RESPONSE

Several factors distinguish the hazardous waste site environment from other occupational situations involving hazardous substances. One factor is the uncontrolled condition of the site. Another factor is the variety of hazardous materials that may be present at the site. It is frequently impossible to accurately assess all chemical hazards at a waste site due to the large number of substances and the potential chemical interactions among materials. Especially in the early stages of the investigation many of the substances may not be identifiable. Although fewer in number and of lesser variety, these same types of hazards may exist during emergency response operations.

There are many potential hazards at a waste site, the most obvious of which is chemical exposure. Most hazardous waste sites contain a variety of gaseous, liquid, and solid chemical substances. On a similar note is the risk of exposure to biological agents. Wastes from hospitals and research facilities can contain biological hazards, disease-causing organisms that can infect personnel working at the site. Routes of exposure to biological agents are similar to those for chemical hazards. Protection may be afforded through the use of dermal, eye, and respiratory protective equipment.

Explosion and fire can occur from the reactions of two or more chemicals that come in contact, from the ignition of explosive or flammable materials, from the ignition of materials due to oxygen enrichment, from agitation of shock- or friction-sensitive materials, or from the sudden release of materials under pressure. At hazardous waste sites or emergency response operations, explosions and fires may not only result in intense heat, open flame, smoke inhalation, and flying objects, but they can also cause the release of toxic chemicals.

Oxygen deficiency can result from the displacement of oxygen by another gas or by the consumption of oxygen by a chemical reaction. The normal concentration of oxygen is approximately 21%. When this concentration drops to 16%, impaired attention, judgment, and coordination and increased breathing and heart rate may result. Air-supplied respirators should be used whenever oxygen levels are measured below 19.5% (by volume) in the work environment.

Radioactive materials in the hazardous waste environment may emit ionizing radiation. Use of protective clothing as well as conscientious personal hygiene and decontamination procedures will provide protection against α- and β-radiation. γ-radiation passes easily through normal chemical-protective clothing and requires the use of specially designed clothing, or preferably engineering controls, for adequate protection.

Safety hazards such as sharp objects, slippery surfaces, uneven terrain, and unstable surfaces and objects (such as drums, walls, or flooring) contribute to the hazardous nature of this work environment. Not only may accidents directly injure workers, but they can damage protective equipment, or result in a dangerous mixture of chemicals. Electrical hazards can result from power lines, electrical wires, buried cables, or from electrical equipment used on site. Special protective clothing may be required for protection against electrical hazards.

Temperature extremes may also be of concern at the hazardous waste site. The PPE worn to protect employees against other exposure conditions can contribute to a heat stress hazard. For this reason, it is important to not "overprotect" employees by requiring the use of unnecessary equipment, especially when work activities are conducted under warm environmental conditions. Cold exposure can be a hazard when working in low temperatures or where the wind chill factor is low. Appropriate clothing as well as a provision for warm shelter can minimize the risk of hypothermia and frostbite under these conditions.

Excessive noise at the hazardous waste site may lead to physical damage to the auditory system ("noise-induced hearing loss"). Loud noises may also startle or distract workers or can interfere with communication among personnel. If noise levels cannot be maintained at a safe level, hearing protectors should be worn. Some devices are available which include communication capability in hearing protection equipment which allows for conversation between workers.

Levels of protection for workers at hazardous waste sites or those conducting emergency response activities are categorized by the Environmental Protection Agency (EPA) into four levels: A, B, C, and D. These levels provide basic guidelines for the selection of PPE in a given environment, but flexibility may be applied based upon the existing circumstances. A summary of the equipment, protection provided, applications, and limitations for each of these levels is presented in Table 7. Further information on PPE requirements may be found in the OSHA Standard for Hazardous Waste Operations and Emergency Response (29 CFR 1910.120). Selection of PPE for emergency response, particularly in the planning stages, can vary greatly from one work environment to another.

SUMMARY

PPE is a useful and often necessary measure to reduce employee exposures to hazardous agents in the work environment. PPE can protect employees from the inhalation of airborne contaminants, from damage or disease resulting from contact of a hazardous substance with the eyes, and from skin absorption of chemical or physical agents. As suggested in this chapter, it should not be used as the only control measure, except in cases where engineering and work practice controls are totally infeasible. When relying upon PPE to isolate an individual from a hazard, it is impossible to ensure that the equipment always will be used properly. Whereas the efficiency of an engineered system can be monitored to determine whether it is operating within acceptable standards, it is difficult to ascertain that PPE is being conscientiously used, maintained, and inspected. PPEs have limited capabilities and equipment failure is not always immediately detectable. Therefore, it is always preferable, from a health and safety standpoint, to remove the hazard altogether rather than to rely exclusively on PPEs for employee

Table 7. EPA Levels of Protection for Hazardous Waste and Emergency Response Operations

Level of Protection: A

Equipment	Protection Provided	Application	Limitations
Fully encapsulating chemically impervious suit	Highest level of respiratory, eye, and dermal protection available	Chemical has been identified as highly toxic in the concentration present or in concentrations that may be potentially generated	Encapsulating suit must be impervious to agents in the work environment
Chemically impervious boots and gloves			
SCBA or air line respirator with escape SCBA		Materials which are very hazardous to the skin	
Two-way radio		Operations in confined or poorly ventilated areas	

Level of Protection: B

Limitations	Equipment	Protection Provided	Application
Chemically impervious clothing (not fully encapsulating)	Highest level of respiratory and eye protection available, lower level of dermal protection	Chemical has been identified as highly toxic via inhalation, but requiring less dermal protection	Must not be used in the presence of gases or vapors that may be harmful to the skin or capable of skin absorption
Chemically impervious gloves and boots		Maybe some gases or vapors have not been identified, but are not suspected of being highly toxic upon skin contact	Unlikely potential of generation of high concentrations of hazardous materials or of splashes of hazardous materials that affect the skin
SCBA or air line respirator with escape SCBA			
Two-way radio		Less than 19.5% oxygen	

Table 7 (continued). EPA Levels of Protection for Haxardous Waste and Emergency Response Operations

Equipment	Protection Provided	Application	Limitations
Level of Protection: C			
Chemically impervious boots and gloves		Air contaminants have been identified and can be removed with a respirator canister	
Full-face air-purifying respirator with canister		Criteria for the use of air-purifying respirators are met (see Table 2)	
Two-way radio			
Level of Protection: D			
Coveralls	No respiratory protection, minimal dermal protection	The work environment has been determined to be nonhazardous	Work environment must contain at least 19.5% oxygen
Boots			
Safety glasses or goggles		No potential for generation of hazardous gases or vapors; no potential for chemical splashes	

protection. Table 8 contains a summary of applicable PPE standards and reference documents.

APPENDIX

The following reference materials can provide additional information for the interested reader:

Patty's Industrial Hygiene and Toxicology, 3rd ed., Clayton, G. P., and F. E. Clayton, Eds., John Wiley & Sons, New York, 1978.

The Industrial Environment — Its Evaluation & Control, U.S. Department of Health, Education & Welfare, Public Health Service, National Institute for Occupational Safety and Health, Washington, D. C. 1973.

Table 8. Applicable Standards and Reference Materials for PPE

The following published standards provide additional information on requirements and/or guidelines for the use of PPE:

Occupational Safety and Health Administration (OSHA) Standards

1. Hazardous Waste Operations and Emergency Response
   29 CFR Part 1910.120
2. General Requirements for PPE
   29 CFR Part 1910.132
3. Eye and Face Protection
   29 CFR Part 1910.133(a)
4. Standard Practice for Respiratory Protection
   29 CFR Part 1910.134
5. Clothing and Equipment for Protection against Electrical Hazards
   29 CFR Part 1910.137

American National Standards Institute (ANSI) Standards

1. Practices for Respiratory Protection
   Z88.2-1980
2. Practices for Respiratory Protection for the Fire Service
   Z88.5-1981
3. Respiratory Protection - Respirator Use - Physical Qualifications for Personnel
   Z88.6-1984
4. Identification of Air-Purifying Respirator Canisters and Cartridges
   K13.1-1973

Guide to Industrial Respiratory Protection, DHHS (NIOSH) Publ. No. 87-116, U.S. Department of Health, Education & Welfare, Public Health Service, National Institute for Occupational Safety and Health, Washington, D.C., 1987.

Occupational Safety and Health Guidance Manual for Hazardous Waste Site Activities, DHHS (NIOSH) Publ. No. 85-115, U.S. Department of Health and Human Services, Public Health Service, National Institute for Occupational Safety and Health, Washington, D.C., 1985.

CHAPTER 15

Personal Protective Equipment (PPE) for Preventing Physical Injury

Leo DeBobes

INTRODUCTION

Occupational safety and health risks have existed since humans first began "careers" as hunter-gatherers. As knowledge and abilities expanded, so did the potential risks. Various occupations and endeavors presented their own peculiar hazards as technology developed. For example, in 19th century England, adolescent males were employed as chimney sweeps (their smaller size enabled them to be lowered down a chimney in order to clean it). Sir Percival Pott documented an increased incidence of scrotal cancer in these youths as a result of their exposure to the creosote-containing soot that caked the inside of the chimneys.[1] Construction workers in the U.S. recognized early on the hazards of being struck by falling materials or striking against fixed objects while engaged in building erection or demolition. In order to minimize their potential for injury, these craftworkers often wore derbies as a form of makeshift head protection.

The publicity surrounding the mounting toll of occupational casualties during the Industrial Revolution was the first large-scale acknowledgment of concern for those in working life. Poorly designed equipment often presented serious machine-guarding risks and an often-uneducated and management-intimidated workforce was unwilling to complain or to refuse hazardous assignments. This flood of protest resulted in the passage of the first Workers' Compensation law in the U.S. in Wisconsin in 1911. Soon, similar legislation was being enacted in other states, spurred by a citizenry outraged at the prevalent barbaric industrial health and safety conditions. Increased agitation by organized labor unions as well as the emergence of occupational safety and health professionals also helped stimulate a trend toward regulating and promoting safety in the workplace. The National Safety Council (NSC), formed in 1912, the American Society of Safety Engineers (ASSE), formed in 1911, and the American Industrial Hygiene Association (AIHA), formed in 1939, all lent considerable effort to developing a heightened awareness of occupational hazards, while making recommendations geared to providing safer work.

With engineering and administrative controls often dismissed by owners and managers as overly expensive, frivolous, or as constraints to productivity, a key means of preventing injury and illness in the workplace was through the introduction of personal protective equipment (PPE). Even today, despite an increased emphasis on the use of engineering controls for hazard abatement in industry, the Occupational Safety and Health Administration (OSHA) suggests that some 22 million workers in the U. S. who are not already so supplied need to wear PPE (specifically hard hats, safety shoes, and eye and face protection) on the job. In addition, government figures approximate that nearly 3 million of these workers are not wearing appropriate PPE on the job.[1a] OSHA estimates that about 411,000 injuries sustained annually by U.S. workers are linked to improper usage or failure to use PPE.[1a]

This chapter provides an overview of the various types of PPE, specifically addressing those devices noted in Table 1.

HEAD PROTECTION

In OSHA's *General Industry Standards* (29 CFR 1910), specific requirements are established under OSHA 1910.135, "Occupational Head Protection," wherein head protection must be provided for workers who may be at risk of head injury from impact and penetration from falling or flying objects and from limited electrical shock and burns.[2] The OSHA standard stipulates that head protection shall meet the specifications outlined in the 1969 revision of the American National Standards Institute's (ANSI) "Safety Requirements for Industrial Head Protection" (Z89.1-1969).[3] While Z89.1, as is the case with most ANSI standards, has subsequently been revised, the 1969 recommendations remain incorporated into the OSHA rule as formal requirements. A prudent employer, however, would

Table 1. Types of Personal Protective Equipment (PPE)

Head protection
Eye and face protection
Foot protection
Body protection
Hand protection
Respiratory protection
Fall protection
Musculoskeletal protection

be well advised to adhere to the more conservative current ANSI standard (Z89.1-1986).

A large variety of different types of head protection are currently available. The most common example of such devices would be the ubiquitous hardhat. Other forms of head protection include bump caps, miner's caps, welder's hoods, firefighter's helmets, military helmets, or riot helmets. Even devices utilized by professional or amateur athletes would constitute head protection. Most people may not automatically categorize football helmets as occupational head protection, yet that is their precise purpose.

There are several forms of devices that are generally called "hardhats." ANSI places industrial head protection into two subcategories: type I helmets, which have a continuous 11/4 in.-wide brim around the circumference of the hat, and type II helmets, which have no surrounding brim, but do have a visor-like peak projecting from the front. Both type I and II helmets are further delineated into three classifications: classes A, B, and C. Factors to be considered in selecting appropriately classed protectors are insulation resistance against electrical shock, impact resistance, penetration resistance, flammability, and water absorption. Both classes A and B cannot be pierced by more than 3/8 in. in testing, while class C helmets cannot be pierced by more than 7/16 in. Another significant difference between the classes is insulation resistance. Class A, which is the most commonly used head protection, is intended to protect against up to 2200 V AC at 60 Hz for 1 min with no current leakage in excess of 3 mA. Class B affords maximum protection against electrical shock, being rated for up to 20,000 V AC at 60 Hz for 3 min, with no current leakage in excess of 9 mA. Class C gear provides no voltage protection at all and may not be worn where there is the potential for contact with energized electrical equipment.

Each helmet type is equipped with a suspension system intended to maintain some clearance between the helmet and the user's head. Each should have a replaceable headband that adjusts to fit the size of the wearer's head and a sweatband at the front. Helmets must have the manufacturer's name and the date of manufacture imprinted on the inside. Helmets have a 5-year life span and should be replaced after that time. Users should be cautioned not to paint or modify the shell as this tends to weaken the device, and not to store anything in the clear space between the head and the inside of the shell.

Typical applications for helmet wear include construction, demolition, rigging, logging, tree trimming, work on overhead electrical lines, mining, and shipbuilding. Of course, there may be many other instances for which head protection is advisable, such as in warehousing or chemical manufacture. Each individual task must be independently examined in determining the need for and the type of head protection required, as some employees may only require head protection for a fraction of their workday.

Bump caps are often used to guard against injury caused by impact with a stationary object such as low ceilings, low piping, platforms, mezzanines, or projections such as pipe supports. Bump caps are lighter and lower profiled than traditional head protection; however, they are *never* substitutes for those operations where class A, B, or C helmets are required. In contrast to hardhats, bump caps are not intended as protection against falling objects.

Bump caps may be utilized in plant maintenance, particularly in servicing equipment in difficult to reach areas, or meat processing, where the cap will help protect against inadvertent contact with hand tools or suspended meat. Where a worker might have occasion to require an appropriately classed helmet yet normally wears a bump cap for other activities, it is essential that the worker be made aware that the bump cap may not be substituted for a required helmet.

The selection of other forms of head protection is based on specific hazards and may be unique to one occupation. A primary example is the firefighter's helmet. This specialized type of PPE is intended to afford the firefighter protection against impact from falling, flying, or stationary objects, and from the possibility of contact with energized electrical equipment and heat. The firefighter's helmet also has a 1 1/4 in. brim around the circumference of the helmet and an elongated brim in the rear to allow water to run off and to prevent embers or debris from falling onto the back of the user's head and neck.

Career firefighters are not the only employees who would normally be required to wear such a helmet. Employees who are utilized on call as plant structural fire brigade members are also required to wear the appropriate protection, as mandated under OSHA 1910.156. Specifics on firefighter's helmets may be found in the National Fire Protection Association's (NFPA) *Standard on Structural Firefighter's Helmets* (NFPA 1972-1979).[4] These helmets should have provisions for ear flaps and for flip-down faceshields.

The miner's helmet is intended to protect against injury associated with impact from falling or stationary objects, as well as against electrical shock. The unique feature of the miner's helmet is that in addition to its low profile, which enables the user to be unencumbered in tight confines, it holds a portable lamp to allow the miner to illuminate the area directly in front of him/her while keeping both hands free.

Most forms of head protection are designed to accommodate the attachment of accessories, which can range from hearing protectors to faceshields to ear flaps or facecovers for cold weather to hair nets preventing long hair from becoming entangled in machinery. For example, an employee working with a chainsaw in the logging industry might wear a type II, class A hardhat equipped with an attached faceshield and hearing protectors rather than trying to wear three not necessarily compatible devices simultaneously. It would be cumbersome for such a worker to otherwise seek to combine a hardhat, faceshield, and muff-style hearing protector since each has an individual independent suspension system. By utilizing a single device possessing the features of each protector, the PPE is made more "user friendly."

EYE AND FACE PROTECTION

Specific references to required eye and face protection are made in several areas of OSHA 1910, 1926, and 1915. Each of the OSHA requirements refer the user to ANSI's "Standard for Occupational and Educational Eye and Face Protection" (Z87.1-1968)[5] as a mandatory reference. ANSI has repeatedly revised its document, but the OSHA requirements remain aligned solely with the 1969 revision. The law may be outdated when compared to the more recent ANSI standard. Among the primary differences between the newer and older versions of the standards are the recent acceptance of alternate designs in protective eyewear, most notably the introduction of polycarbonates or the use of single-piece frame and lens, rather than separate lenses inserted into frames. In addition, the use of safety glasses with sideshields is now more strongly encouraged over those without sideshields. Figures 1 and 2 show examples of different types of eye protection.

During the Industrial Revolution, traumatic loss of eyesight was common because little eye protection was available to workers. Tasks such as blacksmithing and railroad track laying caused frequent eye injuries since metal impacting upon metal yielded flying projectiles. As early as 1913, the Workmen's Compensation Service Bureau gave "approval only of the principle" of an early welding mask. It is interesting to note that even then safety practitioners recommended double protection, i.e., in the form of what passed as safety glasses at that time being worn simultaneously underneath the welding mask. A listing of the various types of eye and face protection can be found in Table 2.

Impact goggles are intended for protection against flying particles or projectiles. They may be worn as a substitute for safety spectacles with sideshields and are useful in that they may be worn over a user's personal prescription eyewear. A principal advantage of impact goggles over most types of safety spectacles is that since they fit flush against the wearer's face, they protect against floating or flying particles that may otherwise invade traditional spectacle-type protectors. The most frequently cited disadvantage is that they are warm, although they are manufactured with vents, and may tend to fog. Most goggles are now designed to minimize potential fogging.

SELECTION CHART				PROTECTORS		
		ASSESSMENT SEE NOTE (1)	PROTECTOR TYPE	PROTECTORS	LIMITATIONS	NOT RECOMMENDED
IMPACT	Chipping, grinding, machining, masonry work, riveting, and sanding.	Flying fragments, objects, large chips, particles, sand, dirt, etc.	B,C,D, E,F,G, H,I,J, K,L,N	Spectacles, goggles faceshields SEE NOTES (1) (3) (5) (6) (10) For severe exposure add N	Protective devices do not provide unlimited protection. SEE NOTE (7)	Protectors that do not provide protection from side exposure. SEE NOTE (10) Filter or tinted lenses that restrict light transmittance, unless it is determined that a glare hazard exists. Refer to OPTICAL RADIATION.
HEAT	Furnace operations, pouring, casting, hot dipping, gas cutting, and welding.	Hot sparks	B,C,D, E,F,G, H,I,J, K,L,*N	Faceshields, goggles, spectacles *For severe exposure add N SEE NOTE (2) (3)	Spectacles, cup and cover type goggles do not provide unlimited facial protection. SEE NOTE (2)	Protectors that do not provide protection from side exposure.
		Splash from molten metals	*N	*Faceshields worn over goggles H,K SEE NOTE (2) (3)		
		High temperature exposure	N	Screen faceshields, Reflective faceshields. SEE NOTE (2) (3)	SEE NOTE (3)	
CHEMICAL	Acid and chemicals handling, degreasing, plating	Splash	G,H,K *N	Goggles, eyecup and cover types. *For severe exposure, add N	Ventilation should be adequate but well protected from splash entry	Spectacles, welding helmets, handshields
		Irritating mists	G	Special purpose goggles	SEE NOTE (3)	
DUST	Woodworking, buffing, general dusty conditions.	Nuisance dust	G,H,K	Goggles, eyecup and cover types	Atmospheric conditions and the restricted ventilation of the protector can cause lenses to fog. Frequent cleaning may be required.	
OPTICAL RADIATION				TYPICAL FILTER LENS SHADE / PROTECTORS		
	WELDING: Electric Arc		O,P,Q	SEE NOTE (9) 10-14 Welding Helmets or Welding Shields	Protection from optical radiation is directly related to filter lens density. SEE NOTE (4). Select the darkest shade that allows adequate task performance.	Protectors that do not provide protection from optical radiation. SEE NOTE (4)
	WELDING: Gas		J,K,L, M,N,O, P,Q	SEE NOTE (9) 4-8 Welding Goggles or Welding Faceshield		
	CUTTING			3-6		
	TORCH BRAZING			3-4	SEE NOTE (3)	
	TORCH SOLDERING		B,C,D, E,F,N	1.5-3 Spectacles or Welding Faceshield		
	GLARE		A,B	Spectacle SEE NOTE (9) (10)	Shaded or Special Purpose lenses, as suitable. SEE NOTE (8)	

Figure 1. ANSIZ87.1-1989 Selection Chart for Personal Protective Eyewear. Protection types (A, B, C) are pictured in Figure 1. Notes are explained in Figure 2. This material is reproduced with permission from American National Standard "ANSI Practice for Occupational and Educational Eye and Face Protection," Z87.1-1989, Copyright 1989 by the American National Standards Institute. Copies of this standard may be purchased from the American National Standards Institute at 1430 Broadway, New York, NY 10018.

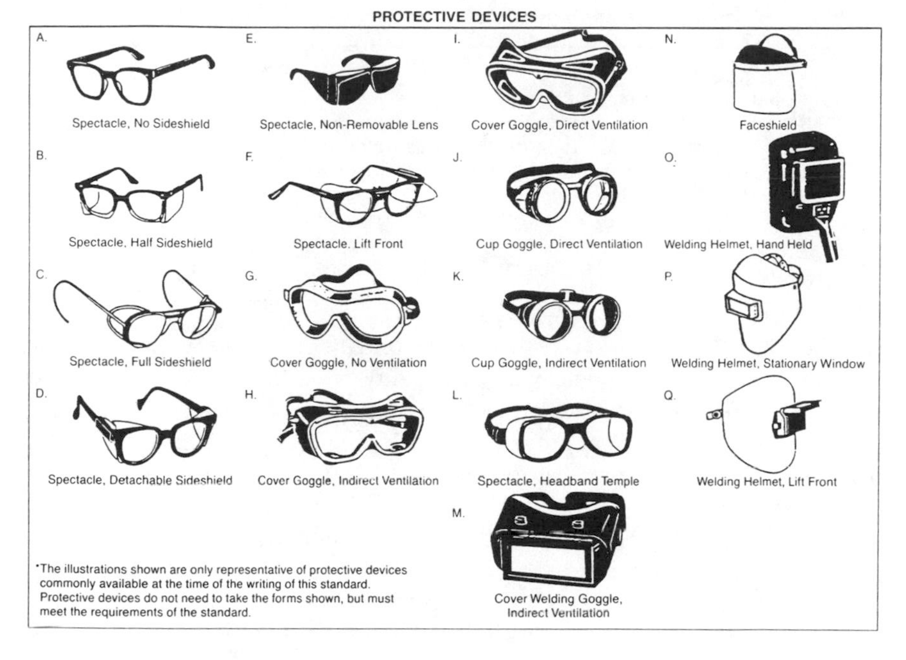

NOTES:

(1) Care shall be taken to recognize the possibility of multiple and simultaneous exposure to a variety of hazards. Adequate protection against the highest level of each of the hazards must be provided.

(2) Operations involving heat may also involve optical radiation. Protection from both hazards shall be provided.

(3) Faceshields shall only be worn over primary eye protection.

(4) Filter lenses shall meet the requirements for shade designations in Table 1.

(5) Persons whose vision requires the use of prescription (Rx) lenses shall wear either protective devices fitted with prescription (Rx) lenses or protective devices designed to be worn over regular prescription (Rx) eyewear.

(6) Wearers of contact lenses shall also be required to wear appropriate covering eye and face protection devices in a hazardous environment. It should be recognized that dusty and/or chemical environments may represent an additional hazard to contact lens wearers.

(7) Caution should be exercised in the use of metal frame protective devices in electrical hazard areas.

(8) Refer to Section 6.5, Special Purpose Lenses.

(9) Welding helmets or handshields shall be used only over primary eye protection.

(10) Non-sideshield spectacles are available for frontal protection only.

Figure 2. Personal protective eyewear devices, ANSI Z87.1-1989. These protective devices are discussed in Figure 1. This material is reproduced with permission from American National Standard "ANSI Practice for Occupational and Educational Eye and Face Protection," Z87.1-1989, Copyright 1989 by the American National Standards Institute. Copies of this standard may be purchased from the American National Standards Institute at 1430 Broadway, New York, NY 10018.

Chemical splash goggles are similar to impact goggles, but with an important difference: they are either nonventilated to prevent irritation from vapors or are indirectly vented through baffles to prevent splashed liquid from entering the device via ventilation ports.

Safety spectacles are manufactured with plano (or nonprescription) lenses, although optical companies or ophthalmic dispensers may furnish prescription safety lenses in a safety frame as well. Spectacles are primarily intended for protection against flying particles and projectiles; however, tinted lenses may also guard against injury from intense light as found in ultraviolet (UV) or infrared (IR)

Table 2. Types of Eye and Face Protection

Types of Eye and Face Protection
Impact goggles
Chemical splash goggles
Safety spectacles with or without sideshields (sideshields may be cup style or flat-fold)
Safety spectacles without sideshields
Welding goggles
Chipping goggles
Welding helmets
Plastic or mesh faceshields

radiation. For obvious reasons, either eyecup or flatfold sideshields should always be utilized in the appropriate workplace. Any circumstance requiring frontal protection also necessitates side protection, even when not dictated specifically by statute. Some manufacturers provide browguards to prevent material from entering the eye area from above the frame.

It is important to recognize that nonsafety eyewear is never a substitute for safety eyewear. Many people share the misconception that ordinary heat-treated lenses intended for improved vision are "safety glasses," although they do not meet the stringent requirements demanded for industrial safety spectacles. Similarly, safety lenses in a nonsafety frame do not constitute eye protection. An employee wearing safety lenses in a nonsafety frame might experience breakage more readily than with safety frames in the event of being struck by or against a large object. The lenses might also be easily forced back out of a nonsafety frame and into the wearer's face if subjected to direct frontal impact. Where spectacles may dislodge easily, elastic straps should be provided to retain the frames on the user's face.

Welding goggles may be found in the form of eyecup types (a separate eyecup over each eye) or coverspec types (shaped similarly to impact goggles, but also have heat-resistant frames). Both types require the use of one of 16 different filter lens shades to guard against UV, visible or IR radiation. Selection of appropriate filter shades is dependent upon the type of welding, cutting, soldering, or brazing to be done. Filter shades should be chosen in accordance with the guidelines established by the National Bureau of Standards (NBS).

Welding helmets also require properly selected filter lenses. The advantage of the helmet is that it affords the welder protection against injury to the face from radiation, heat, or particles. Safety spectacles should always be worn in conjunction with a welding helmet. Additionally, set-up personnel or welder's helpers (i.e., nonwelders who "work" directly with or in close proximity to welders) also need to wear appropriately shaded lenses to guard against flash burns or keratitis.

Chipping goggles are usually of the eyecup variety, but have clear lenses. Their purpose is to seal off the eyes in order to preclude dust from entering the eye, as well as shielding against projectiles or injury from hand tools.

Faceshields are worn expressly to guard against injury to the face, although serendipitously, they also may reduce the possibility of eye injury. It must be emphasized that faceshields in and of themselves are *never* to be considered eye protection. Plastic faceshields may be worn in conjunction with safety spectacles having sideshields or with impact goggles as protection against flying particles, particularly where the exposure is gross, as in heavy welding. Faceshields should be worn over chemical splash goggles during chemical handling operations, particularly when dealing with strong corrosives. Mesh faceshields are comprised of a wire screen that may be used to block splashes of molten metals as in pouring hot metals. The mesh allows increased ventilation, which is important when exposed to elevated temperatures. Tinted faceshields may also be utilized to guard against injurious nonionizing radiation, such as UV or IR. Faceshields may include an independent suspension system for use where head protection is not a consideration, or as an attachment to a helmet where overhead or impact risks are present.

Specialty eye protection, such as for use in laser operations, also exists, and should be selected in accordance with the American National Standard, "Safe Use of Lasers," ANSI Z136.1-1986.[6]

FOOT PROTECTION

Current OSHA guidelines are less than conservative in specifying foot protection, thus leaving much to the discretion of the employer. For example, in stating that abrasive blasting operators must wear safety shoes, OSHA 1910.94 (a)(5)(v) states that this should be done "wherever heavy pieces of work are handled." OSHA 1910.136 (a) notes that where safety-toe footwear is to be worn, it must meet the requirements of ANSI Z41.1-1981, "Men's Safety-Toe Footwear."[7] Z41.1 divides safety toe footwear into three groups based on the severity of the impact. This standard recognizes that workers may face varying risks of foot injury which are largely dependent on the weight and construction of the materials that they may handle or work around. For this reason, ANSI provides for classes 30, 50, and 75 safety toes that are of varying strengths. The market for safety-toe footwear is large with a significant variety of styles available in ANSI classes 30, 50, or 75. Shoes may include slip-resistant soles, impervious uppers, high tops, steel innersoles, or be static electricity-dissipating conductive footwear for grounding static electricity built up in the body of the worker.

Other options available for specific risks include metatarsal guards (extended guards covering the instep and that portion of the upper foot from the toe cap to the front of the ankle), which may be designed as a component of the shoe or as a detachable accessory. Similarly, strap-on leg, shin, and metatarsal guards may be utilized. In the foundry industry, where rapid and easy removal of footwear may be necessary in the event of molten splash, elastic tops or quick-release buckles are used. Wooden soles may be indicated in order to reduce the transfer of heat from hot floors to a worker's feet. For temperature extremes at the other end of the scale, safety footwear expressly intended for cold weather application may be obtained. For example, with loggers requiring protection against cold weather and against chainsaw lacerations or punctures from other tools, safety footwear may include uppers having ballistic nylon construction (in addition to the safety toe-box). Where workers may be exposed to energized electrical components, electrical hazard footwear offers protection against up to 600 V. For protection against chemical absorption due to splashing or immersion, impervious overshoes or chemical spats should be selected based on the specific hazards of the chemicals being handled.

BODY PROTECTION

Body protection other than for those areas covered separately in this chapter makes up a broad category. Encapsulating and impervious protective apparel is addressed in Chapter 14. Other forms of body protection may include any of the items listed in Table 3.

Since the various requirements of different types of operations will dictate much in the way of unique PPE, each activity must be addressed individually. Often a good resource in determining appropriate PPE for a specific application will be a safety equipment manufacturer's or distributor's representative; however, comparisons should be made to minimize the possibility of steering by a salesperson biased in favor of his/her product.

HAND PROTECTION

The risk of soft tissue injuries to the hands are present in most occupational activities. Protection of the hands from chemical exposure is covered elsewhere in this book (Chapter 14), for example, with individual gloves being selected based on permeation and degradation of the protective devices when exposed to a given chemical. This chapter treats gloves as protection from nonchemical occupational exposures, as summarized in Table 4.

Note that gloves should never be worn while working with or around moving machinery, i.e., mills or lathes, as the potential for amputation or avulsion is sizable. A good reference person will likely be a glove manufacturer's or distributor's representative. Various types of gloves for similar risks should be experimented with in order to determine what types provide good protection and find worker acceptance.

RESPIRATORY PROTECTION

Workers have long attempted to develop their own makeshift respiratory protection. Agricola documented primitive respirators in *De Re Metallica.*

Table 3. Body-Protecting Types of Personal Protective Equipment

Wire mesh aprons such as in the meat cutting industry to guard against knife cuts to the torso. The mesh aprons resist lacerations, yet breathe easily and may be readily sanitized. Mesh gloves and arm protectors are also available.

Heat-resistant aprons, bibs, capes or snoods (a hood-like device providing coverage of the head, hair, throat, and nape of the neck) for welding, cutting, and brazing. These will generally be leather or other heat-resisting fabric (i.e., Nomex®) and serve to guard against sparks, slag, and radiant heat or light.

Aprons, coats, or jackets for abrasive blasting operations. These are intended to minimize punctures, abrasions, or mechanical impingement caused by blasting shot or abrasive grit.

U. S. Coast Guard-approved lifejackets or buoyant work vests for working over or near water. These are flotation devices that will keep an unconscious, injured, or nonswimming wearer afloat so as to allow rescue where an employee cannot help himself/herself, as well as to aid an individual in self-rescue.

Red or orange reflecting warning garments for flagmen or others exposed to hazards from vehicular traffic. Open weave materials are also available for this purpose when high temperatures might otherwise cause increased heat stress due to an added garment on an already overheated worker.

Insulated clothing for low temperature extremes. This is particularly important for crews working outdoors or in artificially created cold environments, such as walk-in freezers or refrigerated warehouses, for extended periods of time.

Water, air, or chemically cooled clothing for high temperature extremes. These will help to minimize heat stress resulting from hot environments. A typical example might be for a welder inside a tank, pipe, or other confined space.

Fire entry turnout gear for fire suppression personnel to protect the wearer from chemical hazards, punctures and lacerations as well as protecting against high temperature hazards. This is especially necessary since the firefighter's work environment is often navigated by crawling and feeling with limited visibility. At such times, unseen hazards present an added threat since they are not readily observed.

Foul weather suits for workers exposed to rain, sleet, or other inclement weather. This will help to minimize skin irritation and chafing caused by wet clothing, as well as allowing for increased employee comfort.

Leggings or chaps for protection against chainsaw cuts or abrasive blasting. These devices will resist laceration caused by inadvertent contact with an operating saw or with blasting grit sprayed under pressure.

Waders for immersion in water may be encountered by fishermen, dock construction, or waterworks personnel. This alleviates the discomfort of working in saturated clothing.

Kickback aprons to guard against impact and lacerations as might occur to saw operators. The apron is padded to resist impact such as that caused by lumber being kicked back by a running saw as well as affording some protection against cuts and punctures.

Flame-retardant work clothing such as might be worn by firefighters underneath their turnout gear. This is important since high temperatures or sparks encountered during firefighting might otherwise cause ordinary street wear to smolder or flame even though shielded by turnout gear. Nomex® is often used in constructing such "station uniforms."

Table 3 (continued). Body-Protecting Types of Personal Protective Equipment

Electromagnetic radiation suits for workers in high level radar fields. The suits are designed to prevent the deep body heating associated with exposure to nonionizing radiation.

Conductive clothing for tower linemen working on high voltage equipment. These devices serve to prevent the worker's becoming part of the circuit in the event of inadvertent contact with energized electrical equipment.

Table 4. Hand Protection Equipment

Rubber gloves for protection against energized electrical circuits (OSHA references ANSI J6.6-1971 which has since been superseded by ASTM D120-87).[8] This would include gloves intended for use by tower linemen who may handle live electrical equipment. The rubber gloves are protected by a second outer glove to prevent punctures, tears, or premature wear.

Heavy rubber or metal-studded leather gloves for abrasive blasting. Since the hands are quite vulnerable to abrasion or impingement of particles during abrasive blasting, gloves must be durable and capable of resisting grit under pressure.

Cut-resistant gloves, such as steel mesh, or Kevlar®, for protection against lacerations as in handling sheet metal or knives. The mesh gloves are readily sanitized and afford good dexterity while guarding against lacerations. Mesh sleeves may also be added to protect the arms.

Leather or aluminized fabric gloves for protection against heat, sparks, or molten metal in welding, cutting, soldering, or brazing; thin leather gloves might be utilized for tasks requiring considerable manual dexterity as in TIG welding.

Insulated gloves or mittens for low temperature extremes; mittens should not be used where dexterity is important, since the absence of unencumbered fingers will impede manual dexterity.

High intensity heat-resistant gloves or mittens for heat, sparks, and molten metal splashes as might be experienced in foundry or furnace operation, or in fire suppression activities.

Fingerless gloves for protecting a worker's palms and lower fingers where dexterity is critical and the risks are not excessive. An example of such use might be for a carpenter who wishes to avoid splinters caused by handling lumber, yet needs to have fingers free to position nails while hammering.

Rubber surgical gloves for use in potential exposure to blood and body fluids. In cases of heavy exposure or where the possibility of glove breakage is high, as in emergency medical care, double gloving is recommended.

Finger and wrist tape for prevention of abrasions and lacerations to finger tips where gloves cannot be used such as in machine operation. This is common in assembly line workers or production machine operators who may be at increased risk of injury if gloves are to be worn near rotating mechanisms.

Padded palm or padded thumbed gloves to absorb vibration such as might be experienced in manual grinding or jackhammering. This will minimize the potential for vibration disorders which may become disabling over periods of chronic exposure.

American cowboys of the 19th century utilized bandannas to ward off the choking dust churned up by their animals. Early firefighters used to place their beards in their mouths in order to increase their ability to breathe. Today's respiratory protection might include self-contained breathing apparatus (SCBA); powered air-purifying respirators with chemical canisters; air line respirators; hose mask with blower; vapor and gas removing respirators; particulate removing respirators; and combination particulate, gas, and vapor removing respirators. Respiratory protection is covered at length in Chapter 14.

FALL PROTECTION

Of course, falls should be prevented with vigor; however, in high risk activities such as bridge painting, window washing, or structural steel erection, personal fall protection devices (i.e., safety belts with lanyards or automatically braking fall-arrestor devices) are a wise contingency. Where a worker cannot be adequately protected through the use of guardrails, midrails, and other formidable restraints, safety belts, lanyards, and lifelines may be suitable. The prescribed standard for safety belts and lanyards is ANSI A10.14, which offers minimum construction and testing parameters. Safety belts will generally be a webbed belt with D-rings for attaching lanyards. The lanyards are connected between the belt and a fixed point such as a structural member or lifeline. Lanyards are not intended to allow a fall in excess of 6 ft. Since even a fall of 6 ft might cause serious injury, belts and lanyards are a last resort rather than a casually accepted means of restraint. Rescue harnesses are designed for personnel who must work in confined spaces. The harnesses are connected to a lifeline manned by a standby observer. In the event of an emergency, the standby can extricate the entry employee without entering the danger area himself/herself.

MUSCULOSKELETAL PROTECTION

While no specific OSHA standards presently exist with regard to musculoskeletal PPE, such devices have long been in use in industry. The most commonly seen device is the lumbosacral support belt. The design is such that air or rigid splints support the lower back during lifting activities, much like a weightlifting belt. The intent is that the devices will prevent lower back injury by allowing the worker to minimize stress on the spine during strenuous physical exertion.

Other musculoskeletal PPE would include wrist supports geared toward lessening the incidence of repetitive stress which might ultimately result in carpal tunnel syndrome or tenosynovitis. Repetitive stress-related wrist disorders are often seen in meat cutters, assembly line workers, and video display terminal (VDT) operators. The wrist supports would reduce the stress while enabling the worker to continue normal activities. Such a device is not a substitute for proper ergonomic workstation design, but may help as an interim measure. Since ergonomically designed workstations are not always available, as in the case of motor vehicle operators, lumbar supports such as rolls or wedges are advisable to prevent drivers from suffering continuous low back pain.

HEARING PROTECTION

Earmuff-style protectors and disposable or reusable earplugs may help reduce unacceptable noise levels where engineering controls are not feasible, i.e., temporary exposures during construction operations, or as an interim measure while awaiting the installation of noise abatement equipment. Where protectors are required, specific training is necessary and workers should be permitted to select from a variety of protectors, as discussed in Chapter 11. Protective devices are particularly important where audiometric testing has shown a significant threshold shift.

HUMAN FACTORS AFFECTING PERSONAL PROTECTIVE EQUIPMENT SELECTION

There is definitely a difference in sizing of PPE based on sex, physical size, physical handicaps, or deformities. There may be ethnic and racial characteristics that affect respirator sizing.[9,10] As an example, women's gloves and boots are generally significantly smaller than men's. Normally a small man's shoe, boot, or glove will still not properly fit the average woman. Physical size, prominent facial features (such as high cheekbones), and facial deformities (such as the absence of teeth) will greatly affect respirator fit. Consideration of worker's physical characteristics is critical in ensuring worker acceptance of PPE.

Personal preferences should be accommodated wherever possible. A reasonable variety of devices should be available to ensure proper fit, and there is a greater likelihood that the worker will find a style that he/she prefers, as not every worker wants to look like everyone else. If all of the available safety eyewear is identical, some resistance may be anticipated. Many workers will also express apprehension about PPE due to myths and misconceptions. It seems that in every group, someone will share some anecdotal "true story" of someone who was injured while wearing protective equipment, such as the fabled toes amputated by a safety-toe box. Appropriate training will go a long way in assuaging employee aversions. This training should cover the strengths and weaknesses of the equipment required, discuss how the equipment should be worn, and point out how equipment may prevent the overwhelming majority of injuries.

Some medical contraindications may preclude the use of some form of PPE. For example, an employee with a history of chronic obstructed pulmonary disease should not be assigned to a task that requires the use of a respirator. Here, the determination of a qualified medical professional must be relied on and is required under OSHA's respiratory protection standard.

PPE is of no value if it is not used properly and consistently. An employee who wears protectors only while being observed must be made aware of managerial expectations with regard to safety. Some of the keys to motivating worker acceptance of PPE are summarized in Table 5.

Table 5. Keys to Motivating Workers Acceptance of PPE

Communicate the reason why the equipment is needed; do not let the workers feel it is a punishment or they will continue to resist it as a means of saving face.
If possible, allow a group of employees to participate in the selection process; this should include experimentation to ensure that the devices selected do function as expected.
Have several styles available; people have different tastes and want to feel like individuals.
Do not force the worker to furnish his/her own safety and health equipment; when forced to choose between essentials or personal goods and job-related materials, most workers will place personal needs first.
Listen objectively to employee criticisms of specific devices; sometimes an employee will find that a protector is so difficult to wear as to be counterproductive to proper usage; there may be better devices on the market.
Provide training for all safety equipment; do not simply fit equipment, but also explain the reasons for use, the advantages, and the disadvantages.
Remember — if the worker is dissatisfied with the equipment, productivity will deteriorate; help the employee to be able to work comfortably and efficiently

SUMMARY

While engineering controls and administrative controls are always preferred over requiring employees to wear PPE, personal safety devices may still be a tremendous aid in providing for the safety and health of people at work. Not every process may be designed so as to virtually eliminate all possible risks, so PPE may be needed as a backup or redundant control. Often, temporary projects, such as construction or demolition, cannot feasibly engineer out all risks, so PPE is used to adapt to the varying risks of a dynamic environment. Where PPE is warranted, it must be readily available and its use must be enforced.

REFERENCES

1. Lippmann, M., and Schlesinger, R. *Chemical Contamination in the Human Environment* (New York: Oxford University Press, 1979), p. 9.
1a. "A Proposal to Revise General Industry Safety Standards Governing Personal Protective Equipment" (54 FR33832, August 16, 1989, Sect. V).
2. *General Industry Standard (29 CFR1910)* (Washington, D.C.: Occupational Safety and Health Administration, 1988)
3. "Safety Requirements for Industrial Head Protection." Z89.1-1969, American National Standards Institute, New York; this is superseded by the 1986 revision entitled, "Protective Headwear for Industrial Workers. Z89.1-1986."
4. "*Standard on Structural Firefighter's Helmets. NFPA 1972-1979,*" National Fire Protection Association, Quincy, MA; this is superseded by the 1987 revision entitled, "Helmets for Structural Firefighting, NFPA 1972-1987."

5. "Standard for Occupational and Educational Eye and Face Protection. Z87.1-1968," American National Standards Institute, New York; this is superseded by the 1989 revision of the same title, numbered ANSI Z87.1-1989.
6. "Standard for the Safe Use of Lasers, Z136.1-1986" American National Standards Institute, New York.
7. "Standard for Men's Safety Toe Footwear, Z41.1-1981," American National Standards Institute, New York; this is superseded by the 1983 revision of the same title, numbered ANSI Z41.1-1983.
8. *ASTM Standards on Electrical Protective Equipment for Workers*. 7th ed. (Philadelphia: American Society for Testing and Materials, 1988).
9. "Half Mask Respirator Selection for a Mixed Worker Group," *Appl. Occup. Environ. Hyg.*, 5(4):234 (1990).
10. "Distribution of Face Seal Leak Sites on a Half-Mask Repair and Their Association with Facial Dimensions," *Am. Ind. Hyg. Assoc. J.*, 51 (1990).

PART V

Training to Optimize Occupational Health

Training workers (with respect to occupational health and safety) is sometimes required by law and frequently is done as a good business practice. This section of the book, without reducing management's own responsibility in making the workplace safe, deals with the techniques of training managers and workers to think and act safely:

Designing Workers' Health and Safety Programs provides an overview of health education methods for application to the design of occupational health and safety programs.

Worksite Health Promotion (WHP) discusses the adoption of WHP (programs such as smoking cessation and stress management) to enhance employee health.

CHAPTER 16

Designing Educational Programs for Workers' Health and Safety

Lisa Lieberman

INTRODUCTION

Over the past 2 decades, attention has been increasingly focused on the roles of workplace design and of workers' behavior in improving workers' health and safety. As a result of the Occupational Safety and Health Act (OSHAct) of 1970 and enactment of subsequent standards by the Occupational Safety and Health Administration (OSHA), the training and education of workers in reducing the hazards associated with their jobs has become a well-recognized responsibility of employers. Many employers, however, are themselves not trained in educational and training techniques.[1] Thus, employee training in the use of safety equipment and in safeguarding themselves from potentially toxic substances may not have the desired effect of improving workers' health and safety.

The principles and practices of health education can be applied to the occupational setting for improving the effectiveness of health and safety education and training efforts. The field of health education has evolved over the years from simply providing information to one in which the desired outcomes are generally behavioral in nature. A working definition of health education serves to illustrate this thinking in the field, "Health education is any combination of learning experiences designed to facilitate voluntary adaptation of behavior conducive to health."[2]

Particularly relevant to occupational health and safety is the further delineation of "health promotion," which broadens the scope of health education to include organizational, economic, and structural support for behaviors conducive to health.[3] In this context it is recognized that workers' health and safety may be improved both through the behavior of workers and through what OSHA refers to as "engineering controls."

When the use of personal protective equipment (PPE) is required, worker acceptance and use of it may be increased by the provision of safety equipment that is comfortable and convenient to wear. Similarly, a worker's compliance in wearing a measuring device for toxic substances is likely to improve if the worker perceives that the company is providing adequate protection through the use of up-to-date, well-maintained equipment, proper ventilation, and other such safeguards.

This chapter serves as an overview of health education theories, methods, and media and their applications to the design of occupational health and safety education programs. Such programs include three different classes of behaviors, discussed below.

Behaviors Relevant to the Work Environment

The first class of health and safety related behaviors are those specific to the work environment itself. These may include increasing the use of PPE and putting into effect engineering and/or work practices that reduce the risk of exposure to toxic substances. Such behaviors readily lend themselves to on-the-job training, use of incentives for maintaining appropriate behaviors, and positive peer pressure among workers.

OSHA requires the "establishment and supervision of programs for the education and training of employers and employees in the recognition, avoidance, and prevention of unsafe or unhealthful working conditions."[1] Such educational efforts include teaching the worker to avoid behaviors that may reduce the effectiveness of engineering controls and PPEs, to engage in behaviors that use PPEs and existing engineering controls to the worker's advantage, to reduce contact with hazardous substances, and to limit opportunities for illness and injury.

Behaviors Indirectly Related to the Work Environment

The second class of behaviors are those indirectly related to the work environment through their potential contribution to occupational hazards. Examples include smoking, which may increase the risk of cancer for workers exposed to certain carcinogenic substances, or alcohol abuse, which increases the risk of work-related accidents. Recognition of the significance of these behaviors to overall worker health and safety, and specifically to issues of worker productivity and economics, has led many companies to offer assistance to workers in these areas. Such assistance may take the form of formal on-site programs, financial incentives to enter programs outside of the workplace, employee assistance programs (EAPs) that direct employees into drug and alcohol treatment, often under the threat of job loss, or improved health care benefits that cover treatment programs.

Health Promoting Behaviors

The third class of behaviors are those related to the general health of employees, such as nutrition, exercise, and stress management. These behaviors are not directly related to occupational health and safety, but have the potential to reduce absenteeism, increase productivity, and reduce employer health care costs. In a national survey of worksites with 50 or more employees, some 65% of the respondents reported offering some type of "health promotion" program to employees (i.e., nutrition, stress, blood pressure, fitness, weight control, smoking cessation).[4] Programs of this type are discussed in detail elsewhere in the volume.

It is important to note that this focus on worker behavior is not intended to imply that the responsibility for work-related illness or injury lies solely or even predominantly with the worker. Poor or unsafe workplace design, lack of available PPEs, and stressful conditions on the job are likely to have a significant influence on worker health, illness, and behavior. Thus, employers should consider these issues prior to and simultaneously with the development of health and safety education programs for workers.

THEORETICAL BACKGROUND AND APPLICATIONS

Several theories having their origins in the social sciences and in communications help to explain and predict health-related behaviors. The application of these theories has been critical to improving educational strategies for facilitating behavior changes.

Health Belief Model

The health belief model,[5,6] which was developed in the early 1970s and continues to be the focus of much health education research to date, attempts to explain health-related behavior on the basis of three factors: (1) the existence of sufficient concern or motivation for health; (2) the perceived threat of a particular condition to the individual, i.e., the belief that one is susceptible or vulnerable to a particular condition and its perceived severity; and (3) the belief or perception that a particular action will be beneficial in reducing the perceived threat at a reasonable cost.

Suppose a worker were at risk of developing lung cancer as a result of combined smoking and exposure to asbestos. The model postulates that in order to make behavior changes that will reduce the risk of lung cancer:

1. Health must be of concern (i.e. salient) to the individual
2. The individual must perceive that he/she is personally at risk for (i.e., susceptible to) lung cancer and that the consequences of lung cancer are sufficiently serious to warrant preventive action
3. That quitting smoking and/or regularly wearing a protective mask would actually reduce the risk of lung cancer.
4. That the barriers or difficulties (i.e., costs) involved in taking these actions do not outweigh their benefits.

Health Belief Model Applied

Programs and educational messages that utilize the factors identified by the health belief model would be designed such that:

- Learners accurately perceive their own risks or susceptibility to the condition
- Learners accurately perceive the consequences of a particular condition
- Learners perceive that the preventive action is an effective method for reducing the threat of the condition
- Learners are assisted in reducing the barriers, difficulties, financial costs, or discomforts of taking the preventive action

Finally, the efficacy of the health message can be increased through "cues to action" which bring the health message and motivation to the conscious mind of the individual. Table 1 describes the various aspects of learner and message characteristics relevant to the health belief model.

Self-Efficacy

A related theory of health behavior is the social cognitive theory,[7] based on the premise that behavior is determined by an individual's expectations about the behavior and incentives for performing that behavior. Expectations include those

Table 1. Using the Health Belief Model

Learners should:	Message should:
Accurately perceive their own risks	Be personalized to relate directly to the workers' concerns
Accurately perceive the consequences of taking or failing to take a particular action	Portray the severity of the consequences of a particular action in a believable manner
Perceive that a specific action will reduce the perceived threat	Focus on the benefits of taking the particular action
Perceive that the barriers to and costs of the action are small	Discuss ways to reduce the barriers, costs, or difficulties of taking the action
Receive cues and reminders to take the action	Appear often and at the appropriate times and places in order to remind the learner to take action

of the learners concerning the outcomes and consequences of one's own behavior, as well as expectations about one's competence to perform the behavior, called "self-efficacy." In particular, the likelihood of performing a behavior is increased by one's belief in the ability to adequately perform the behavior, i.e., self-efficacy.[6]

Self-Efficacy Applied

Self-efficacy can be enhanced in several ways. Personal mastery of the behavior, or evidence that one can perform the behavior, increases one's sense of self-efficacy. Thus, practicing and successfully performing the behavior or steps leading to the behavior under supervised conditions may lead to mastery of the behavior and increases in self-efficacy. Vicarious experience can enhance self-efficacy through observation of the successful performance of others, providing the sense that, "if he can do it, so can I." Verbal or written persuasion ("you CAN do it!") is a less powerful, but useful enhancement to self-efficacy. Finally, physiological states, such as increased anxiety, may reduce one's sense of ability and conversely, relaxed states may improve self-efficacy.[6]

Health and safety education programs should be designed that identify reasonable behavioral goals that the individual feels capable of achieving and that build in activities that serve to enhance self-efficacy. For example, activities should include opportunities to personally master specific skills through supervised practice, should break down tasks into manageable parts, and should provide opportunities for viewing and learning of the success of others. In addition, social support provides the worker with the sense that he can achieve the behavior and reduce stress. Table 2 summarizes and provides examples of these methods of enhancing self-efficacy.

Table 2. Enhancing Self Efficacy

Method	Specific Strategies
Personal experience	Supervised practice of the behavior; break down the behavior into manageable parts
Vicarious experience	Observing peers performing the behavior; hearing the experiences of others
Verbal persuasion	Positive messages that support the learner
Physiological states	Relaxation and anxiety reduction exercises during the learning process

The PRECEDE Model

The PRECEDE model[2] is a framework for the design and planning of health education programs. It is particularly well known for its focus on three factors that influence the performance of health-enhancing behaviors. Predisposing factors include knowledge, attitudes, beliefs, values, and perceptions. Enabling factors are the resources and skills necessary to perform the behavior. Reinforcing factors are the support for the behavior, and include such things as social norms, social support from family, friends, co-workers, and supervisors, and reminders to perform the behavior.

PRECEDE Applied

Occupational health and safety programs should focus on all of these factors for facilitating positive behavior changes. Educational programs may be aimed at the predisposing factors by increasing knowledge and supporting and enhancing appropriate beliefs, attitudes, and perceptions. It has become clear that increased knowledge about a behavior and its health effects does not necessarily lead to a change in behavior. Certain types of knowledge, however, such as how to use a particular piece of safety equipment, are crucial to the performance of the behavior. Such knowledge, sometimes called "utility knowledge," is necessary, but not sufficient to facilitate behavior change.[2]

Educational programs must also focus on the enabling value of skills training and practice. In addition, the organization should make every effort to encourage the behavior through proper workplace design, convenient and comfortable equipment, and consistent and fair safety and health-related policies.

Finally, reinforcement can be built into the program by encouraging positive peer pressure and support among co-workers, and supportive relationships between supervisors and staff. As some workers change their safety and health-related behaviors, changes in the social norms of the work environment may occur, i.e., workers' expectations about what is appropriate and acceptable behavior will reflect the new, desired behavior.[1,8]

Table 3. Designing Persuasive Messages

Feature	Characteristics	Examples
Source	Credible Attractive	Union leaders; well-respected co-workers; well-respected management
Message	Clarity Salience Repetition	Simple message with a single, specific focus; draw on existing attitudes and beliefs that are salient to the worker
Channel	Multiple modes appropriate for the task potential to review message	Visual and auditory modes of message delivery; use visual modes for tasks needing demonstration; step-by-step reminder materials for complex tasks
Receiver	Knowledge Attitudes Beliefs	Message consistent with existing level of knowledge; message should reinforce existing positive attitudes toward behavior; message should call into action existing beliefs about behavior's usefulness or benefits

Persuasive Communication

Communication theories are also relevant in educating workers, since health education programs often depend on use of persuasive messages to encourage behavior change. The characteristics of the source, the message itself, the channel, and the receiver all contribute to the persuasiveness of the message.[9] Source refers to the characteristics of the individual or organization delivering the message. *Channel* refers to the medium through which the message is communicated. The message itself is the content of the communication, and the receiver is the object of the message, in this case, the worker.

Communication Theory Applied

Table 3 reviews examples of the four characteristics of persuasive messages. Source may influence persuasiveness of the message and includes the credibility and attractiveness of the source. A credible and attractive source is one whom the workers believe has their true, best interests in mind, one whom workers feel confident in, can identify with, and whom they believe qualified to provide the information. Knowledgeable and well-respected co-workers (such as a union leader) may be considered credible and attractive sources and may be recruited to assist in the program. Furthermore, programs that are designed and delivered jointly by labor and management may have significantly enhanced credibility.[10]

Message may increase persuasiveness and includes the clarity and simplicity of the message. The message must also be salient to the learner, i.e., it must draw on attitudes and beliefs of which the learner is already aware and concerned. Repeated messages may improve retention of the message, but once a certain threshold of repetition is reached, the receiver tends to "turn off," or cease reception. The most important aspects of the message should be presented first and then reviewed.

Another important characteristic is the level of fear aroused by the message. Health education programs and messages have often relied on the use of fear-inducing messages to attempt to motivate behavioral changes. A great deal of communications research has concluded, however, that messages that invoke high levels of fear are not likely to lead to the desired behaviors.[11] High fear messages may produce reactions such as denial (e.g., "It won't happen to me") or high anxiety, which might actually induce an undesired behavior, such as drug or alcohol use. This is particularly relevant to occupational health and safety. The consequences of the failure to use PPE could be presented very graphically and may provide great fear or shock value. Such approaches may be ineffective, however, because consequences that are too gruesome or only a remote possibility, may be perceived by the worker as not personally relevant. For example, the worker who has not used safety equipment properly can reduce the message's relevance with the logic that, "I must not be vulnerable to this problem because I've worked without the safety device for six months and none of those things have happened to me."

Some studies have found that low to moderate levels of fear arousal can positively influence behavior change, but only when the following guidelines are considered:

- The level of fear employed should be relatively low, so that inappropriate responses (such as denial or anxiety which illicits the behavior itself) are not implemented to deal with the fear
- The message should be accompanied by a prescribed action perceived as being effective for removing the threat.[11]

The *channel*, or medium, in which the message is "sent" is equally important to the persuasiveness of the message. Adults learn in a variety of ways; thus, modes of message delivery should be both visual and auditory whenever possible. The medium should be appropriate for the task; for example, messages that require complicated procedures should be demonstrated using a visual mode or channel. In addition, learners should have the opportunity to review the message; thus it may be important to provide written instructions for complex behaviors that learners can refer to at an individual pace, or to provide opportunities to review films or videos as needed.

The predisposing factors, i.e., knowledge, attitudes, and beliefs, of the learner constitute important *receiver* characteristics. Such factors may determine how a receiver perceives the message and which messages are likely to be accepted.

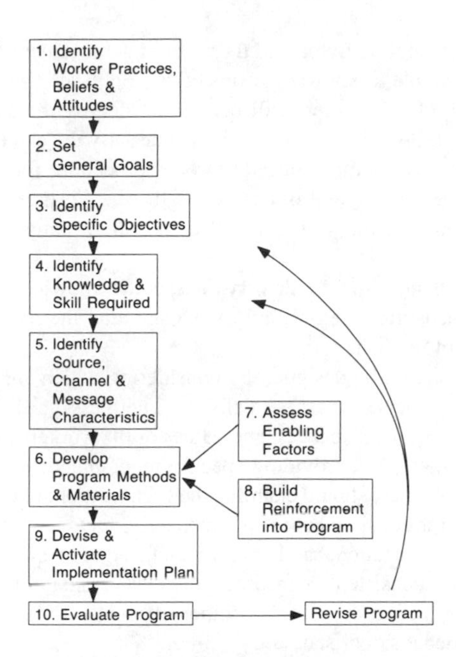

Figure 1. Ten-step model of program design for workers' health and safety education.

Understanding the beliefs and attitudes that are most salient or of concern to the receiver, aids in designing messages that can call these salient beliefs and attitudes into play. Research has shown that receivers who are expected to serve as message senders (e.g., workers who are being trained to teach other workers) may be more easily persuaded.[12]

PROGRAM DESIGN FOR WORKER HEALTH AND SAFETY EDUCATION

Designing programs to improve the health and safety of workers can be conducted using a step-by step procedure that takes into account many of the factors addressed in these theories. A model for health and safety education program planning is depicted in Figure 1. This ten-step model is described below.

1. The model requires that the program planner be knowledgable about the current behaviors or practices of the workers, and their relevant beliefs, attitudes, and values. Such information can be gathered through observation, formal and informal surveys and interviews, and company records.
2. The program developer sets general goals that identify the desired health status changes for the workers.
3. The general goals are furthered delineated into specific objectives, which are stated in terms of what the workers will achieve by the end of the program. Objectives

should specify the time frame and the level of achievement desired in measurable terms. For example, a knowledge objective might be stated, "By the end of the program, 90% of the workers will be able to identify three safety features of the new equipment." A skill objective might state, "By the end of the program, 100% of the workers will demonstrate the proper procedure for using the new equipment." Finally, a behavioral objective might state, "Three months after program completion, 90% of the workers will demonstrate daily use of the new safety device."

4. The program planner should identify the specific knowledge that workers need to perform the behaviors, i.e., utility knowledge, and the required skills for proper performance of the behavior.
5. The program should be designed by considering which are the most appropriate sources and channels of message delivery. The message should be designed with attention to the attitudes, beliefs, and values of the workers, using low to moderate fear, if appropriate, and providing specific recommendations for reducing risks.
6. The program planners should examine the organizational factors that may contribute to or hinder the desired behavior and identify factors which enable the behavior through such organizational and environmental adjustments. These include, but are not limited to, consistently enforced policies, availability of comfortable and convenient protective devices, and other efforts on the part of the employer to maintain adequate safety standards.
7. Program methods and materials, such as print materials, films, computer programs, group activities, etc. (described in a subsequent section) should be chosen from existing available sources or designed specifically for the program. Such methods and materials should be appropriate for the required tasks (i.e., knowledge gain, skill attainment, motivation, etc.) and should include varied media and methods to reach workers with different learning styles and abilities.
8. The program should include reinforcements such as incentives, social support, reminders, and other messages to provide incentive, feedback, and reminders of the desired behavior.
9. A program implementation plan should be designed to identify all of the necessary financial, staff, and physical resources needed to conduct the program; a method for pretesting or pilot testing program materials; a plan for recruitment to the program (if workers must attend or sign up for program activities); and the time(s) and place(s) for the program.
10. A plan for evaluating the effectiveness of the program should be designed. The evaluation plan should include methods of assessing the program process, i.e., how the program is actually conducted, and the program outcome, i.e., how well the specific objectives have been met. Evaluation measures may include program attendance records, surveys, interviews, focus groups, direct observation, and biomedical testing, and whenever possible should be compared to an appropriate comparison or control group who have not received the program.[13] A detailed discussion of health education program evaluation is beyond the scope of this volume. The author suggests that the program planner consult any of a variety of available texts on the subject, as presented in the resource guide at the end of this chapter. The program planner should be prepared to modify the program on the basis of evaluation results.

In general, it is critical that the program designer be cognizant of the particular organizational dynamics and relationships between management and staff in order to facilitate adoption of the program. For example, in most settings, workers who are being asked to take responsibility for their health and safety on the job, must take an active role in shaping the programs, rather than perceiving that it is simply the mandate of management.[14]

Table 4 provides two examples of how this ten-step model can be applied to safety and health education programs in an occupational setting. Example A is aimed at increasing the use of protective eyewear, a behavior that is directly related to the work environment. The program focuses on the proper use and fit of protective eyewear, and reducing perceptions of discomfort or inconvenience. Example B is aimed at reducing smoking among asbestos workers, an example of a behavior that is indirectly related to the work environment. The program is focused on increasing motivation to quit smoking and providing the skills to overcome addiction to cigarettes.

HEALTH EDUCATION METHODS

As the field of health education has changed its focus from providing information to behavior, so have its various strategies. While providing information that is accurate, understandable, and relevant to the behavior itself is still a necessary strategy for change, it is clearly insufficient. Effective health education programs, therefore, should include both a knowledge component and specific techniques aimed at training workers in the necessary skills for the practice of the desired behavior. In addition, efforts to reduce the "costs" or difficulties involved in practicing the behavior and to positively reinforce the desired behavior need to be built into the program.

The following section describes a variety of educational strategies that can be used in programs aimed at improving workers' health and safety-related behavior.

Skills Training

Gaining and practicing the skill necessary to engage in a behavior is one of the most critical factors in performance of a behavior. Regardless of a worker's beliefs, attitudes, knowledge, and intentions, a behavior will not be performed correctly without the requisite skill. Proper handling of hazardous materials, correct use of safety features of machinery, proper positoning and use of PPEs, and breaking a habit or addiction are all types of skills that can be learned and practiced within the occupational setting.

Adults learn best by doing. Skills training, therefore, should entail both demonstration and opportunities to practice the skill. Practice of the skill under realistic circumstances provides the learner with a sense of self-efficacy, or mastery of the skill. In addition, perceived barriers to the desired behavior may be

Table 4. Program Planning Examples for Worker Health and Safety Education

Planning Step	Example A: Protective Eyewear	Example B: Smoking Cessation
(1) Current practices, attitudes, knowledge, etc.	50% of workers use protective eyewear at appropriate times; workers perceive eyewear to be uncomfortable, hot, and to reduce visibility	35% of asbestos workers smoke cigarettes; smokers perceive that their asbestos exposure is riskier than their smoking; they enjoy smoking; smoking is an addiction; smokers believe that difficulties and discomforts of quitting outweigh benefits
(2) Program goals	Decreased eye injuries	Reduced incidence of lung cancer
(3) Specific objectives	90% of workers will wear protective eyewear at all appropriate times	50% of workers who smoke will quit smoking
(4) Utility knowledge and requisite skills	Knowledge regarding proper use, appropriate times, specific hazards, potential consequences of not using eyewear Skills are proper fitting, proper wear	Knowledge regarding synergistic effect of smoking and exposure to asbestos; resources available for aid in quitting Skills are smoking cessation techniques
(5) Source, message, channel factors	Good sources are peers and supervisors Message factors: low to moderate fear; focus on personal susceptibility to eye injury; focus on high preventive benefit of eyewear Channel: pamphlets, visual reminders, group "eyewear fitting session"	Good sources are company health educator; representatives from local Lung Association or Cancer Society Message factors: focus on personal susceptibility to lung problems; focus on benefits of quitting; decrease perceived barriers to quitting Channel: group clinic; self-help materials
(6) Enabling factors	Improve eyewear design for better fit and ventilation; company policy regarding use should be enforced	Nonsmoking policy in work areas should be enforced; healthy snacks and exercise breaks can be offered in place of smoking during break time; off-site smoking cessation programs can be covered by employee benefit plan

Table 4 (continued). Program Planning Examples for Worker Health and Safety Education

Planning Step	Example A: Protective Eyewear	Example B: Smoking Cessation
(7) Program methods and material development	Develop written materials to provide utility knowledge; group meetings can be used to provide demonstration of proper fit and use, demonstrations of "visibility" with and without protective eyewear; "What is my Eye-Q?" self-test	Assess available existing cessation resources from local Lung Association, Cancer Society, or Heart Association; hire a clinic leader to offer on-site cessation clinics; provide self-help cessation materials for those who prefer them
(8) Reinforcements	Engage workers who regularly use protective eyewear to encourage co-workers; provide incentives such as contests between shifts or different areas of the plant for highest percentage of use	Begin a smoking crisis "buddy system" for smokers trying to quit; enlist ex-smokers to assist those trying to quit; use self-monitoring materials that allow smokers to see progress in cutting down; use biomedical tests such as heart rate and lung function to demonstrate benefits of quitting
(9) Program Implementation	Provide written materials in employee paychecks and in union newsletters; post visual reminders in the work area; conduct group sessions during work hours for a pilot group of 20 workers; follow with full-scale implementation with modifications based on the pilot test	Offer group clinic on a "first come/first served" basis to 20 workers as a pilot test; conduct clinic during work hours or other convenient times such as lunch hours or break times; conduct full-scale implementation after assessment of pilot test phase
(10) Program evaluation	Conduct survey of workers' perceptions of benefits of eyewear, susceptibility to injury both before and after program; monitor actual use by direct observation in the work area	Use self-reports, biomedical methods to assess quitting and cutting down rates; conduct surveys to assess workers' attitudes and beliefs about the benefits of quitting, their perceived ability to quit, and intentions to quit before and after program

reduced if learners have the opportunity to practice the behavior without difficulty or discomfort. Furthermore, supervised practice of the behavior in a simulated setting provides the opportunity for correcting improper procedures and mastery of the behavior with little or no risk.

Self-Assessment

Self-assessment, or monitoring of personal behaviors, can be an effective tool for facilitating behavior change. Used prior to the educational program, monitoring and keeping records of a specific behavior, such as how often a piece of safety equipment is worn and under what circumstances, can assist in increasing personal relevance of the program and in setting realistic goals for change. Used throughout the educational program, self-assessment serves as an important self-evaluation tool that can guide the worker and the educator to determine needs for additional practice or can provide a sense of mastery and reinforcement.

The health risk appraisal (HRA) is a unique form of self-assessment that is often used in comprehensive employee health promotion programs. HRAs can be self-scored, computer-based, or commercially scored. They are based on the premise that there are statistically measurable risks associated with particular behaviors. By answering questions about one's health-related practices and medical history, the computer generates information regarding life expectancy, risks for injury and illness, etc., that are individualized. Most computer-scored HRAs are accompanied by specific computer-generated suggestions for behavior change and the resultant statistical risk reduction. HRAs, like other methods of self-assessment, can increase personal relevance of the educational program, and provide evidence of personal susceptibility to specific conditions. Self-scoring HRAs and those available for interactive use on microcomputers have the added advantage of providing immediate feedback to the learner. HRAs can be used as part of a health education intervention, but should be followed up with specific recommendations and skills training. A listing of HRAs available for corporate or personal use can be obtained from the National Health Information Clearinghouse, as identified in the "Additional Resources" following the References to this chapter.

Incentives

Many occupational health and safety education programs have successfully utilized incentive programs for changing worker behavior. Incentives are generally small, but desirable items or privileges that are offered in exchange for meeting specified behavioral goals. These goals may include attendance at a meeting, participation in a program, as well as performance of the desired behavior.

Incentive-based educational programs have been effective in the occupational setting. For example, small cash incentives were offered to workers in a plastic manufacturing plant as part of a successful program aimed at reducing exposure to styrene.[1] Other workplace incentive programs have offered t-shirts, gift certificates, extra pay, time off, or trophies.[3] In many cases, incentives are tied to group performance; thus, there is an added advantage of peer pressure and social support for the desired behaviors. Furthermore, as behaviors change among a group of workers the workplace norms may be effectively changed to reflect generalized expectation of the new behavior.

The contingency contract that has been utilized in many patient education programs to improve compliance with medical regimens[15] is a specific, negotiated agreement between two parties to meet a specific behavioral goal. The agreement identifies rewards and/or penalties associated with those goals. The rewards may be tangible items such as money, lottery tickets, gift certificates, or extra vacation time. Less tangible rewards such as social approval may also be effective. Often, programs require participants to deposit money which is then returned contingent on meeting specific performance requirements. Such programs have been effective for reducing drug, alcohol, and cigarette use, as well as in weight control programs and in improving compliance with medical regimens. The potential of contingency contracts for use in occupational settings is unexplored. Such contracts may be made between co-workers or between supervisors or health educators and workers.

Social Support

Health-related behavior can be heavily influenced by the support, encouragement, and expectations of significant others. In the workplace, those most likely to influence behaviors are co-workers, supervisors, or union leaders. Programs can build supportive relationships through "buddy systems" or contracts/agreements between two or more individuals. Social support for behaviors could include reminding one another, or providing positive feedback for performing the desired behavior, and providing emotional or crisis support as requested. A formal agreement to serve a mutually supportive role should be based on the identified needs of the individuals; i.e., what is supportive to one individual may be perceived as nagging to another. It is therefore important to determine the form of support ahead of time and revise it as needed.

It has been suggested that the worksite offers a particularly unique opportunity for health education/promotion programs because of the built-in, cohesive support network and the potential for corporate reinforcements.[3] This has been particularly true of smoking control efforts, in which cessation efforts accompanied by consistent worksite smoking control policies have been very effective in reducing tobacco smoking among workers.[3]

Biomedical Feedback

Similar to self-assessment, biomedical tests that measure levels of hazardous substances absorbed or heart and respiratory capacities can provide evidence to increase personal relevance of the educational program and positive feedback to reinforce behavior changes. When the desire to improve health status itself is salient to the individual, biomedical feedback can serve as a powerful incentive to behavior change. Biomedical feedback was used successfully in a program in which lead levels in blood and urine were monitored as part of a safety education program.[1] The information was transmitted to the participants and lead levels were significantly lower in the safety-trained group than in the control group. The design of this study did not enable evaluators to separate the effects of the feedback from the program as a whole. In another setting, however, workers who received feedback from audiometric tests were more likely to wear hearing protectors than workers who were instructed to wear the protectors, but received no biomedical feedback.[8]

Biomedical feedback provides information to the learner about the effectiveness of the preventive actions they are taking or about their personal susceptibility to specific problems. In the case of the noise protection program, it was concluded that the relief from noise, once workers began wearing the protectors, provided its own positive reinforcement for continuing the behavior. In fact, in this case, the program was successful in changing the norms and expectations of the social environment, as use of hearing protectors eventually increased in workers who had not received the educational intervention and among new workers in the department.[8]

Environmental Supports and Engineering Control

Critical to the performance of desired safety and health-related behaviors is the availability of comfortable and convenient safety equipment and other aspects of the work environment which are conducive to the performance of the behavior. Uniformly enforced consistent and fair policies can be used to facilitate behavior change. New policies should be accompanied by programs that assist the worker in achieving the desired behavior. For example, a new company policy that prohibits on-the-job smoking (due to the recognition of particular risks of exposure to a substance for workers who smoke) is likely to be met with resistance by smokers. Offering assistance for those who wish to quit through formal on-site group programs, free self-help materials, or financial incentives to attend other formal programs will increase the likelihood that the policy will be successfully implemented. Furthermore, penalties for violations of the policy should be fair, consistently applied, and consistent with behavioral change efforts, rather than punitive.

Employee Assistance Programs (EAPs) to aid workers with drug, alcohol, and other addictive or psychological problems have become more prevalent in the workplace. Generally part of personnel departments, some programs offer counseling and treatment while others require that employees who show evidence of such problems seek outside help for the problem. Many are predicated on the threat of job loss and are reserved for employees who have demonstrated significant problems with absenteeism, productivity, and on-the-job injuries as a result of substance use. With the social climate changing in favor of earlier recognition of potential drug and alcohol-related problems, EAPs have expanded their services to provide counseling and resources for employees who are demonstrating the "early warning" signs of problems. EAPs are an example of the potential for employers to "enable" certain kinds of behavior change. The National Institute on Drug Abuse (NIDA) publishes a directory of educational resources for EAPs, as identified in the "Additional Resources" following the References to this chapter.

HEALTH EDUCATION MEDIA

A variety of media can be utilized in health and safety education programs. Since adults learn in different ways, occupational health and safety programs that encorporate several types of media can more effectively reach a wider audience.[16] The following section discusses specific types of media for use in occupational safety and health education programs. Their key features, advantages, and disadvantages are summarized in Table 5.

Pamphlets

The most traditional of health education media, pamphlets are relatively inexpensive to produce, especially with the use of desktop publishing programs available in most worksites. Pamphlets are best utilized to provide information or to serve as "cues" or reminders. Pamphlets should be straightforward, attractive, and simple to read. Pamphlets should not be written to include lengthy or complex information, as this will tend to decrease attention to the most important message. Only the information or reminder relevant to the behaviors should be included (and in such a way as to emphasize the most important information).

A good pamphlet begins with a "hook", i.e., something to make the reader recognize that it is he/she that this pamphlet addresses. This should be followed by the desired message presented in a simple, clear, and easily readable manner. If the information presented is a suggestion to perform a particular behavior, then the behavior should be described in a way that makes it seem convenient and beneficial, even fun. Suggestions can be made that may reduce specific barriers or difficulties in performing the behavior. Readers should be directed to an available company resource for additional information and assistance. Finally, the pamphlet should end on a positive note, with a reminder or memorable "tag line."

Pamphlets should be distributed at a time and place in which workers' attention spans are most keen. Some examples of distribution methods include paycheck envelopes, company newsletters, union correspondence, or during meetings or formal health education programs.

Table 5. Health Education Media for Worker Health and Safety Programs

Medium	Uses	Key Features	Disadvantages
Pamphlets	Reminders; cues; limited information	Message can be reviewed as needed	Not likely to change complex behaviors; can be easily discarded
Posters	Cues; reminders; limited information	Visual display; reinforces behavior	May be expensive; not likely to change complex behaviors
Audio-visual media (videos, films, etc.)	Information; demonstration of skills	Visual and audio display; complex behaviors can be demonstrated; real situations can be depicted; maintains interest	Expensive to produce; commercially available films or videos may not be appropriate for specific needs; should not be used without additional staff intervention
Interactive media computers, interactive videodisks	Demonstration and rehearsal of complex skills	Provides immediate feedback; learner can work at own pace	May be expensive; some learners may experience anxiety due to the use of the technology itself
Biomedical tests	Self-assessment; evaluation of programs; incentive and personalization of message	Provides feedback; individualized for each worker; actual demonstration of behavioral consequences	Can be very expensive: confidentiality must be preserved

Posters

Posters are most appropriate as a type of reminder or cue. Rarely do posters receive more than a passing glance. Thus, a good poster has a single message that is clearly stated and illustrated. These would include reminders to do a specific behavior (e.g., "Wear your protective goggles!") or to attend a meeting, seminar, or program (e.g., "Want to quit smoking for good? Meet here on Thursday at noon").

Posters may be slightly more expensive to produce than pamphlets since they generally require more sophisticated artwork. Many posters that may meet the needs of a particular company are already available from voluntary and private agencies free of charge or for a nominal fee. Resources committed to in-house poster development should probably be limited and used only for reminders of the most critical behaviors.

Audiovisual Media

Videos, films, and other audiovisual (A-V) media are the most expensive to produce or to purchase. They have several clear advantages over print media in that viewers can receive both visual and audio stimuli. In addition, complicated behaviors can be demonstrated and specific on-the-job situations and settings can be depicted.

A-V media tend to maintain the attention of workers better than print media. The addition of color, music, and other visual effects inherently make these media more interesting. Participants may perceive video learning to be easier than learning through print media and thus be more motivated to participate. Video, TV, or films, however, should not stand alone as a program since viewers need to be able to process, discuss, and clarify what they have seen and heard.

The group setting for viewing offers an opportunity both before and after the viewing to discuss questions, perceived difficulties in performing the behaviors, and suggestions that build on the information provided in the program. Such groups can be formal, with an "instructor" or leader providing the structure for discussion or for practice of behaviors. Alternately, group viewing can be informal, without a leader. If viewing is done individually (e.g., a new employee views a safety videotape in the company library before beginning the job), some follow-up should be included, such as a self-test, or an opportunity to discuss questions or comments with another employee or administrator. This will ensure that what was viewed passively was mentally reviewed and processed.

Videos, films, and TV programming can be developed by the company to meet specialized program needs. Although this is an expensive option, it is often difficult to find the commercially available program that meets the needs of a specific organization. Nevertheless, there are many excellent existing programs that can be used in whole or in part. Appropriate follow-up is particularly important in this case, so that the relevance of the material to the specific health and safety needs of the viewers is made clear. Ready-made programs are likely to be most effective if they deal with the same industry.[16]

Films and videos that rely heavily on extensive footage of an expert sitting at a desk discussing the topic or being interviewed (colloquially termed "talking heads") tend to be less interesting. On-site demonstration, illustration, and activity are more likely to engage the attention of the viewers. The demonstration and setting, however, should be closely related to the specific worksite situation. Commercially available films that appear dated or take place in a vastly different setting are not likely to engage the learners.

Interactive Media

Interactive computer programs, such as self-tests, and interactive videodiscs, which add sophisticated graphic displays to the screen, can be used very effectively for training and the practice of safety and health-related behaviors. Such media are especially useful for learning complicated behaviors, as the learner has an opportunity to progress at his/her own pace, to master each step before continuing, and to focus on the material that is most relevant to the worker.

Some studies have found computerized health education programs to be equal to face-to-face educational formats in participant satisfaction and better than face-to-face programs in terms of information recall.[3] This may be due to the fact that computerized learning allows for flexibility and tailoring to individual needs while still maintaining quality control and consistency. This is not to suggest that computerized health education programs can replace the motivational and supportive value of group programs and personalized approaches, but instead can be used as powerful program enhancements. Furthermore, individual reactions to computer-based formats, such as increased anxiety, must be considered in designing and using computers in the health education program.

Costs for setting up interactive, computer-based programs depend on what facilities are available in-house and the level of complication, i.e., branching, of the program. Interactive videodiscs are much more complicated to design, but have the advantage of sophisticated graphics to support the learning process.

Biomedical Tests

Through the medium of medicine, occupational health and safety programs can be greatly enhanced. Biomedical tests are therefore included in the category of media because medical test results serve to personalize information, increase belief in personal susceptibility to the consequences of failing to take preventive action or increase outcome expectancies, and increase self-efficacy; thus, they may warrant the usually high expense.

Biomedical test results should always be confidential and workers must have confidence in this confidentiality. Furthermore, results should be provided in such a way as to minimize fear by coupling them with specific recommendations that will effectively reduce risks. Biomedical tests can serve as a self-assessment tool and as an instrument to evaluate the effectiveness of the educational program.

SUMMARY

It is critical that occupational safety and health be considered the shared responsibility of the employer and employee, such that each perceives a significiant contribution by the other. Behavioral programs can never take the place of safe engineering and design of the worksite. Similarly, workers have the responsibility of utilizing the safe devices and practices that such design demands. Worksites that capitalize on this shared responsibility are likely to be more effective in reducing work-related injuries and health problems and in improving workers' health overall.

As discussed in the program planning model presented here, the following components are critical to the success of occupational health and safety education programs:

- Utility knowledge
- Skills training
- Skills rehearsal
- Environmental reminders or "cues to action"
- Contributing engineering and design factors that enable the behavior
- Efforts to achieve social normative change in expectations about worker behavior.

Program designers should utilize a variety of learning and behavior change strategies including skills training, self-assessment of behavior, provision of tangible incentives, social support, and continuous feedback both of the learning process and of outcomes. Furthermore, attention to policies and conditions that provide environmental support for safe and healthy work environments will further facilitate behavior changes.

Finally, a variety of media are available for program support, including print, video, audio, and interactive formats. Using several different media in a given program will enhance the ability to reach learners with differing needs and abilities.

This discussion of program design for workers' health and safety is intended to provide employers with a basic overview of the principles and practices of health and safety education. The resource guide at the end of this chapter provides several additional sources for more detailed discussions of program planning, evaluation methods, and available materials.

ADDITIONAL RESOURCES

Several useful books and government resources are listed below as potential sources for additional information on planning health and safety education programs for workers.

Texts

Evaluation of Health Promotion and Education Programs, R. A. Windsor, T. Baranowski, N. Clark, G. Cutter, Eds., Mayfield Publishing, Palo Alto, CA, 1984.

Handbook of Health Education, 2nd ed., P. M. Lazes, L. Hollander Kaplan, K. A. Gordon, Eds., Aspen Publishers, Rockville, MD, 1987.

Health Education Planning: A Diagnostic Approach, L. W. Green, M. W. Kreuter, S. G. Deeds, K. B. Partridge, Eds., Mayfield Publishing, Palo Alto, CA, 1980.

The Future of Work and Health, C. Bezold, R. J. Carlson, J. C. Peck, Eds., Auburn House, Dover, MA, 1986.

Directory of Academic Institutions and Organizations Offering Drug, Alcohol, and Employee Assistance Program Educational Resources, U.S. Department of Health and Human Services, National Institute on Drug Abuse, Office of Workplace Initiatives, DHHS No. ADM 88-1579, 1988.

Information Clearinghouses

Office of Disease Prevention and Health Promotion
National Health Information Clearinghouse
P. O. Box 1133
Washington, D.C. 20013-1133
1-800-336-4797
301-565-4167 (in Maryland)

The National Health Information Center provides a variety of free resources. These include the Healthfinder series of resource lists and the Health Information Resources in the Federal Government directory, as well as an online directory of over 1000 health-related organizations providing health information and assistance.

National Clearinghouse for Alcohol and Drug Information
P. O. Box 2345
Rockville, MD 20852
301-468-2600

Clearinghouse of the Handicapped
Switzer Building, Room 3132
330 C Street, SW
Washington, D.C. 20202
202-732-1244

National Injury Information Clearinghouse
5401 Westbard Avenue, Room 625
Washington, D.C. 20207
301-492-6424

REFERENCES

1. Vojtecky, M. A. "Education for Job Safety and Health," *Health Educ. Q.* 15(3):289-298 (1988).
2. Green, L. W., M. W. Krueter, S. G. Deeds, and K. B. Partridge. *Health Education Planning: A Diagnostic Approach* (Palo Alto, CA: Mayfield Publishing Company, 1980).

3. Lazes, P. M., L. H. Kaplan, and K. A. Gordon. *The Handbook of Health Education* (Rockville, MD: Aspen Publishers, 1987).
4. Fielding, J. E., and P. V. Piserchia. "Frequency of Worksite Health Promotion Activities," *Am. J. Publ. Health,* 79(1):16-38 (1989).
5. Becker, M. H. Ed. "The Health Belief Model and Personal Health Behavior," *Health Educ. Monogr.* 2:324-473 (1974).
6. Rosenstock, I. M., V. J. Strecher, and M. H. Becker. "Social Learning Theory and the Health Belief Model," *Health Educ. Q.* 15(2):175-183 (1988).
7. Bandura, A. *Social Foundations of Thought and Action* (Englewood Cliffs, NJ: Prentice-Hall, 1986).
8. Zohar, D., A. Cohen, and N. Azar. "Promoting Increased Use of Ear Protectors in Noise through Information Feedback," *Hum. Factors* 22:69-79 (1980).
9. McGuire, W. J. "The Probabilogical Model of Cognitive Structure and Attitude Change," in *Cognitive Responses in Persuasion,* R. E. Petty, T. M. Ostrom, and T. C. Brock, Eds. (Hillsdale, NJ: Lawrence Erlbaum Associates, 1981) pp. 291-307.
10. Robins, T. G., and S. Klitzman, "Hazard Communication in a Large U.S. Manufacturing Firm: The Ecology of Health Education in the Workplace," *Health Educ. Q.* 15(4): 451-472 (1988).
11. Soames Job, R. F. "Effective and Ineffective Use of Fear in Health Promotion Campaigns," *Am. J. Publ. Health* 78(2):163-167 (1988).
12. Fishbein, M., and I. Ajzen. "Acceptance, Yielding and Impact: Cognitive Processes in Persuasion," in *Cognitive Responses in Persuasion,* R. E. Petty, T. M. Ostrom, and T. C. Brock, Eds. (Hillsdale, NJ: Lawrence Erlbaum Associates, 1981), pp. 339-359.
13. Windsor, R. A., T. Baranowski, N. Clark, and G. Cutter, *Evaluation of Health Promotion and Education Programs* (Palo Alto, CA: Mayfield Publishing, 1984).
14. Green, K. L. "Issues of Control and Responsibility in Workers' Health," *Health Educ. Q.* 15(4):473-486 (1988).
15. Janz, N. K., M. H. Becker, and P. E. Hartman. "Contingency Contracting to Enhance Patient Compliance: A Review," *Patient Educ. Couns.* 5(4): 165-178 (1984).
16. LaBar, G. "Flexibility: Key Tactic in Safety Training," *Occup. Haz.* (February 1989), pp. 69-70

CHAPTER 17

Worksite Health Promotion

Mary Beth Love

INTRODUCTION

Corporate America is becoming increasingly interested in promoting employees' health and fitness. Nationally, about two out of three large companies now have at least one type of health promotion program.[1] Although large corporations have led the way in this area, smaller businesses have also invested in employee well-being. A survey of small- to medium-sized companies in California found that one third had sponsored at least one worksite health promotion program during the previous year.[2]

In general, the goal of Worksite Health Promotion (WHP) is to educate employees about the health-promoting/disease preventing rewards of adopting positive life habits (proper diet, nonsmoking, regular exercise, stress management) and to change organizational policies and the social and physical environment so that they support employee health. This chapter provides an overview of WHP by discussing the rationale behind corporate interest in health promotion, the advantages of promoting health at work, the three levels of intervention for WHP, and the characteristics of successful WHP programs.

WHY HEALTH PROMOTION?

The interest of business in developing programs to enhance employee health has been stimulated predominantly by two factors: (1) the escalating cost of health care, and (2) the growing consensus in the epidemiologic literature for the relationship of individual behavior to disability and disease.[3] The employer's role in employee health care has emerged only in the last several decades as employee benefits expanded to include health care coverage. As a result, health care expenditures and the demand for health care have risen dramatically along with the burden of health care cost to employers. Employers now pay one third of the nation's health care bill.[2] According to the U.S. Chambers of Commerce's most recent survey of employee benefits, American companies were paying an average of $2560 per employee for health care, an amount that constitutes up to 11.8% of the payroll.[4] Company costs for health care are rising so quickly that they threaten to consume a larger and larger portion of corporate profits if left unchecked.[5] In fact, over the last 3 decades, employer health care costs have been the most rapidly rising cost item associated with operating a business in the U.S. Between 1972 and 1982 the increase in health care cost averaged 15%/year; these costs are currently increasing at a rate of 15 to 40%/year.[2] Most important is that more than half of all the health insurance claims responsible for these health care costs are due to potentially preventable conditions.[6]

The second catalyst for this movement toward promoting health and preventing disease emerged as a result of "The First Public Health Revolution." This "revolution" involved the institution of public health measures at the turn of the century (including the provision of clean water and the disposal of waste) that resulted in a shift in the disease pattern among Americans.[7] Whereas our ancestors died mostly from acute infections (such as pneumonia, influenza, and tuberculosis), today, chronic diseases (such as heart disease, cancer, and stroke) are responsible for the majority of deaths. Unlike infectious disease, many of the risk factors for chronic disease are modifiable. Experts believe that over half of the deaths from chronic illness and a large portion of chronic disease can be prevented or postponed with a change in personal behavior.[6]

The economic burden of preventable diseases has motivated employers to invest in programs aimed at reducing the risk factors for chronic disease. Table 1 lists today's major killers and the modifiable risk factors associated with them. Epidemiological studies of population-wide risk factors have found that adults who do not smoke, exercise regularly, maintain an average weight for their height and age, and who drink alcohol moderately, have one third the mortality rate of those with high-risk behaviors.[2] Mortality is just part of the cost of these risk behaviors. A recent study of 15,000 Control Data Corporation employees found

Table 1. Major Causes of Death and Associated Risk Factors, U. S., 1986

Cause	Annual Rates per 100,000 Population	Behavioral Risk Factor
1. Heart disease	873.2	Smoking[a], hypertension,[a] elevated serum cholesterol,[a] diet, lack of exercise, diabetes, stress, family history
2. Malignant neoplasms	317.5	Smoking,[a] worksite carcinogens,[a] environmental carcinogens, alcohol, diet
3. Stroke	62.1	Hypertension,[a] smoking,[a] elevates serum cholesterol,[a] stress
4. Accidents other than automobile	19.7	Alcohol,[a] drug abuse, smoking; motor vehicle accidents(fires), product design, handgun availability
Motor vehicle	19.1	Alcohol,[a] no seat belts,[a] speed,[a] roadway design, vehicle engineering
5. Chronic obstructive pulmonary condition	31.8	Smoking
6. Influenza and Pneumonia	29.0	Smoking, vaccination status[a]
7. Diabetes	15.4	Obesity[a]
8. Suicide	12.8	Stress,[a] alcohol and drug abuse, gun availability
9. Chronic liver disease/cirrhosis	10.9	Alcohol abuse[a]
10. Arteriosclerosis	9.4	Elevated serum cholesterol[a]

[a] Major risk factors.

From *Morbidity and Mortality Weekly Report,* Centers for Disease Control, U.S. Department of Health and Human Services, Public Health Service, March 3, 1989:38 (8).

that those with high-risk health habits cost more in health insurance. Conversely, workers with healthy lifestyles spend fewer days in the hospital, have lower medical claims costs, and have fewer claims of $5000 or more.[8] Table 2 summarizes relevant data on the effect of the illnesses associated with risk behaviors on the bottom line for employers.

WHY THE WORKSITE?

The worksite has been identified as a promising arena for health promotion programs for a number of reasons. First, since the majority of Americans work, the worksite provides easy access to the adults at risk for preventable disease. Since the adult population spends about half of their waking hours at work, on-site educational programs are convenient for employees as well.

Second, the worksite is viewed as providing the health promotor with the opportunity to influence the group's social norms. By virtue of employee proxim-

Table 2. Employer Costs of Preventable Illness: The Facts[a]

Absenteeism
Costs an employer 11/2 times the absent employee's daily wage.

Turnover
Costs an employer an average of $4595 per employee turnover.

Back Pain
In 1984, back and spine injuries accounted for 21% of total work-related injury/illness cases reported in California.
$715.5 million was paid out in medical claims by California Workers' Compensation carriers in 1982 for back and spine-related injuries.
Nationwide, backaches accounted for 93 million lost workdays, $1.0 billion in lost output, and $250 million for Worker's Compensation claims.

Cigarette Smoking
Between 7 to 16% of the workday is lost to the smoking ritual — up to 30 min per smoker per day.
About 81 million workdays are lost each year due to smoking; smokers account for approximately a 50% higher rate of absenteeism than nonsmokers.
The average pack-a-day smoker costs an employer an excess of between $432 to $775, plus another $53 to $105 in annual side-stream smoke-related costs.

Lack of Exercise
Persons who do not exercise vigorously and regularly (equal to climbing 15 flights of stairs or walking 1.5 miles 4 times per week) have 14% higher health care claim costs and 30% more hospital inpatient days than those who do.
A sedentary employee has a 30% higher absenteeism rate as compared to the average active employee.

Hypertension
About 29 million workdays and $16.0 billion are lost each year due to hypertension-related cardiovascular diseases.
Persons at high risk of hypertension (BP above 159/94 mmHg) have health care claim costs 11% higher than those with normal blood pressure.

Obesity
Employees who are 40% or more overweight visit the doctor and miss work twice as often as the average employee — at an excess cost to employers of $1000 per year per overweight employee.
Persons 30% or more above their desirable weight have 11% higher health care claim costs and 45% more hospital inpatient days than those less than 20% above their desirable weight.

Excessive Stress
Between 1974 and 1979, cumulative stress-related injury claims rose threefold in California.
In 1980, more than 3000 psychiatric injury claims were filed in California, half of which were awarded monies.

As expensive as employer-paid expenditures for medical care are, they represent only a portion of the true costs of illness to business and industry. Employers must also absorb the indirect costs of lost productivity, i.e., absenteeism, turnover, reduced job performance, as well as losses from early retirement and premature death. These costs appear to run two to three times those of direct medical care costs. The President's Council on Physical Fitness places the annual cost to American business of premature death alone at more than $2.5 billion and 132 million workdays of lost production.

[a]From Kizer, K. California Department of Health Services, Sacramento, 1988.

Table 3. Benefits of Worksite Health Promotion Programs

Reduction of costs: decreases in life, health, and workers' compensation insurance costs
Reduction in employee absenteeism and turnover
Increased productivity
Increased morale
Improved corporate image
Improved employee health and well-being

ity, common schedules, and work functions, social networks already exist and can be mobilized to reinforce and encourage positive health behaviors. The stability of the workplace can provide consistency and the opportunity to offer a wide range of programs. Finally, the corporate communications system (newsletters, bulletins, etc.) and the employee training sessions can be utilized.[2]

Third, the potential direct and indirect benefits to the employer are an incentive to management to invest in health promotion. The benefits typically cited are presented in Table 3.[9]

CONCEPTUAL FRAMEWORK FOR WORKSITE HEALTH PROMOTION INTERVENTIONS

The goal of worksite health promotion programs is to improve employee health by reducing the prevalence of behavioral risk factors among employees. These risk factors are patterns of behavior that have proved to be deleterious to health, such as smoking, overeating, poor dietary practices, and alcohol abuse. In order to develop effective health promotion interventions, one need first understand the determinants of such behavior. A conceptual model of the determinants of patterned behavior is useful in planning and evaluating health promotion programs because it draws attention to different levels and types of influences on behavior and serves as a guide to the development of appropriate interventions.

The conceptual model the author finds most informative for designing effective WHP programs views patterned behavior as being determined by the following:

1. *Individual factors* — characteristics of the individual such as knowledge, attitudes, behavior, self-concept, skills, etc.
2. *Social factors* — formal and informal social networks and social support systems, including family, work group, and friendship network
3. *Organizational factors* — organizational characteristics and formal and informal rules and regulations for operation[10,11]

The strength of this framework is its recognition of the importance of targeting not only individual behavior, but the social surroundings that can support or sabotage that behavior and the organizational environment that is increasingly recognized as having a profound impact on health. Historically, individually oriented behavior change strategies have been the primary mode of intervention in WHP.

Program failures and the high relapse rate experienced in many programs have forced a recognition of the limitations of such an approach.[10] The advantage of targeting all three levels is the resulting complementarity: the limitations inherent in an intervention focusing on one level is the strength of the intervention that focuses on another. For example, a social intervention can provide the motivational climate for change but without the skill-building experience provided by the individual level intervention, this motivation would result in frustration rather than behavior change. The strength of such an organizational level intervention as the provision of flexible work time would serve to further enhance the likelihood of successful behavioral changes.

The three levels outlined above reflect the range of strategies currently available for worksite health promotion programing. Table 4 provides examples of the worksite health promotion interventions that can be implemented at each of the three levels. At the individual level the principles of social learning theory (SLT),[11] and especially self-efficacy, guide practice. Research has demonstrated consistently that an individual's self-efficacy beliefs mediate change and the maintenance of change. Self-efficacy is the belief of an individual that he/she has mastered the skills necessary to engage in a new behavior.[12] Perceptions of one's self-efficacy are best increased by incorporating modeling, participant practice, demonstration of new behaviors, reinforcement, and feedback into the various educational programs tailored for individual level interventions. One-on-one education, small group workshops, lectures, mass media campaigns, physical screening, and health risk assessments are all categorized as individual level change strategies.

At the social level, the major objective is to change the normative influences within the worksite. No matter how good the skill training is at increasing self-efficacy, or how strongly motivated the individuals are, if they are left to fend for themselves in a social environment filled with negative health practices, the strength of peer pressure and the lack of external support will erode commitment in the majority of people. Interpersonal relationships are an important influence in the health-related behavior of individuals.[14] Soliciting the involvement of socially popular opinion leaders within the workforce (in program planning and implementation) can be an effective strategy to confer social prestige to a worksite program. Additionally, management's strong verbal and financial support for the health promotion program and encouragement of employee participation will help to establish positive health expectations. Health competitions are a fun and effective strategy for mobilizing peer support for positive health behavior. Peer support groups and health-oriented clubs such as a lowfat gourmet cooking club or a corporate running team will create alternative networks in which healthy behavior is a recognized value receiving strong social support.

The important ingredient in interventions at the social level is that they create a milieu within the workplace that makes it desirable for employees to be associated with co-workers who practice health behaviors and who publicly affirm their

Table 4. Potential Health Promotion Interventions

	Individual Level Factors
Object of Change:	Individual's behavior, knowledge, attitudes, intentions
Interventions:	Educational programs, e.g., lectures Skill training workshops Audio-visual library Mass media campaigns Peer support groups and peer counseling Screening and assessment instruments Incentive systems
	Social Level Factors
Object of Change:	Modification of interpersonal social influences — group norms
Interventions:	Involvement of social leaders in WHP planning and implementation Strong commitment and verbal support from upper and middle management Creating desirable social networks based on healthy values, e.g., corporate running club Self-help support groups Health competitions
	Organizational Level Factors
Object of Change:	Organizational structures and processes
Interventions:	Changes in health-related policies, e.g., smoking policy Programs to improve worker-supervisor relationships Healthy food options in cafeteria Changes in the structure of work, e.g., flex time Changes in employee benefits On-site exercise facilities

value for health. This requires that the interventions be socially rewarding and incorporate fun, prestige, and recognition.

The structure and process of the worksite can have a substantial influence on health and health-related behavior. Occupational health services protect the worker from many of the environmental risks to health that the worksite may impose, such as hazardous chemicals and risks from injuries and accidents; however, there are other hazards of the workplace environment that tend to be psychologically stressful and thus have negative health effects. Excessive work pace or workload, ineffective management style, poor employee/employer relations, and job monotony and/or complexity, are examples of organizational level characteristics which, if ignored, will undermine the health promotion efforts being made at the individual and social level.[14]

Organization level interventions can capitalize on the characteristics of the worksite to promote health. Stress reduction interventions have included workshops to improve management styles and worker/supervisor relationships. Policies to ban smoking or to establish no smoking areas have been developed. Other examples of health promotion policies include time off from work to participate in health promoting activities, flexible hours to enable participation on-site day care for working parents, and/or the introduction of lowfat food alternatives in the work cafeteria.[4] The use of incentives has been a recent addition to the health promotion manager's strategies for motivating behavior change. Although typically used for individual level interventions, e.g., to reinforce the adoption of healthy behaviors, they actually can and should be applied across all three levels. An incentive system that reinforces not only the adoption of healthy habits but also health leadership activities in the workplace, such as organizing a lunchtime walking group or working on a health policy committee, could also earn the employee incentive points.[15]

CHARACTERISTICS OF QUALITY WORKSITE HEALTH PROMOTION PROGRAMMING

The scope of WHP programs varies among workplaces. There is no "package" of services or mode of delivery that will work for every organization. Each program must be carefully tailored to the specific needs and interests of an organization and its employees; however, Table 5 lists some of the general strategies believed to increase the probability of program success.

Management Support: Commitment from all levels of management is very important to program success. Management should not only provide the needed financial resources but also a strong social influence within the organization in motivating employee participation. The role of gate keeper, played by middle management, makes their support critical, especially in programs aimed at staff and line workers.

Careful Planning: Due to the idiosyncratic nature of every workplace, planning must be tailored to the specific needs and organizational realities of that workplace and its employees. Program management issues such as budget and lines of authority must be negotiated up front along with the expectations regarding program objectives. To develop a sense of ownership among employees and to minimize the problems of coercion and paternalism, employees' active participation should be sought in every stage of the planning and implementation process.

Employee Participation: Participation in WHP programs must be voluntary and the health information gathered kept confidential. Employees should be assured of this confidentiality whenever personal data are collected.

Program Implementation: As outlined earlier, the WHP program should incorporate strategies that address all three levels of the system within an organization (see Table 4). Programs should be implemented so that they are convenient and the educational methods varied, participatory, and fun.

Table 5. Characteristics of a Successful Workplace Health Promotion Program

Management Support
- Commitment from all management levels, beginning with the CEO and other senior managers.

Careful Planning
- Involvement of employees from all organizational levels
- Responsibility for the program assigned to one key manager
- Realistic measurable program goals
- Budget identified
- Health promotion integrated into company's policies, benefit structure, recognition system, and occupational health programs

Employee Participation
- Emphasis on confidentiality
- Participation voluntary
- Program designed to meet employee needs

Program Implementation
- Programs scheduled at convenient times and in locations that are accessible and pleasant
- A multifaceted approach that offers a variety of educational techniques and opportunities to be utilized
- Program activities participatory, relevant to the lifestyles of the employee, and fun

Maintenance of Program Support
- Market plan in place
- Data systems that provide clear, meaningful information on existing programs
- Development of support networks among employees
- Involve employees' families
- Measure and capitalize on program successes; generate momentum

From Kizer, K. California Department of Health Services, Sacramento, 1988.

Maintenance of Program Support: Program evaluation is essential to the ongoing support of the WHP program. Management information systems must be established and methods of measuring program effectiveness designed. This can often be easily accomplished if the organization has such a system already in place. Although in other cases it will require a considerable investment of staff energy, its importance to program justification and continued support should not be underestimated.

Another component important to maintaining support is a program marketing plan. This plan outlines how the program will continue to interest employees in its services. It requires ongoing assessments of employee needs and the tailoring and publicizing of the health promotion services in a manner that will attract participation.

CONCLUSION

Motivated by escalating health care costs, employers are increasingly looking toward WHP programs to enable and motivate the employees to adopt healthy behaviors and thus lower the risk of ill health. The implementation of health promotion at the worksite has many direct and indirect benefits both for the employee and the employer. This chapter presented a conceptual framework for understanding the three levels of influence within the workplace (individual, social, and organizational) that determine the lifestyle practices known to result in an increase in the employees' risk of premature death and disability. Health promotion strategies available for intervention at each of the three levels were outlined and the characteristics of successful WHP programs were presented.

REFERENCES

1. Fielding, J. K., and P. Piserchia. "Frequency of Worksite Health Promotion Activities," *Am. J. Public Health* 79(1):16-38 (1989).
2. Kitzer, K. W. "Health Promotion at Work in California: An Employer's Guide to Assessing Quality in Worksite Health Promotion Programs," California Department of Health Services (1988), pp. 1-46.
3. Herzlinger, R., and D. Calkin. "How Companies Tackle Health Care Costs: Part III," *Harvard Bus. Rev.* (January-February 1986), pp. 70-80.
4. Sloan, R., J. Gruman, and J. Allegrante. *Investing in Employee Health: A Guide to Effective Health Promotion in the Workplace* (San Francisco: Jossey Bass, Inc., 1987), pp. 55-77.
5. Herzlinger, R., and J. Schwartz. "How Companies Tackle Health Care Costs: Part I," *Harvard Bus. Rev.* (July-August 1985), pp. 69-81.
6. Matarazzo, J. "Behavioral Health: A 1990 Challenge for the Health Professions," in *Behavioral Health: A Handbook of Health Enhancement and Disease Prevention,* Matarazzo, J. D., C. Weiss, J. Herd, N. Miller, and S. Weiss, Eds. (New York: John Wiley & Sons, 1984), pp. 3-40.
7. Bates I., and A. Winder. *Introduction to Health Education* (Palo Alto, CA: Mayfield Publishers, 1984), pp. 15-37.
8. Brink, S. "Health Risks and Behavior: The Impact on Medical Costs," Preliminary Study by Milliman & Robertson, Inc. and Control Data (1988), pp. 1-16.
9. O'Donnell, M. and T. Ainsworth. *Health Promotion in the Workplace* (New York: John Wiley & Sons, 1984), pp. 1-50.
10. McLeroy, K., D. Bibeau, A. Steckler, and K. Glanz. "An Ecological Perspective on Health Promotion Programs," *Health Educ. Q.* 15(4):351-377 (1988).
11. Abrams, D. "SLT Principles for Organizational Health Promotion: An Integrated Approach," in *Health and Industry: A Behavioral Medicine Perspective,* Cataldo, M. F., and T. L. Coates, Eds. (New York: John Wiley & Sons, 1986).
12. Steckler, V., B. McEvoy DeVellis, M. Becker and I. Rosenstock, "The Role of Self-Efficacy in Achieving Health Behavior Change," *Health Educ. Q.* 13(1): 73-92 (1986).

13. Levine, S., and J. Sorenson, "Social and Cultural Factors in Health Promotion," in *Behavioral Health: A Handbook of Health Enhancement and Disease Prevention,* Matarazzo, J. D., C. Weiss, J. Herd, N. Miller, and S. Weiss, Eds. (New York: John Wiley & Sons, 1984), pp. 222-229.
14. Rosen, R. "Developing Healthy Companies through Human Resources Management," Worksite Wellness Series, Washington Business Group on Health, (1986), pp. 1-26.
15. Yenney, S. "Using Incentives to Promote Employee Health," Worksite Wellness Series; Washington Business Group on Health (1986), pp. 1-38.

GLOSSARY

GLOSSARY

For any chapter in this book, terminology specific to that chapter can also be found explained in the references for that chapter. Glossary terms can have multiple meanings but are defined here only in the context of occupational health and safety. Terms listed in **boldface** are defined elsewhere.

A

AAIH. American Academy of Industrial Hygiene, a professional society of **CIH**s (Certified Industrial Hygienists; see **ABIH**).

Abatement. The elimination of a potential or actual hazard.

ABIH. American Board of Industrial Hygiene, a professional industrial hygiene accreditation organization, administering the **CIH** and **IHIT** program. **ABIH**, 4600 W. Saginaw, Suite 101, Lansing, MI, 48917 (517) 321-2638.

Absorbed Dose. The amount of a substance entering the worker's body after exposure.

Absorbents, Sound. Materials that absorb sound readily. These are usually construction materials designed specifically for the purpose of absorbing acoustic energy.

Absorption. To suck up or drink in; contrast with **adsorption.**

Absorption Coefficient (α)**.** For a surface, the ratio of the sound energy absorbed by a surface of a medium (or material) exposed to a sound field (or to a sound radiation) to the sound energy incident on the surface. Tables of absorption coefficients usually list the absorption coefficients at various frequencies.

Absorption Loss. That part of the transmission loss due to the dissipation or conversion of sound energy into other forms of energy (for example, heat), either within the medium or attendant upon a reflection.

Acclimatization. The physiological changes which occur when an individual spends several successive days exposed to a thermally stressful environment.

Accuracy. The measure of the correctness of data, often defined as the difference between the measured value and the "true" value.

ACGIH. The American Conference of Governmental Industrial Hygienists. An organization of individuals employed by governmental agencies or educational institutions and engaged in occupational safety and health programs. The ACGIH develops recommended occupational exposure limits for chemical and physical **agents** (see **TLV**). ACGIH, 6500 Glenway Avenue, Building D-7, Cincinnati, OH, 45211 (513) 661-7881.

Acid. An inorganic or organic compound that reacts with metals to yield hydrogen, reacts with a base to form a salt, dissociates in water to yield hydrogen or hydronium ions, has a pH of less than 7.0, and neutralizes bases or alkaline media. Acids turn litmus paper red and are corrosive to human tissue and are to be handled with care. See also **Base, pH**.

Acoustic. Acoustical. The qualifying adjectives acoustic and acoustical mean containing, producing, arising from, actuated by related to, or associated with sound.

Acoustics. The science of sound, including its production, transmission, and effects. The acoustics of a room are those qualities that together determine its character with respect to sound and hearing.

Action Level. The exposure level at which OSHA regulations to protect employees take effect (29 CFR 1910.1001-1047). Generally, once an action level has been reached, the employer must initiate workplace air or noise analysis, employee training, medical monitoring, and record keeping.

Acute. In this context, anything having a sudden onset and/or a short duration.

Acute Exposure. An exposure that occurs over a short period of time and often at high levels.

Acute Health Effect. An adverse health effect resulting from (usually) one high exposure, with severe symptoms developing rapidly and coming quickly to a crisis. Contrast with **Chronic Health Effect**.

Acute Toxicity. The adverse toxic effects resulting from a single exposure to a material.

Administrative Controls. Approaches for reducing worker exposure other than through the use of engineering controls or personal protective equipment (PPE), i.e., training or rotation of workers from job to job. Contrast with **Engineering Controls, Personal Hygiene** and **PPE**.

Adsorption. The attachment of molecules (or atoms if elemental) to the surface of another substance.

Aerosol. A fine suspension in air (or any other gas) of liquid (**mist, fog**) or solid (**dust, fume, smoke**) particles which are sufficiently small in size to be stable and remain airborne (and frequently respirable).

Agent. In this context, any chemical substance, physical force, ionizing or nonionizing radiation, biological organism, or other influence that affects the body.

AHERA. Asbestos Hazard Emergency Response Act; legislation to regulate asbestos in schools.

AIHA. The American Industrial Hygiene Association, a professional industrial hygiene society of over 7000 persons practicing in the science of industrial hygiene. AIHA, P.O. Box 8390, 345 White Pond Dr., Akron, OH, 44320 (216) 873-2442.

Air Bone Gap. The difference (in decibels) between the hearing levels for sound, at a particular frequency, determined by air conduction and bone conduction-threshold measurement.

Air Conduction. The process by which sound is conducted to the inner ear through the air in the outer ear canal utilizing the tympanic membrane and the ossicles as part of the pathway.

Air Flow Rate. See **Flow Rate.**

Air-Line Respirators. A respiratory protection system that supplies fresh air to the user through airlines or hoses which tie into an air compressor located in a clean environment.

Air Sampling. In general, air is pumped through various media (Filter, Sorbent and Impinger are examples) such that the airborne contaminants are captured for chemical analysis. If the volume of air pumped is known, the amount of chemical contaminant captured can be expressed as a concentration (**ppm or mg/m^3**).

ALARA. An acronym for As Low As Reasonably Achievable.

Alkali. Also see **Base**. In general, any compound having highly basic properties (one that readily ionizes in aqueous solution to yield OH anions, with a pH above 7, and that turns litmus paper blue). Alkalies are caustic. Common commercial alkalies are sodium carbonate (soda ash), NaOH, lye, potash, caustic soda, KOH, water glass, and bicarbonate of soda. See also **Acid**; **pH**.

Allergic Reaction. An abnormal physiological response to a chemical or physical stimulus.

Ambient Noise. The "background" noise associated with a given environment, being usually a composite of sounds from many sources near and far.

Ambient Temperature. The temperature of the air around an individual's body.

American Board of Industrial Hygiene. See **ABIH**.

American Conference of Governmental Industrial Hygienists. See **ACGIH**.

American Industrial Hygiene Association. See **AIHA.**

American Public Health Association. See **APHA.**

American Society of Heating, Refrigerating, and Air Conditioning. See **ASHRAE.**

Amps. Amperage is the unit of current flow in an electrical circuit.

Analytical Methods. The chemical or physical analytical techniques used to measure the presence or amount of an **agent** in the workplace, usually in the air.

Anemometer. An instrument used to measure air flow velocity.

Anesthetic. Any substance that causes a loss of sensation.

Anhydrous. Without water. A substance in which no water molecules are present in the form of a hydrate or as water of crystallization.

Annoyance. With respect to noise, high-pitched (approximately 1500 to 10,000 Hz) noise is more annoying than low-pitched noise. Also, intermittent or irregular noise may be considerably more annoying than a steady noise of the same intensity.

ANSI. American National Standards Institute. A privately funded, voluntary membership organization that identifies industrial and public need for national consensus standards and coordinates their development. Many ANSI standards relate to safe design/performance of equipment and safe practices or procedures. ANSI, 1430 Broadway, New York, NY, 10018 (212) 642-4900.

Antidote. A substance that can counter the effects of a poison.

Antislip Index. A scale of reference measured by the Horizontal Pull Slipmeter (HPS) which may be converted to (**COF**) readings by moving the decimal point one place to the right (for example, a 0.515 slip index = 5.15 COF).

APCA. The former Air Pollution Control Association; see **AWMA**.

APHA. American Public Health Association. APHA, 1015 15th Street, Washington, D.C., 20005 (202) 789-5600.

Aqueous, aq. Describes a water-based solution or suspension or a gaseous compound dissolved in water. Contrast with **anhydrous**.

Area Sample. A sample collected at a fixed point in the workplace. The data from the area sample may or may not correlate with an individual's personal sample results (due to the often high degree of variability in exposures).

Article. Any manufactured item whose function is dependent on its shape or design (for example, pencils or pens). Unless they give off dust or fumes, articles are excluded from hazard laws. An article does not release or result in exposure to a hazardous material in normal use.

ASHRAE. American Society of Heating, Refrigerating and Air Conditioning Engineers, a professional society interested in all aspects of heating and cooling of occupied spaces. ASHRAE, 1791 Tullie Circle NE, Atlanta, GA, 30329 (404) 636-8400.

ASP. Associate Safety Professional, indicating partial certification toward the **CSP**.

Asphyxiant. A vapor or gas that can cause unconsciousness or death through suffocation (lack of oxygen). Most *simple asphyxiants* are harmful to the body only when they become so concentrated that they displace the available oxygen in the air (normally about 21%) to dangerous levels (19.5% or lower). Examples of simple asphyxiants are CO_2, N_2, H_2, and He. Other asphyxiants are *chemical asphyxiants* like carbon monoxide (CO) or cyanide, which reduce the blood's ability to carry oxygen.

ASSE. American Society of Safety Engineers, a professional society of nearly 23,000 members dedicated to the advancement of the safety profession. ASSE, 1800 E. Oakton St., Des Plaines, IL, 60018-2187 (312) 692-4121.

ASTM. The American Society for Testing and Materials. The ASTM is a voluntary membership organization whose members devise consensus standards for materials characterization and use. ASTM, 1916 Race Street, Philadelphia, PA, 19103 (215) 299-5400.

Asymptomatic. The lack of identifiable symptoms.

atm. Atmosphere. 1 atm = 14.7 $lb/in.^2$ 1 atm represents the pressure exerted by the air at sea level that will support a column of mercury 760 mm high (29.92 in. Hg) and expressed as 760 mmHg. 1 torr = 1 mmHg.

Attitude. A relatively constant or stable feeling directed toward an object or phenomenon in the workplace that includes an evaluative aspect. Attitudes inherently imply a good/bad or positive/negative dimension. See also **Belief**.

Audiogram. A chart or table relating hearing level (for pure tones) to frequency. A record of hearing loss or hearing level measured at several different frequencies, usually 500 to 6000 Hz. The audiogram may be presented graphically or numerically.

Auditory Nerve. The eighth cranial nerve.

Aural Temperature. Temperature pertaining to the ear; usually the temperature of the external acoustic canal.

Autoignition Temperature. In general, the minimum temperature to which a substance (usually in an enclosure) must be heated, without the application of a flame or spark, which will cause that substance to ignite.

Avulsion. Removal of a body part by accident.

AWMA. Air and Waste Management Association, new name for the **APCA**, a professional society interested in environmental pollution issues. AWMA, P.O. Box 2861, Pittsburgh, PA, 15230 (412) 232-3444

B

Background Noise. The total of all sources of interference in a system used for the production, detection, measurement, or recording of a signal, independent of the presence of the signal.

Baluster. A vertical support member that supports a handrail or guardrail.

Balustrade. A railing with supporting balusters.

Band-center Frequency. The designated (geometric) mean frequency of a band of noise or other signal. For example, 1000 Hz is the band center frequency for the octave band that extends from 707 to 1414 Hz, or for the third octave band that extends from 891 to 1123 Hz.

Base. A substance that liberates OH anions when dissolved in water, receives a hydrogenation from a strong acid to form a weaker acid, and gives up two electrons to an acid. Bases react with acids to form salts and water. Bases have a pH >7 and turn litmus paper blue. They may be corrosive to human tissue and are to be handled with care. See also **Alkali**, **Acid**, **pH**.

BCSP. Board of Certified Safety Professionals, a professional safety accreditation organization, administering the **CSP** and **OHST** program. BCSP, 208 Burwash Ave., Savoy, IL, 61874-9510 (217) 359-9263.

BEI, Biological Exposure Indexes. Procedures to determine the amount of material absorbed into the human body by measuring it (or its metabolic products) in tissue, fluid, or exhaled air. See the **ACGIH** publication, "Documentation of the Threshold Limit Values and Biological Exposure Indices," for a full explanation.

Belief. A conviction that a phenomenon or object is true; a belief does not necessarily include a good/bad or positive/negative dimension. See also **Attitude**.

Bias. A systematic error inherent in a sampling or analytical method.

Bioassay. A test to determine the potency of a substance at producing some adverse health effect on a biological system.

Biological Agent. Disease-causing organisms which can cause infection if inhaled, ingested, or absorbed through mucous membranes or impaired skin.

Biological Monitoring. Also see **BEI**. Certain chemicals, if absorbed by the human body, can be detected as metabolites in blood, urine, etc. Periodic examination of body substances can determine the extent of a worker's hazardous material absorption or exposure. Occupational exposures to lead can be monitored by performing blood analysis.

Biological Time Constant. The time required for a portion of an absorbed dose to undergo metabolic changes in the body, for example, the biological half-life would be the time required for one half of an absorbed dose to be biotransformed.

Blast Gate. A sliding valve located in **ductwork** to regulate flow.

Body Burden. The total amount of a chemical that a person has ingested or inhaled, from all sources, over time. This includes any source of exposure whether occupational or nonoccupational, and is the net sum of all exposures and excretions, etc.

BOHS. British Occupational Hygiene Society, 1 St. Andrew's Place, Regent's Park, London NW1 4LB, England (01-486-4860).

Boiling Point, BP. The temperature at which the vapor pressure of a liquid is equal to the surrounding atmospheric pressure, i.e., the temperature at which it boils. If a chemical is a mixture of different liquids a range of boiling points can be given.

Bonding. A fire safety practice where two objects (tanks, cylinders, etc.) are fastened together with clamps and bare wire. Electrically connecting two metal containers or surfaces with an electrical strap. This equalizes the electrical potential between the metal containers, etc., and helps prevent static sparks that could ignite flammable materials. See also **Grounding**.

Bone Conduction. The process by which sound is conducted to the inner ear through the cranial bones.

BP. See **Boiling Point.**

Breakthrough. The situation in which contaminated air is passed through a sorbent system, yet a portion of the contamination is not captured and passes out of the sorbent system. When this term is used to refer to chemical cartridge respirators, the implication is that contaminated air is allowed into the facepiece of the respirator. Breakthrough may occur when the sorbent system of the respirator has become depleted, or if the type of sorbent system used is not appropriate for protection against the hazardous **agent** in the workroom **air**.

Breathing Zone. The zone of air from which the worker breathes; not rigidly defined but usually understood to be the volume in front of the worker's face, chest, etc.

British Occupational Hygiene Society. See **BOHS.**

Btu. British thermal unit. The quantity of heat required to raise the temperature of 1 lb of H_2O 1°F at 39.2°F. See also **Calorie**.

Buffer. A substance that reduces the change in hydrogen ion concentration (i.e., change in **pH**) that otherwise would be produced by adding acids or bases to a solution.

C

C. is used by **OSHA** and **ACGIH** to designate ceiling exposure limit. Less commonly, "C" can also mean continuous exposure if used in the expression of toxicological data, for example, "LD_{50} >10 g/kg, 24 H-C" would mean that 10 g/kg was the LD_{50} resulting from continuous exposure for 24 hr. See also **Ceiling Limit**; **TLV**.

°C. Degrees centigrade. Degrees Fahrenheit equals (°C × 1.8) + 32. Degrees centigrade equals (°F = 32) × 5/9. See also **°F**.

CAA. The Clean Air Act (EPA). Public Law PL 91-604, 40 CFR 5080. The CAA sets and monitors airborne pollution hazardous to public health or natural resources. Enforcement and issuance of discharge permits are carried out by the states and are called state implementation plans.

CAER. Community Awareness and Emergency Response program of the **Chemical Manufacturers Association.**

Calorie. A standard unit of heat. A calorie is the amount of heat required to raise 1 g of water 1°C. See also **Btu**.

Capture Velocity. Airflow velocity needed to overcome contaminant movement and cross-drafts in order to draw airborne contaminants into an exhaust hood.

Carcinogen. Used in general to refer to a material that has been found to cause cancer. There are different distinctions in types of carcinogens, for example, a chemical could be an **animal carcinogen,** a **suspect human carcinogen,** or a **human carcinogen.** The designations are a function of the available exposure and laboratory test data. In general, a material is considered to be a carcinogen if it has been evaluated by the International Agency for Research on Cancer (**IARC**) and found to be a carcinogen or potential carcinogen, if it is listed as a carcinogen or potential carcinogen in the "Annual Report on Carcinogens," published by the National Toxicology Program (**NTP**) latest edition, if it is regulated by **OSHA** as a carcinogen, or if one positive study has been published.

Carotid Artery. Principal artery of the neck.

Carpal Tunnel Syndrome. A wrist injury caused by repetitive motion, such as frequent, rapid or forceful movements.

CAS Registration Number, CAS, CAS#. CAS stands for Chemical Abstracts Service. CAS numbers identify "specific" chemicals and are assigned sequentially. Chemicals may have more than one name but only one CAS number. For example, methyl ethyl ketone and 2-butanone are different names for the same chemical (and thus both have a CAS# of 78-93-3). The CAS number is useful because it is a concise, unique means of material identification. Chemical Abstracts Service, Division of the American Chemical Society, Box 3012, Columbus, OH, 43210 (614) 421-3600.

Case. An injured or infected, etc., worker, test subject, or test animal.

Case-Control Study. A type of epidemiologic study design in which a group of individuals with a disease is identified (**cases**) and an appropriate group of individuals without the disease is identified (**controls**) so that the past exposure experiences of the two groups can be compared.

Case Fatality Rate. A measure of the frequency of deaths, due to a particular cause, among all **cases.**

Case-History. See **Case-Control.**

Caustic. See Alkali.

cc, cm^3. Cubic centimeter.

Ceiling. See **Ceiling Limit.**

Ceiling Limit, C. The concentration that should not be exceeded during any part of the working exposure. See also **PEL-C** and **TLV-C**.

Centigrade. See °C.

CERCLA. The Comprehensive Environmental Response, Compensation, and Liability Act, a.k.a. The Superfund Law, Public Law PL 96-510, 40 CFR 300 (EPA). CERCLA is a major environmental legislative action to control hazardous waste sites. It provides for the identification and the cleanup of the hazardous materials that have been released over the land and into the air, waterways, and groundwater. It covers areas affected by newly released materials and older leaking or abandoned dump sites. CERCLA established the Superfund, a trust fund to help pay for the cleanup and sites where hazardous materials have been released. Report releases of hazardous materials to the National Response Center (800) 424-8802; the EPA Superfund Information Hotline is (800) 535-0202.

CFR. Code of Federal Regulations. Regulations established by law. Contact the agency that issued the regulation for details, interpretations, etc. Copies are sold by the Superintendent of Documents, Government Printing Office, Washington, D.C., 20402 (202) 783-3238.

CGA. The Compressed Gas Association, Inc., 500 Fifth Ave., New York, NY, 10036 (212) 354-1130.

Charcoal Tube. A type of **Solid Sorbent Air Sampling** media.

Chemical Asphyxiant. See **Asphyxiant.**

Chemical Cartridge Respirator. A respirator able to purify inhaled air of certain gases and vapors. Contrast with **HEPA**.

Chemical Family. A group of compounds with a common general name; e. g., acetone, methyl ethyl ketone (MEK), and methyl isobutyl ketone (MIBK) are members of the ketone family.

Chemical Formula. States formula for a molecule, i.e., the number and kind of atoms that comprise a molecule of a material. The chemical formula of water is H_2O (each molecule of water is made up of 2 atoms of hydrogen and 1 of oxygen).

Chemical Manufacturers Association. See **CMA.**

Chemical Name. The scientific designation of a chemical or a name that will clearly identify the chemical for hazard evaluation purposes. Also see **CAS number**.

Chemical Reactivity. Refers to the ability of a material to chemically change. Undesirable and dangerous effects such as heat, explosions or the production of noxious substances can result.

Chemical Transportation Emergency Center. See **CHEMTREC**.

CHEMTREC. Chemical Transportation Emergency Center. Established by the Chemical Manufacturers Association (CMA) to provide emergency information on materials involved in transportation accidents. Twenty-four-hour number: (800) 424-9300. In Washington, D.C., Alaska, and Hawaii call (202) 483-7616.

Chronic. In this context, anything of long duration. Also usually of frequent occurrence and at low levels.

Chronic Health Effect. An adverse health effect as a result of a long or continuous exposure (usually a low level exposure). Contrast with Acute Health Effect.

Chronic Toxicity. Chronic effects result from repeated exposures to a material over a relatively prolonged period of time.

CIH. Certified Industrial Hygienist. An individual who has passed the certification requirements of the American Board of Industrial Hygiene (see **ABIH**).

Clean Air Act. See CAA.

CMA. The Chemical Manufacturers Association; an industry group. Chemical Manufacturers Association, 2501 M Street, N.W., Washington, D.C., 20037 (202) 887-1216; (202) 887-1100.

CNS. Central Nervous System. "CNS" on an **MSDS** indicates effects on the CNS by the material and can include effects such as headache, tremors, drowsiness, convulsions, hypnosis, anesthesia, nervousness, irritability, narcosis, dizziness, fatigue, lethargy, peripheral myopathy, memory loss, impaired concentration, sleep disturbance, etc.

Cochlea. The auditory part of the internal ear, shaped like a snail shell, containing the basilar membrane on which are distributed the end organs of the acoustic or eighth cranial nerve.

Code of Federal Regulations. See **CFR**.

Coefficient of Friction (COF). The ratio of horizontal force required to move one level surface over the other to the total vertical force perperdicularly applied to the two surfaces. COF = horizontal force/vertical force. The COF serves as an indicator of the relative friction between two level surfaces in contact.

Coefficient of Variation, CV. The sample standard deviation divided by the mean, often expressed as a percent.

Coefficient of Water/Oil Distribution. Also called the partition coefficient, this is the ratio of the solubility of a chemical in oil to its solubility in water. The coefficient is used to indicate how easily a material can be absorbed into or stored by the human body.

Cohort. A group of persons selected for a study.

Cohort Study. A type of epidemiologic study design in which disease-free individuals are followed over time and comparisons are made between the exposure experiences of those who subsequently develop disease and those who do not.

Combustible. Used by the **NFPA**, **DOT**, and others to classify on the basis of flash points liquids that will burn. Both NFPA and OSHA define combustible liquids as having a flash point of 100°F (38°C) or higher. Nonliquid materials such as wood and paper are classified as ordinary combustibles by the NFPA. OSHA defines combustible liquid within the Hazard Communication Standard as any liquid having a flash point at or above 100°F (38°C) but below 200°F (93.3°C). Also, any mixture having components with flash points of 200°F (93.3°C) or higher (the total volumes of which make up 99% or more of the total volume of the mixture).

Combustible Liquid. A liquid that has a flash point at or above 140°F.

Common Name. A designation for a material other than its chemical name, such as code name or code number or trade, brand, or generic name.

Common Wire. The electrical return wire (usually has white insulation) that is used with the hot electrical wires. Also known as the system ground.

Communicable Disease. A disease or illness caused by the transmission of an infectious agent, or its toxic products, from a reservoir to a susceptible host.

Community Noise Equivalent Level (CNEL). Community noise equivalent level is a scale that takes account of all the A-weighted acoustic energy received at a point, from all noise events causing noise levels above some prescribed value. Weighting factors place greater importance upon noise events occurring during the evening hours (7 to 10 p.m.) and even greater importance upon noise events at night (10 p.m. to 6 a.m.).

Compliance Monitoring. Usually, measuring worker exposures for the purpose of evaluating an employer's compliance with governmental standards.

Compressed Gas. A gas (or mixture of gases) in a closed container having an absolute pressure exceeding 40 psi at 70°F (21. 2°C) or having an absolute pressure exceeding 104 psi at 130°F (54.4°C), regardless of the pressure at 70°F (21.2°C); or, a liquid having a vapor pressure exceeding 40 psi or 100°F (38°C) as determined by **ASTM** D-323-72.

Compressed Gas Association. See **CGA**.

conc. Concentration.

Concentration. The quantity of a contaminant per volume of air; see also **ppm** and **mg/m^3**.

Confidence Interval. For sample data, the range of values about a given number with a probability (for example, 95%) of including the true population value in an underlying distribution.

Confidence Level. The probability that a given confidence interval will include the real population parameter.

Confidence Limits. The upper and lower limits of a confidence interval.

Consumer Product. Any product such as medicines packaged and distributed for retail sale and for consumption by individuals or households. Consumer products are exempt from **HCS** if used in the home or if used in the workplace in the same manner as in normal consumer use; repeated or extensive use would not be exempt.

Consumer Products Safety Commission. See **CPSC**.

Continuous Air Monitors, CAMs. Instrumentation that can be used to continuously measure air concentrations in the workplace (as opposed to area samples and personal samples which are often only taken for a day or part of a day at a time).

Contraindications. Something, such as a symptom or condition, which makes a specific treatment or procedure inadvisable.

Controls. In general, equipment to reduce the amount of contaminant released into the workplace air (or to remove it from the workplace air), etc. See **Administrative Controls** and **Engineering Controls.**

Core Temperature. Temperature of the central part of the body.

Corrosive. A chemical that causes visible destruction of living tissue or irreversible alterations in living tissue by chemical action at the site of contact. Any liquid that causes a severe corrosion rate in steel. Any waste that exhibits a characteristic of corrosivity (40 CFR 261.22) as defined by **RCRA**, may be regulated by EPA as a hazardous waste.

CPSC. Consumer Products Safety Commission, a federal agency responsible for regulating hazardous materials when they are used in consumer goods (per the Hazardous Substances Act and Poison Prevention Packaging Act of 1970).

Criteria Document. In-depth toxicological review of a hazardous chemical or physical **agent,** published by **NIOSH**.

Critical Pressure/Critical Temperature. A temperature above which a gas cannot be liquefied by pressure. The critical pressure is that pressure required to liquefy a gas at its critical temperature.

Cross-Sectional Study. A type of epidemiologic study design in which a group of individuals are selected at a single point in time and comparisons are made between the exposure experiences of those who have the disease and those who do not.

Cryogenic. Extremely low temperature (such as for refrigerated gases).

CSP. A Certified Safety Professional, as certified by the **BCSP**.

Cutaneous. Pertaining to the skin.

CVS. Cardiovascular effects.

Cycle Per Second (CPS). A unit of frequency. The preferred terminology is hertz, abbreviated **Hz.**

Cyclone. An **Air Sampling** device used to collect the **Respirable Fraction** of an airborne particulate.

Cylinder. In this context, a container of various sizes used for he storage of compressed gas. See **CGA.**

D

Damping. Any means of dissipating or attenuating vibrational energy within a vibrating medium. Usually the energy is converted to hear. Any influence which extracts energy from a vibrating system is known as damping.

Dangerously Reactive Material. A material that can react by itself or with air or water to produce a hazardous condition. Preventive measures can be taken if the conditions that can cause the dangerous reaction are known.

Dec, Decomp; Decompose, Decomposition. The breakdown of a material (by heat, chemical reaction, electrolysis, decay, or other processes).

Decibel (dB). A nondimensional unit used to express sound levels. It is a logarithmic expression of the ratio of a measured quantity to a reference quantity. In audiometry, a level of zero decibels represents roughly the weakest sound that can be heard by a person with good hearing. dBA — A sound level reading in decibels made on the A-weighted network of a sound level meter. dBC — A sound level reading in decibels made on the C-weighted network of a sound level meter. See also **Weighting network.**

Delivered Dose. Refers to the amount of a substance available for interaction in a particular organ or cell.

Demulcent. A material capable of soothing or protecting inflamed, irritated mucous membranes.

Density. Ratio of weight (mass) to volume of a material, usually expressed in grams per cubic centimeter. One cc of H_2O weighs 1 g/cm^3. See also **Specific Gravity**.

Dermal. Pertaining to the skin.

Dermal Toxicity. Adverse effects resulting from the skin's exposure to a material.

Dermatitis. Inflammation of the skin.

Descriptive Statistics. Simple statistical ways to characterize the exposure data obtained in a sample (for example, mean, standard deviation, etc.).

Dewpoint Temperature. The temperature at which water vapor in the air first starts to condense.

Die. With respect to stamping presses, the piece of the press that cuts the material into the shape of the final part.

Diffusion. Dispersion of sound within a space so that there is uniform energy density throughout the space.

Dilution Ventilation. See also **General Ventilation;** ventilation designed to control airborne contaminants by diluting concentrations to acceptable levels.

Disconnect Boxes. Switch boxes used to disconnect machinery from the electrical power source.

Disease. A deterioration of the normal, healthy health status. Also see **Occupational Disease**.

Dispensing Drum. Large drum containing chemicals used as a reservoir to fill portable containers. Usually 55 gal drums are fitted with spigots and set on stands horizontally to fill smaller containers.

Dose. The amount of a substance that enters the worker's body; see **absorbed dose** and **delivered dose.** Frequently, **dose** means the same as absorbed dose.

Dose Rate. The dose per unit time.

Dose-Response. The relationship between dose and health effect (sometimes between exposure and health effect). In general, the curve that results when dose (X-axis) is plotted against response (Y-axis).

Dosimeter. A device worn on the person for determining the accumulated sound exposure with regard to level and time. In general, a dosimeter is any individually worn device for measuring exposure to chemical or physical **agents.**

DOT. U.S. Department of Transportation. Regulates the transportation of materials in order to protect the public as well as fire, law, and other emergency-response personnel. DOT classifications specify appropriate warnings and labels that must be used. DOT Locator, 400 7th Street, SW, Washington, D.C., 20590 (202) 366-4000.

Double Insulated. Electrical appliances and tools that have an additional layer of insulation in the form of plastic coating of the tool or an internal insulated layer.

Dry Bulb Temperature. See **Ambient Temperature.**

Ductwork. A closed pathway for ventilation airflow that connects system components from intake to discharge.

Dust. Solid particles suspended in air (usually produced by some mechanical process such as crushing, grinding, abrading, or blasting). Most dusts are hazardous with respect to inhalation, fire, and dust explosion.

Dynamic friction. The resistance to movement, once sliding has begun, between the two sliding surfaces in contact.

E

EAP. Employee Assistance Programs.

Ear Protector. A device worn to reduce the passage of ambient noise into the auditory system. Earplugs are inserted in the external ear canal. Earmuffs fit over the entire ear and snug against the head.

EC_{50}. Median effective concentration. The concentration of a material in water that has been observed to cause a biological effect on 50% of a group of test animals.

Effective Exposure. With respect to respirators, the concentration of a contaminant inside the **respirator**.

Effective Temperature. An index used for estimating the effect of temperature, humidity, and air movement on the subjective sensation of warmth.

EHS. A substance determined by the U.S. EPA to be so potentially hazardous to life and health that local planning committees are required to be notified if an EHS is present in a **facility.** See **SARA.**

Electrolyte. A nonmetallic substance that conducts an electric current in solution by the movement of ions.

Embryotoxin. A material harmful to a developing embryo.

Emergency Response. Mitigation of a spill or leak of a hazardous material.

Emergency Response Planning Guideline. See **ERPG**.

Emetic. An **agent** that induces vomiting.

Endemic. The presence of a disease or infectious agent within a specific geographic area.

Engineering Control. Environmental, mechanical, or structural factors that serve to encourage, facilitate, or complement safe and healthful behaviors in the worksite. Engineering controls work by removing the worker from the hazard or by removing the hazard from the work environment (using ventilation in particular). Compare with **Administrative Control, Personal Hygiene, PPE, Ventilation.**

EPA; U.S. Environmental Protection Agency. A federal agency with environmental protection authority (regulatory and enforcement). EPA, 400 M Street, SW, Washington, D.C., 20460, EPA Locator (202) 382-2090. Public Information (202) 382-2080, EPA main library (202) 382-5921.

EPCRA. Emergency Planning and Community Right-to-Know Act. The specific title for **SARA Title III.**

Epidemic. An illness or illnesses that occur at a greater rate than normally expected and prevalent in a particular community or region.

Epidemiology. The study of disease occurrence in populations with the aim of relating disease occurrence to characteristics of people and their environment.

Equipment Ground. The return wire (usually white colored insulation) to electrical systems that are with the hot electrical wires. Also known as the system ground.

ERC. Educational Resource Center. ERCs are **NIOSH**-funded interdisciplinary graduate school occupational safety and health training programs.

Ergonomics. The concept that human bodies are components of the workplace, with their own limitations. The study of human characteristics for the appropriate design of living and work environments.

ERPG. Emergency Response Planning Guidelines. An **AIHA** exposure guideline dealing with nonoccupational community exposures to chemicals (i.e., resulting from chemical spills due to accidents, etc.). ERPGs are short-term guidelines for exposure to chemicals during spills or emergencies. ERPGs provide concentration levels at which adverse health effects are expected for the general public. ERPGs are defined as follows:

ERPG-1 is the maximum airborne concentration below which it is believed that nearly all individuals could be exposed for up to one hour without experiencing other than mild transient health effects or perceiving a clearly defined objectionable odor.

ERPG-2 is the maximum airborne concentration below which it is believed that nearly all individuals could be exposed for up to 1 hr without experiencing or developing irreversible or other serious health effects or symptoms that could impair an individual's ability to take corrective action.

ERPG-3 is the maximum airborne concentration below which it is believed that nearly all individuals could be exposed for up to 1 hr without experiencing or developing life-threatening effects.

Etiology. The causes and origins of disease(s).

Evaporation Rate. The rate at which a material will vaporize, volatilize, or evaporate from the liquid or solid state (compared to the rate of vaporization of a known, baseline material). The known material is usually normal butyl acetate, with an evaporization rate of 1.0. Faster evaporization rates are >1 (slower, <1).

Evaporative Heat Loss. The loss of body heat to the environment through the evaporation of sweat.

Excursion Limit. Usually short duration exposure limits (such as **STEL**s or **Cs**).

Explosive. A material that explodes when subjected to abrupt shock, pressure, or high temperature.

Explosive Limits. See **Flammable Limits**.

Exposure. A worker's contact with agents in the workplace. Exposure can be reduced if engineering controls are effectively in place and the worker wears the proper **PPE**.

Exposure Assessment. The study and measurement of exposure, usually including its magnitude, frequency, duration, and route.

Exposure Group. See **Homogeneous Exposure Group**.

Exposure Limits. The concentration of a chemical in the workplace air that is thought to be safe. See **PEL** and **TLV**.

Exposure Route. The manner by which an **agent** enters the body, for example, inhalation, ingestion, or skin absorption.

Extension Ladders. Nonself-supporting ladder consisting of two sections with one sliding within the other.

Extinguishing Media, Agents. The type of fire extinguisher or extinguishing method appropriate for use on a specific material. Some chemicals react violently to the presence of water, so other methods such as the use of foam or CO_2 should be followed.

Extremely Hazardous Substance List. See also **EHS**. Any of approximately 400 chemicals listed under **SARA Title III** as extremely hazardous.

Eye Protection. In general, anything worn to protect the eyes (safety glasses, goggles, face shields, etc.).

F

°F or F. Degrees Fahrenheit. 32 °F (0°C) equals the freezing point of water. 212°F (100°C) equals the boiling point of water. Degrees Fahrenheit equals (°C × 1.8) + 32. Degrees centigrade equals (°F – 32) × 0.56.

Face Velocity. The average velocity across an opening (face) of a local exhaust hood.

Facility. As defined for **SARA Title III,** any building, equipment, structure, and/or other stationary items located on single, adjacent, or contiguous sites.

Fan. A mechanical device that rotates to cause a pressure differential and resulting air movement.

Fast Response. A selectable mode of operation of a sound level meter or analyzer in which the needle indicator has minimum damping and can therefore respond rapidly to changes in sound level.

Federal Register. See **FR**, also **CFR**.

Feed Opening. The opening in a guard that is used to put the part or material into the press to be formed or cut.

Federal Emergency Management Agency. See **FEMA**.

FEMA. The Federal Emergency Management Agency, 500 C Street S.W., Washington, D.C., 20472 (202) 646-4600.

Fibrillation. A condition of the heart when the two halves are beating out of synchronization and are not pumping blood. The heart can be shocked into this condition from contact with electrical current.

Fibrosis. The formation of fibrous tissue in excess of amounts normally present in the lung tissue walls. This reduces the oxygen and CO_2 exchange efficiency.

Field, Far. That part of a sound field for which spherical divergence occurs; i.e., SPL decreases by –6 dB for each doubling of distance. As a general rule, it is also considered as that part of a sound field that is beyond a distance of 3 to 4 times the largest dimension of the source or greater than the maximum wavelength of sound for the lowest frequency of interest.

Filter or Filter Collection. In this context, using a filter to collect airborne particulates so that they can be chemically analyzed.

Fire Diamond (NFPA). The fire diamond has four classes of entries by position:

Position A — Health Hazard (Blue):

0	=	Ordinary Combustible Hazards in a Fire
1	=	Slightly Hazardous
2	=	Hazardous
3	=	Extreme Danger
4	=	Deadly

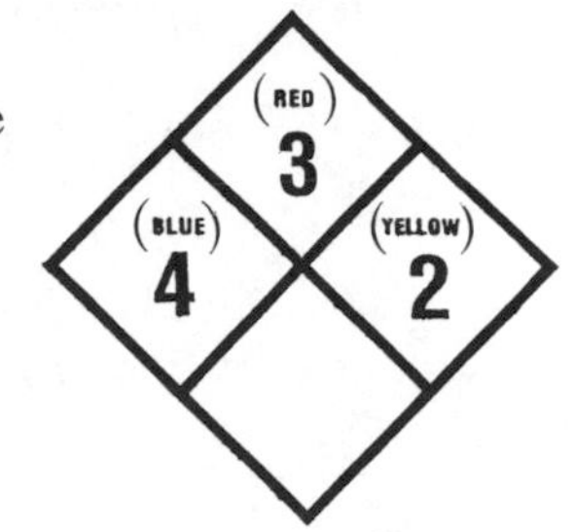

Position B — Flammability (Red):
0 = Will Not Burn
1 = Will Ignite if Preheated
2 = Will Ignite if Moderately Heated
3 = Will Ignite at Most Ambient Conditions
4 = Burns Readily at Ambient Conditions

Position C — Reactivity (Yellow):
0 = Stable and Not Reactive with Water
1 = Unstable if Heated
2 = Violent Chemical Change
3 = Shock and Heat May Detonate
4 = May Detonate

Position D — Specific Hazard (White):
OXY = Oxidizer
ACID = Acid
ALKALI = Alkali
COR = Corrosive
W̶ = Use No Water
☢ = Radiation Hazard

Fit Test. Refers to the testing of respirators (with respect to how a particular respirator fits particular worker's face). See **Qualitative** and **Quantitative Fit Testing**.

Flammable. Describes any solid, liquid, vapor, or gas that will ignite easily and burn rapidly. See also **Combustible**.

Flammable Aerosol. An aerosol is considered to be a flammable aerosol if it is packaged in an aerosol container and can release a flammable material.

Flammable Gas. A gas which, at ambient temperature and pressure, forms a flammable mixture in air at a concentration of 13% by volume or less, or a gas which at ambient temperature and pressure forms a range of flammable mixtures in air greater than 12% by volume, regardless of the lower limit.

Flammable Limits (Flammability Limits). The minimum and maximum concentrations of a flammable gas or vapor at which ignition can occur. Concentrations below the lower flammable limit (LFL) are too lean to burn, while concentrations above the upper flammable limit (UFL) are too rich. All concentrations between LFL and UFL are in the flammable range, and special precautions are needed to prevent ignition or explosion.

Flammable Liquid. A liquid that gives off vapors that can be readily ignited at room temperature. Defined by the **NFPA** and **DOT** as a liquid with a flash point below 100°F (38°C).

Flammable Solid. A solid that will ignite readily and continue to burn.

Flanking Paths. Transmission paths that transmit acoustic energy around a sound barrier; paths that bypass the intended barrier.

Flash Back. This takes place when a trail of flammable material is ignited by a distant spark or ignition source (the flame then travels along the trail of the material back to its source).

Flash Point, FP. The lowest temperature at which a flammable liquid gives off sufficient vapor to form an ignitable mixture with air near its surface or within a vessel.

Flow Rate. Generally used to mean the rate of air flow through an Air Sampling pump; usually expressed in L/min. Also can refer to the volumetric rate of air flow through a ventilation system.

Flywheel. A large wheel on mechanical punch presses that keeps the ram moving after it has hit the material. It stores mechanical energy in the form of kinetic centrifugal energy of a spinning wheel.

Flux. Rate of transfer of fluid, particles, or energy across a given surface.

Foam. A fire fighting material consisting of small bubbles of air, water, and concentrating agents. Foam can put out a fire by blanketing it, excluding air and blocking the escape of volatile vapor.

Fog. A visible suspension of fine droplets in a gas; e.g., water in air.

Fomite. An inanimate vector of disease (a comb is a fomite).

Form R. An EPA report required under **SARA Title III** for annual reporting of releases of toxic chemicals from an industrial facility.

Formula Weight. See **Molecular Weight**.

FP. See **Flash Point**.

FR. The "Federal Register." A daily publication that lists and discusses daily governmental business including regulations of federal agencies. Compare to **CFR**.

Free Field. A free sound field is a field in a homogeneous, isotropic medium free from boundaries. In practice it is a field in which the effects of the boundaries are negligible over the region of interest.

Freezing Point. The temperature at which a material changes its physical state from liquid to solid.

Frequency. The time rate of repetition of a periodic phenomenon. The frequency is the reciprocal of the period. In sound (as in electricity) the unit of measurement for frequency is the hertz (**Hz**) which equals one complete waveform (cycle) per second.

Frequency Analyzer. Electrical apparatus capable of measuring the acoustic energy present in various frequency bands of a complex sound.

Frequency Bands. A division of the audible range of frequencies into subgroups for detailed analysis of sound

Friction. The resistance to movement between two surfaces in contact, for example, resistance to air movement over surfaces (e.g., **ductwork**) which causes pressure loss.

Full Protective Clothing. Fully protective gear that keeps gases, vapor, liquids, and solids (dusts, etc.) away from any contact with skin and prevents them from being inhaled or ingested. Includes **SCBA** (self-contained breathing apparatus).

Fume. Fume is the particulate that is formed by molten metals or molten plastics. It is made up of minute solid particles suspended in air. This heating is often accompanied by a chemical reaction where the particles react with oxygen to form an oxide.

Fundamental Mode of Vibration. The fundamental mode of vibration of a system is the mode having the lowest natural frequency.

Fuse Pullers. A tool with a long pole that is used to pull fuses out of fuse boxes. Used with electrical currents above 600 V.

G

g or gm. Gram.

Gas. A formless fluid that occupies the extent of the volume which contains it. It can be changed to its liquid or solid state only by increased pressure and decreased temperature.

Gate Guard. A guard that uses a door or gate through which the material or parts are fed to the machine.

General Exhaust. See **General Ventilation**.

General Ventilation. Also known as dilution ventilation. Both terms refer to the removal of contaminated air and its replacement with clean air from the general workplace. Contrast with **local ventilation**.

Generic Name. An appellation such as a code name, code number, trade name, or brand name used to identify a chemical by other than its chemical name.

Globe Temperature. The temperature inside a hollow, blackened copper sphere; used to measure radiant heat.

Grain Auger. The screw conveyor at the bottom of a grain bin that is used to empty the bin.

GRAS. Generally Recognized As Safe (a phrase applied to food additives approved by the FDA).

Ground Fault Circuit Interrupters (GFCI). An electrical device that opens a circuit when current leakage is sensed by the device, thus preventing electrical shocks to workers.

Grounding. A fire safety practice to conduct any electrical charge to the ground, preventing sparks that could ignite flammable materials. Electrically connecting a metal container to the building ground with an electrical strap. See **Bonding**.

Guard Rail. A protective barrier railing that is provided to prevent persons from falling from an open side.

H

Hair Cells. Sensory receptors for sound stimuli located in the basilar membrane in the ear.

Hand Protection. PPE worn to protect the hands (such as gloves).

Handrail. A single support member mounted on brackets of a wall or partition paralleling a stair or ramp slope, installed as a means of support to furnish users with a handhold.

Harmonic Frequency. The frequency of a component of a periodic quantity is an integral multiple of the fundamental frequency. For example, a component whose frequency is twice the fundamental frequency is the second harmonic of that frequency.

Hazard. The probability that an adverse health effect will occur if a substance is used in a specified quantity and under a specified set of conditions. Compare with Safety.

Hazard Assessment (also Hazard Identification). The process of determining what hazards, if any, are associated with exposure to a particular chemical.

Hazard Communication Standard. Requires chemical manufacturers and importers to assess the hazards associated with the materials in their workplace (29 CFR 1910.1200). **Material safety data sheets**, **labeling**, and training are all the result of this law.

Hazard Zone. The point on a machine where injury can be caused if a person comes in contact with it. On calendar rolls, the point where the rolls are 3/8 in. apart is the hazard zone.

Hazardous Decomposition. The breakdown or separation of a substance into its constituent parts (sometimes accompanied by the release of heat, gas, or hazardous materials). Some materials can give off hazardous materials when they burn or decompose.

Hazardous Material. In general, any chemical or mixture having properties capable of producing adverse effects on the health or safety of a human. As defined by **OSHA** with respect to Worker Right-to-Know (Hazard Communication Standard), the term Hazardous Material means a chemical or mixture having one or more of the following characteristics:

Has a flash point below 140°F, closed cup, or is subject to spontaneous heating

Has a threshold limit value below 500 ppm for gases and vapors, below 500 mg/m^3 for fumes, and below 25 million particles per cubic foot for dusts

Has a single dose oral LD_{50} below 50 mg/kg

Is subject to polymerization with the release of large amounts of energy

Is a strong oxidizing or reducing **agent**

Causes first degree burns to skin resulting from a short time exposure, or is systemically toxic by skin contact

In the course of normal operations, may produce dusts, gases, fumes, vapors, mists, or smokes which have one or more of the above characteristics.

Hazardous Waste Number. An identification number assigned by the EPA per the **RCRA** law to identify and track wastes.

Hazmat. Usually used in reference to any hazardous material.

HCS. See **Hazard Communication Standard**.

Health Education. Any combination of learning methods or activities designed to facilitate voluntary adaptation of behavior conducive to health.

Health Effect. Any effect on human or test animal health as a result of exposure to an **agent**. Compare **Epidemiology** with **Toxicology**.

Health Promotion. Any combination of health education activities related to organizational, economic, or environmental supports for behavior conducive to health.

Health Risk Appraisal. See **HRA**.

Healthy Worker Effect. A phenomenon in which employed workers appear healthier than the general population due to the fact that ill and disabled individuals are excluded from the workforce. Because of the healthy worker effect, it may be inappropriate in epidemiologic studies to compare the disease experience of workers with that of the general population.

Hearing. The subjective response to sound, and the nervous and cerebral operations that translate the physical stimuli into meaningful signals.

Hearing Conservation. The prevention or minimizing of noise-induced hearing loss through the use of hearing protection devices and the control of noise through engineering methods or administrative procedures.

Hearing Impairment. A deviation or change for the worse in either structure or function, usually outside the normal hearing range.

Hearing Level. A measured threshold of hearing at a specified frequency, expressed in decibels relative to a specified standard of normal hearing. The deviation in decibels of an individual's threshold from the zero reference of the audiometer.

Hearing Loss. An increase in the threshold of audibility, at specific frequencies, as the result of normal aging, disease, or injury to the hearing organs. It is the symptom of reduced auditory sensitivity, synonymous with auditory impairment, when a specific cause can be ascribed. Also used in a general sense to describe the process of losing auditory sensitivity. Types of hearing loss are:

Conductive — A hearing loss originating in the conductive mechanism of the ear.

Sensori-Neural — A hearing loss originating in the cochlea or the fibers of the auditory nerve. (Formerly called Perceptive Deafness.)

Noise-Induced — A sensori-neural hearing loss attributable to the effects of noise.

Hearing Threshold Level (HTL). The amount (in decibels) by which the threshold of audibility for that ear exceeds a standard audiometric threshold.

Heat Cramp. Contractions of the voluntary muscles as a result of insufficient salt intake and profuse sweating resultant from heat stress.

Heat Exhaustion. Muscular weakness, distress, nausea, vomiting, dizziness, pale clammy skin, fainting, etc., as a result of heat stress.

Heat Stroke. An excessive rise in body temperature, loss of consciousness, resultant from heat stress; an acute medical emergency.

HEG. See **Homogeneous Exposure Group**.

Hematopoietic. Refers to having an effect on the blood cell-forming system of the biological organism.

HEPA. High Efficiency Particulate Air filter. A filter capable of removing very small particles from the airstream.

Hepatic. Pertaining to the liver.

Hertz (Hz). Synonymous term for cycles per second. Most standardizing agencies have adopted hertz as the preferred unit of frequency.

High-Frequency Loss. Usually showing a hearing impairment, starting with 2000 Hz and beyond.

Homogeneous Exposure Group, HEG. In general, a group of employees who are defined as having the same potential for the same type and degree of exposure. For example, in a plant spray painters may make up one HEG while typists would make up another.

Hood. Inlet to a local exhaust ventilation system specifically designed to capture, enclose, or receive airborne contaminants.

Hood Entry Loss. Loss in static pressure associated with acceleration of airflow and resistance by friction and turbulence.

Hostage Control. A guard that is interlocked with a machine physically or by distance such that it keeps the whole person out of a large hazard zone.

Household Products Disposal Council, HPDC. 1201 Connecticut Ave. N.W., Washington, D.C., 20036 (202) 659-5535.

HPDC. See **Household Products Disposal Council**.

HRA. Health Risk Appraisal; self-assessment in employee promotion programs.

Humidity. See **Relative Humidity**.

Hygienic Guides. **AIHA** Hygienic Guides are publications containing useful industrial hygiene information for many common chemicals.

Hygroscopic. Readily adsorbing moisture from the air.

Hyperbolic. Self-igniting upon contact of its components (without a spark or external aid).

Hyperpyrexia. A body core temperature greater than 104°F (40°C).

Hyperthermia. A condition of elevated body core temperature (heat stress).

Hypothalamus. A part of the brain that activates, controls, and integrates the peripheral autonomic mechanisms, endocrine activity, and some somatic functions, e.g., regulation of water balance, body temperature, sleep, and food intake, and the development of secondary sex characteristics.

Hypothermia. A condition of lowered body core temperature (cold stress).

I

I. Intermittent.

IAFF. International Association of Fire Fighters, 1750 New York Ave. N.W., Washington, D.C., 20006 (202) 737-8484.

IARC. International Agency for Research on Cancer. World Health Organization, Geneva, Switzerland; distributed in the U.S. from 49 Sheridan Ave., Albany, NY, 12210 (518) 436-9686.

IDLH. Immediately Dangerous to Life and Health. For certain chemicals, the Standards Completion Program of **NIOSH** has recommended IDLH values. For the chemicals for which they exist, IDLH values are defined only for the purpose of respirator selection and represent the maximum concentration from which, in the event of a respirator failure, one could escape in 30 min without experiencing any escape-impairing symptoms or any irreversible health effects. IDLH values are not to be used for evaluating routine occupational exposures.

IES. Illuminating Engineering Society, 345 E. 47th Street, New York, NY, 10017 (212) 705-7926.

Ignition Temperature. The lowest temperature at which a combustible material will catch on fire in air and will continue to burn (independent of the original source of heat).

IHIT. Industrial Hygienist-in-Training, indicating partial certification toward the **CIH**.

Illuminating Engineering Society. See **IES.**

IMDG Code, IMO Classification. The International Maritime Dangerous Goods code classifies materials in shipment (explosives, flammables, oxidizers, poisons, corrosives, and others). International Maritime Organization, #4 Albert Embankment, London, SE 175R, U.K.

Immediately Dangerous to Life and Health. See **IDLH**.

Impact Noise. The noise resulting from the collision of two masses, such as a jackhammer on concrete.

Impervious. A material that does not allow another substance to penetrate or pass through it.

Impinger or Impinger Sampling. An impinger is a device carrying a liquid and through which the workplace air is bubbled; airborne gases will become dissolved in the liquid if it has been properly selected. Chemical analysis can then be done on the liquid to determine the airborne concentration.

Impulse Noise. Impulse noises are usually considered to be singular noise pulses, each less than 1 sec in duration, or repetitive noise pulses occurring at greater than 1 sec intervals. Also defined as a change of sound pressure of 40 dB or more within 0.5 sec.

Inches of Water. Units of pressure measurement based upon displacement of a water column exposed to a pressure differential.

Incidence. The number of new cases of a disease that occur in a specified population during a specified time period.

Incidence Rate. The number of new cases of disease occurring during a specified period of time divided by the total number of unaffected individuals who are at risk for acquiring the disease at the beginning of the time interval.

Incompatible. Materials that could cause dangerous reactions if placed in direct contact with one another are said to be incompatible.

Industrial Audiometry. The use of audiometric procedures in an industrial hearing-conservation program.

Industrial Hygiene. That science or art that devotes itself to the recognition, evaluation, and control of environmental factors arising in or from the workplace which may cause impaired health or significant discomfort among workers or residents of the community.

Inert Ingredients. Anything other than the active ingredient in a product; an ingredient not having active properties. Inert ingredients may still be hazardous.

Inflammable. Capable of being easily set on fire and continuing to burn.

Inflammation. A series of reactions produced in tissue. Characterized by redness and swelling caused by an influx of blood and fluids.

Ingestion. The taking in of a substance through the mouth.

Inhalation, ihl. Breathing in a substance (in the form of a gas, vapor, fume, mist, or dust).

Inhibitor. A material that is added to another to prevent an unwanted reaction; e.g., polymerization.

Inorganic Materials. Materials that do not contain carbon bonded to hydrogen, as do hydrocarbons (a.k.a. organics).

insol. Insoluble.

Interlocked Device. A guard that is connected to or tied in with the working mechanism of a machine it is used to guard such that when the machine is moving or in a hazardous state the guard prevents the operator from being in the hazard zone.

International Association of Fire Fighters. See **IAFF**.

IRR. Irritant effects.

Irreversible Injury. An injury that is not reversible after the exposure has terminated.

Irritant. A material that is noncorrosive and that causes a reversible inflammatory effect on living tissue by chemical action at the site of contact.

Isomers. Chemical compounds that have the same molecular weight and atomic composition but differ in molecular structure.

K

kg, kilogram. 1000 g; 2.2 lb.

Kilocalorie. Amount of heat required to raise the temperature of 1 kg of water 1 °C.

Kilohertz (kHz). 1000 HZ.

L

l. Liter.

Label. Any written sign or symbol displayed on containers of hazardous chemicals. Among other things a label should contain the identity of the hazardous material, appropriate hazard warnings, and the name and address of the chemical manufacturer, importer, or other responsible party.

Lacrimation. Secretion and discharge of tears.

Ladder. A device with steps or rungs used to descend or ascend.

Ladder Marking. A sign or label attached to a ladder containing either general information or hazard warning.

Landing. The floor (or platform) at the bottom, middle or top of a stair.

Latency Period. The time that elapses between exposure and the manifestations of disease or illness. Latency periods can range from minutes to decades, depending on the hazardous material involved.

LC_{50}. Lethal concentration 50. LC_{50} is the concentration of a material in air that killed 50% of a group of test animals when administered as a single exposure.

LC_{Lo}. Lethal concentration low. LC_{Lo} is the lowest concentration of a substance in air that has been reported to have caused death in test animals.

LD_{50}. Lethal dose 50. The dose of a substance that causes the death of 50% of the animals in a test population. Other lethal-dose percentages such as LD_1, LD_{10}, LD_{30}, and LD_{99} are similar in concept. LD_{50} is usually expressed as milligrams (or grams) of material per kilogram of animal weight.

LD_{Lo}. Lethal dose low. The lowest dose (other than LD_{50}) of a substance reported to have caused death in test animals.

LEL. See **Lower Explosive Limit**.

LEPC. Local Emergency Planning Committee. A committee formed, as required by **SARA Title III**, for the purpose of implementing local planning for chemical emergencies and for implementing requirements of that law.

Let-Go Current. A small current of 6 to 9 mA that is sufficient to cause the arm muscles to contract and not allow a person to release his hold on an electrical wire.

Leukemia. A progressive, malignant disease of the blood-forming organs.

LFL. Lower Flammable Limit; see **LEL**.

LFM or lfm. Linear feet per minute.

Limit of Detection. The smallest concentration or amount of a substance that an analytical method can distinguish as being different from the background.

Limits of Flammability. See **Flammable Limits**.

Lipophilic. Displaying a preference or affinity for fats in the biologic system.

Local Emergency Planning Committee. See **LEPC**.

Local Exhaust Ventilation. Ventilation designed to capture and remove airborne contaminants near their point(s) of emission.

Local Ventilation. The removal of contaminated air directly from its source. This type of ventilation is recommended for controlling most hazardous airborne materials. Contrast with **General Ventilation**.

Lockout. The procedure that requires opening an electrical circuit, fixing a barrier (like a padlock), and testing the circuit prior to work or testing of a circuit.

LOD. See **Limit of Detection**.

Lower Explosive Limit. The lowest concentration of gas or vapor (% by volume in air) that will burn or explode if an ignition source is present at ambient temperatures.

M

m^3 or cu m. Cubic meter. m^3 is preferred.

Manometer. An instrument to measure pressure by displacement of a fluid (e.g., water) column.

Masking. Masking refers to the process by which the threshold of audibility of a sound is raised by the presence of another (masking) sound. For example, the stimulation of one ear of a patient by controlled noise to prevent his/her hearing with that ear the tone or signal given to his/her other ear.

Mass Law. The approximately linear relationship between the sound insulation of a partition, expressed in decibels, and the logarithm of its weight per unit area.

Material Safety Data Sheet. See **MSDS**.

Mean. The arithmetic average of a set of data.

Measurement Error. The difference between a value obtained using a measuring device and the true value.

Median. In this context, if all exposures in a workplace are ranked from lowest to highest, the **median** is the exposure level which is at the midpoint of the ranking.

Melting Point. The temperature at which a solid changes to a liquid.

Ménière's Disease. The combination of deafness, tinnitus, nausea, and vertigo.

Metabolic Rate. The rate at which energy is made available for use by the body.

Metabolism. The transformation of chemical energy into energy within the body; used for performing work and generating heat.

Metatarsal. The part of the foot between the tarsus (the joining of the leg and foot) and the phalanges (the digital bones).

Meter (m). A measure of length; 100 cm; the equivalent of 39.371 in.

mg. Milligram (1/1000 of a gram).

mg/kg. Milligrams per kilogram. A dosage commonly used in toxicology testing to indicate the dose administered per kilogram of body weight.

mg/m^3. Milligrams per cubic meter in air; a unit of airborne concentration. For any substance, the relationship between mg/m^3 and ppm is

$$\text{ppm} \times \text{MW} = 24.45 \times \text{mg/m}^3$$

where ppm = parts per million by volume
MW = molecular weight of the substance in question
24.45 = the number of liters occupied by 1 mol of any gas at **STP**.

Microgram (μg). 1/1,000,000 of a gram.

Micrometer (μm). 1/1,000,000 of a meter, also referred to as a micron.

Micron. See **Micrometer**.

Milliamps. 1/1000 of an ampere.

Millimeter (mm). 1/1000 of a meter.

Mine Safety and Health Administration. See **MSHA**.

Miscible. The extent to which liquids or gases can be mixed or blended.

Mist. Small liquid droplets suspended in the air.

Mixture. A heterogeneous association of materials that cannot be represented by a chemical formula and that does not undergo chemical change as a result of interaction among the mixed materials. The constituent materials may or may not be uniformly dispersed and can usually be separated by mechanical means (as opposed to a chemical reaction). Uniform liquid mixtures are called solutions.

ml. Milliliter. 1/1000 of a liter, equivalent to 1 cubic centimeter (1 cm^3).

MLD. Mild irritation effects.

mmHg. Millimeters of mercury. A measure of pressure. See also **atm**.

MMI. Mucous membrane effects.

MOD. Moderate irritation effects.

Mode of Vibration. In a system undergoing vibration, a mode of vibration is a characteristic pattern assumed by the system. Two or more modes may exist concurrently in a multiple-degree-of-freedom system.

Model. A hypothetical representation of a real phenomenon or process.

Mole. The quantity of a chemical substance that has a weight in a unit (usually grams) numerically equal to the molecular weight. It is that amount of a material that has 6.023×10^{23} molecules (Avogadro's number).

Molecular Weight. The mass in grams per mole of a substance. See also **Mole.**

Monitoring Protocols. The specific procedures used for different types of monitoring.

Morbidity. The rate of disease or proportion of disease in a specified population.

Mortality. A measure of the frequency of deaths within a particular population during a specified time period.

Mounting, Resilient. Any mounting, attachment system, or apparatus that permits room surfaces or machinery to vibrate without transmitting all of the energy of vibration to the structure.

mppcf. Millions of particles per cubic foot of air.

MSDS. Material Safety Data Sheet. As required by the **Hazard Communication Standard**, an MSDS is supposed to be issued by chemical manufacturers, importers, or distributors for each hazardous material. The MSDS should contain information relating to the various health risks associated with use of the chemical.

MSHA. Mine Safety and Health Administration. A federal agency within the U.S. Department of Labor that devises and promulgates mandatory safety and health in mines. MSHA, Ballston Tower #3, 4015 Wilson Blvd., Arlington, VA, 22203 (703) 235-1452.

MSK. Musculoskeletal effects.

Mucous Membrane. The mucous-secreting membrane lining the hollow organs of the body; i.e., nose, mouth, stomach, intestine, bronchial tubes, and urinary tract.

Muffler. A special duct of pipe that impedes the transmission of sound by reducing the velocity of the air or gas flow.

Multiple Lockout Device. A clasp that allows several padlocks to be locked onto one lockout point. All padlocks must be removed to remove the clasp.

MUT, Mutagen. A material that induces genetic changes (mutations).

mw or MW. See Molecular Weight.

N

n-. Normal; used as a prefix in chemical names signifying a straight-chain structure; i.e, no branches.

NA, ND. Not applicable, not available.

Narcosis. Stupor or unconsciousness produced by narcotics or other materials.

Natality Rate. A measure of the frequency of births in a population during a specified time period.

National Clearinghouse for Alcohol and Drug Information. P. O. Box 2345, Rockville, MD, 20852 (301) 468-2600.

National Fire Protection Association. See **NFPA**.

National Health Information Clearinghouse. P. O. Box 1133, Washington, D.C., 20013 (800) 336-4797, in Maryland (301) 565-4167.

National Injury Information Clearinghouse. 5401 Westbard Ave., Room 625, Washington, D.C., 20207 (301) 492-6424.

National Institute for Occupational Safety and Health. See **NIOSH.**

National Safety Council. The largest organization in the world dedicating its efforts exclusively to the prevention of accidents. NSC, 444 North Michigan Ave., Chicago, IL, 60611-9811 (800) 621-7619; training information (800) 848-5588 or (415) 341-5649.

National Toxicology Program. See **NTP**.

National Volunteer Fire Council. See **NVFC**.

Natural Frequency. The frequency of free oscillation of a system. The frequency at which a resiliently mounted mass would vibrate when set into vibration, under the influence of gravity alone with no added force or constraints. Often called Resonance Frequency.

Natural Wet Bulb Temperature. The temperature obtained from a thermometer encased in a wetted wick; when taken with the **Dry-Bulb Temperature**, this measure represents the ability of the air to evaporate perspiration.

NCI. National Cancer Institute. A part of the National Institutes of Health.

ND. Not Determined.

Necrosis. Localized death of tissue.

NEO. Neoplastic effects, i.e., the production of tumors.

Neoplasm. A new or abnormal tissue growth that is uncontrollable and progressive.

Nephrotoxic. Poisonous to the kidney.

Neural. Of, relating to, or affecting a nerve or the nervous system.

Neuritis. Inflammation of the nerves.

Neutralize. To render chemically harmless.

NFPA. National Fire Protection Association. An international voluntary membership organization to promote/improve fire protection and prevention and establish safeguards against loss of life and property by fire. NFPA, Battery March Park, Quincy, MA, 02269 (800) 344-3555, (617) 770-3000.

NFPA Fire Diamond. See **Fire Diamond.**

ng. Nanogram. 1/1,000,000,000 or 10^{-9} or of 1 g.

NIOSH. National Institute for Occupational Safety and Health. The research and training "arm" of OSHA. NIOSH, 4676 Columbia Parkway, Cincinnati, OH, 45226-1998. Technical support (800) 35-NIOSH, the *NIOSH Publications Catalog*. DHHS (NIOSH) Publ. No. 87-115 is free from NIOSH Publications (513) 533-8287.

NIPTS. Noise-induced permanent threshold shift.

NOC. Not otherwise classified.

NOEL. No effect level.

Noise. Any undesired sound.

Noise Dosimeter. See **Personal Dosimeter**.

Noise Exposure. A generic term signifying the total acoustic stimulus applied to the ear over a period of time.

Noise-Induced Hearing Loss. This terminology is usually restricted to mean the slowly progressive inner ear hearing loss that results from exposure to noise over a long period of time as contrasted with acoustic trauma or physical injury to the ear.

Noise Level. A level of noise, the type of which must be indicated by further modification of context.

Noise Reduction (NR). A decrease of the sound pressure level at a specified observation point.

Noise Reduction Coefficient (NRC). A measure of acoustical absorption performance of a material, calculated by averaging its sound absorption coefficients at 250, 500, 1000, and 2000 Hz, expressed to the nearest integral multiple of 0.05.

Nonflammable. Incapable of being easily ignited or burning with extreme rapidity when lighted. Also, a **DOT** hazard class for any compressed gas other than a flammable one.

NOS. Not otherwise specified.

Nosing. The projection of the front portion of tread over the riser wall.

NSC. See **National Safety Council**.

NTP. National Toxicology Program. Overseen by the Department of Health and Human Services (with resources from National Institutes of Health, the Food and Drug Administration, and the Centers for Disease Control). Its goals are to develop tests useful for public health regulations of toxic chemicals, to develop toxicological profiles of materials, to foster testing of materials, and to communicate the results for use by others. NTP Information Office, MD B2-04, Box 12233, Research Triangle Park, NC, 27709 (919) 541-3991.

Nuisance Particulates. Dusts that do not produce significant disease or toxic effects from "reasonable" concentrations and exposures. The TLV for nuisance particulates is 10 mg/m^3.

NVFC. The National Volunteer Fire Council, 1325 Pennsylvania Ave., N.W., Ste. 500, Washington, D.C., 20004 (202) 393-3351.

O

OCAW. Oil, Chemical, and Atomic Workers International Union, AFL-CIO, 255 Union Blvd., Lakewood, CO, 80228 (303) 987-2229.

Occupational Disease. A deterioration in health status resulting from circumstances or conditions in the workplace.

Occupational Exposure. See Action Level.

Occupational Exposure Work Practice Guideline. **AIHA** publication, in-depth toxicological and applied industrial hygiene reviews of current interest.

Occupational Safety and Health Act. See **OSHAct**.

Occupational Safety and Health Administration. See **OSHA**.

Octave Band. A division of the audible range of frequencies into subgroups such that in each division the upper frequency limit is twice the lower limit. The center frequencies used to designate the octaves are twice the center frequency of the preceding octave band.

Octave-Band Analyzer. A device for measuring pressure levels of a sound for a frequency band corresponding to a specific octave.

Odds Ratio. A ratio between two odds as, for example, the odds of exposure among individuals with a disease divided by the odds of exposure among individuals without the disease. The odds ratio is used as a measure of association in **case-control** studies.

Odor Threshold. The lowest concentration of a material's vapor (or a gas) in air that can be detected by odor. Frequently expressed as a percentage of a panel of test individuals.

OEL. Occupational Exposure Limit. See also **Exposure Limits**.

Ohms. A unit of resistance. 1 ohm of resistance produces 1A of current in a 1-V circuit.

OHST. Occupational Health and Safety Technologist.

Oil, Chemical, and Atomic Workers Union, AFL-CIO (OCAW). See **OCAW.**

Opaque. Impervious to light rays.

Organ of Corti. An aggregation of nerve cells lying on the basilar membrane that senses vibrations that are transmitted to the brain, where they are interpreted as sound.

Organic Materials. Compounds composed of carbon, hydrogen, and other elements with chain or ring structures.

ORM. Other Regulated Material. A **DOT** hazard classification for the purpose of labeling hazardous materials in transport. For example, **ORM-A** are materials with an anesthetic, irritating, noxious, toxic, or other property whose leakage can cause extreme discomfort to transportation personnel. **ORM-B** are materials, including solids wet with water, that can cause damage to a vehicle if they leak. **ORM-E** are materials that are not in any other hazard classification but are subject to DOT regulations.

OSHA. The Occupational Safety and Health Administration (Part of the U.S. Department of Labor). The regulatory and enforcement agency for safety and health for most U.S. business and industrial sectors. OSHA Technical Data Center Docket Office, Rm N-3670, 200 Constitution Ave. NW, Washington, D.C., 20210 (202) 523-7894, Public Information (202) 523-8151, Publications Office (202) 523-9667.

OSHA Flammable/Combustible Liquid Classification. (29 CFR 1910.106). Flammable/combustible liquid is a standard classification used to identify the risks of fire or explosion associated with a particular liquid. The distinctions are made on the following basis:

Flammable Liquids (with a flash point below 100°F [38°C]) are class I liquids and are divided into the following classes: class IA — flash point below 73°F (22.8°C), boiling point below 100°F (38°C); class IB — flash point below 73°F (22.8°C), boiling point at or above 100°F (38°C); and class IC — flash point at or above 73°F (22.8°C), boiling point below 100°F (38°C).

Combustible Liquids (with a flash point at or above 100°F) are divided into two classes: class II, with a flash point at or about 100°F (38°C) and below 140°F (60°C), except any mixture having components with flash points of 200°F (93.3°C) or higher, the volume of which makes up 99% or more of the total volume of the mixture; and class III, with flash point at or above 140°F (60°C).

Class III liquids are divided into two subclasses: class IIIA, with a flash point at or above 140°F (60°C) and below 200°F (93.3°C), except any mixture having components with flash points of 200°F (93.3°C) or higher, the volume of which makes up 99% or more of the total volume of the mixture; and class IIIB, with flash point at or above 200°F (93.3°C) .

OSHAct. The Occupational Safety and Health ACt of 1970; Public Law 91-596, effective April 28, 1971. Found at 29 CFR 1910, 1915, 1918, 1926. The regulations intended to ensure the safety and health of workers in firms larger than ten employees by setting standards of safety that will prevent injury and illness among the workers. Regulating employee exposure and informing employees of the dangers of materials are key factors of this act.

OSHRC. Occupational Safety and Health Review Commission, the review board for appealed **OSHA** decisions.

Otologist. A physician or surgeon who specializes in the diagnosis and treatment of the disorders and diseases of the ear.

Otosclerosis. A condition marked by the growth of spongy bone around the delicate structures within the ear, particularly the stapes and oval window, resulting in a gradual loss of hearing.

Overexposure. Any exposure greater than the occupational exposure limit (e.g., TLV-TWA) or any exposure that causes some adverse health effect.

Overhang. In essence, the same as **nosing**, i.e., the projection of the tread past the riser face as measured from the tread surface below.

P

PAH. Polycyclic Aromatic Hydrocarbons. Organic compounds, usually formed from incomplete combustion, that may pose a risk of cancer.

Particulate. Tiny pieces of an airborne material. **Dusts**, **fumes**, **smokes**, **mists**, and **fogs** are examples of particulates.

Partition Coefficient. The quotient of the solubility of compound in oil divided by its solubility in water. See also **Coefficient of Water/Oil Distribution.**

PEL. Permissible Exposure Limit. Exposure limits established by **OSHA**. OSHA PELs have the force of law. Note that **TLVs (ACGIH)** and **RELs (NIOSH)** are recommended exposure limits that may or may not be enacted into law by OSHA. PELs may be expressed as

PEL-TWA, a time-weighted average (TWA) limit

PEL-C, a ceiling exposure limit that legally must never be exceeded instantaneously

PEL-STEL, a short term exposure limit (STEL), a 15-min TWA which may not be exceeded.

Percent Volatile. The percentage by volume of a liquid or solid that will evaporate at an ambient temperature of 70°F (20°C).

Period. The time required to complete one cycle; it is the reciprocal of the frequency.

Permanent Threshold Shift (PTS). The component of threshold shift which shows no progressive reduction with a passage of time when apparent cause has been removed.

Permissible Exposure Limit. See **PEL**.

Personal Dosimeter. With respect to *chemical exposures,* a personal dosimeter is a small device worn by a worker to collect airborne contaminants and used to measure exposure. It differs from other forms of **Air Sampling** in that a **Sampling Pump** is not used. For noise exposures, a personal *noise dosimeter* is a mechanical device worn by a worker which will measure and store the sound level experienced by the worker throughout the day. With regard to heat stress, a *heat stress* dosimeter will monitor the environmental, metabolic, and physiological factors contributing to an individual's **Thermal Stress.**

Personal Hygiene. Precautionary measures taken by the individual worker to maintain good health when exposed to potentially harmful materials. This includes keeping hands, other parts of the body, work clothing, and equipment free of chemicals, dirt and chemical residue. It includes the practice of not eating, drinking, applying makeup, or using toilet facilities where chemicals are in use.

Personal Measurement. See **Personal Sample.**

Personal Protective Equipment. See PPE.

Personal Sample. A measurement collected from an individual's breathing zone.

Personal Sampling Pump. See **Sampling Pump.**

pH. The value that represents the acidity or alkalinity of an aqueous solution. By definition pH is the logarithm to the base 10 of the reciprocal of the hydrogen ion concentration of a solution. The pH scale ranges from 1 to 14; pure water has a pH of 7. The strongest acids have an excess of H^+ ions and a pH of 1 to 3 (HCl, pH = 1). The strongest bases have an excess of OH- ions and a pH of 11 to 13 (NaOH, pH = 12).

Phon. A measure of loudness level (on a logarithmic scale) compares the loudness of a sound to the loudness of a 1000 Hz tone of a given sound pressure level.

Physical Hazard. A hazard from a chemical that results in potential damage, death, or injury to a person. These hazards include corrosive materials, explosives, reactive substances, compressed gasses, oxidizers, flammables, and organic peroxides.

Physical State. The condition of a material; i.e., solid, liquid, or gas, at room temperature.

Pink Noise. Noise whose spectrum level decreases with increasing frequency to yield constant energy per octave of band width.

Pitot Tube. In industrial ventilation, an instrument to measure velocity pressure as the difference between total and static pressures.

Plano. Nonprescription lenses, as in noncorrective spectacles or goggles.

Plenum. A portion of a ventilation system designed to contain relatively uniform static pressure and usually relatively low airflow velocity.

Pneumatic. Pneumatic systems used compressed air or gases as a source of stored energy, for example, a pressure vessel used to operate a pneumatic press ram.

Pneumoconiosis. Pathological reaction of the organism to the presence of dusts in the lung tissue. Involves activation of the macrophage/phagocyte response system.

PNS. Peripheral nervous system effects.

Poison. A substance that produces injury or death. *"All substances are poisons; there is none which is not a poison. The riqht dose differentiates between a poison and a remedy."* — Paracelsus.

Poison, Class A. A **DOT** term for an extremely dangerous poison, for example, a poisonous gas or liquid, such that a very small amount of the gas or vapor of the liquid mixed with air is dangerous to life.

Poison, Class B. A **DOT** term for liquid, solid, paste, or semisolid substances other than class A poisons or irritating materials known (or presumed on the basis of animal tests) to be so toxic to man as to afford a hazard to health during transportation.

Poison Control Center. Provides medical information on a 24-hr basis for accidents involving ingestion of potentially poisonous materials. Call your area's largest hospital to find the one nearest to you.

Polarity. When referring to electrical circuits, one wire delivers the current and the other returns completing the circuit.

Polymerization. A chemical reaction in which many small molecules combine to form large molecules. Sometimes a "hazardous polymerization" can take place at such a rate that large amounts of energy causing fires or explosions or burst containers are released. Materials that can polymerize usually contain inhibitors that can delay the reactions.

Point of Operation. The place on a press or other machine where material is formed or cut.

Population. All members of a group (for example, all males aged 20 to 30). Contrast with **sample**.

Population Attributable Risk Percent. The proportion of disease in the total population that is attributable to a particular exposure.

Portable Ladder. A ladder that can be easily moved or carried.

Potentiation. See **Synergy**.

Pour Point. The temperature at which a liquid ceases or begins to flow.

Power Level. In decibels, 10 times the logarithm to the base 10 of the ratio of a given power to a reference power. The reference power must be indicated.

ppb. Parts per billion. A volumetric unit of concentration; 1 ppb = 0.001 ppm.

PPE. Personal Protective Equipment. Equipment (devices or clothing) worn to help isolate a worker from direct exposure to hazardous materials. Examples include gloves and respirators.

ppm. Parts per million by volume; a unit of airborne concentration. For any substance, the relationship between mg/m^3 and ppm is

$$\text{ppm} \times \text{MW} = 24.45 \times \text{mg/m}^3$$

where mg/m^3 = mg of chemical per cubic meter of air
MW = molecular weight of the substance in question
24.45 = the number of liters occupied by 1 mol of any gas at **STP**

Speaking in very rough terms, one air rifle pellet (a "BB") placed in an empty 10 gal pail is approximately 1 ppm.

ppt. Parts per trillion; 1 part per trillion = 0.000001 ppm

Precision. A measure of the reproducibility of a measured value under a given set of conditions.

Presbycusis. Decline in hearing acuity that normally occurs with the aging process.

Prevalence Rate. The total number of individuals affected with a disease divided by the total number of individuals who are at risk for having the disease at a particular point in time.

Primary Guard. A guard that prevents an operator from being able to put any part of his hands or body in the hazard zone.

Process Change. In this context, any change in manufacturing process (type of equipment, substitution of different chemicals) that can lead to changes in worker exposure.

Prospective Study. See also **Cohort Study**. An epidemiological method for identifying the *future* relationship, if any, between exposure to an **agent** and the increased incidence of some adverse health effect in a population. Contrast with **Retrospective Study**.

Psychotropic, PSY. Acting on the mind.

Psychrometric Wet Bulb Temperature. Similar to the Natural Wet Bulb Temperature, this measurement is taken by twirling the wetted thermometer overhead (or otherwise forcing air past it). The resulting temperature, compared to the **Dry Bulb Temperature**, is a measure of how much perspiration can be evaporated with forced air.

PUL. A chemical that has pulmonary systems effects.

Pulmonary Fibrosis. A reaction by the biologic system, occurring in the lungs, marked by increased formation of interstitial fibrous tissue.

Pyrolysis. A chemical decomposition or breaking apart of molecules produced by heating.

Pyrophoric. Describes materials that ignite spontaneously in air below 130°F (54°C). Occasionally friction will ignite them.

Q

Q. See **Ventilation Volume Flow Rate.**

Qualitative Fit Testing. A method of evaluating the fit of a given respirator to the user's face. Irritant smoke is used to indicate whether the seal between the face and the respirator mask is sufficient to prevent the smoke from entering the facepiece.

Quantitative Fit Testing. A method of evaluating the fit of a given respirator to the user's face which measures the concentration of the contaminant inside and outside the mask of the respirator. The method gives a quantitative measure of the percentage of the contaminated air which is allowed to enter the facepiece.

R

Radiant Heat. Heat exchange by radiation between two radiant surfaces having different temperatures.

Rail. The side members of a ladder joined by steps or rungs.

Random Sampling. Samples collected at random from a population, without bias, such that each sample has an equal probability of being collected. If enough random samples are collected, the data represented in the sample will accurately represent the total population.

Range. The difference between the smallest and largest values in a data set.

Rate of Decay. The time rate at which the sound pressure level (or other stated characteristic) decreases at a given point and at a given time. A commonly used unit is decibels per second.

RBC. Red blood cell effects.

RCRA. Resource Conservation and Recovery Act; an amendment to the Solid Waste Disposal Act, (800) 424-9346.

Reactive Material. In general, a chemical substance or mixture that will vigorously polymerize, decompose, condense, or become self-reactive due to shock, pressure, or temperature.

Reactivity. Describes a substance's tendency to undergo chemical reaction either by itself or with other materials with the release of energy.

Recommended Exposure Limit. See **REL**.

Recovery Heart Rate. The heart rate measured at a prescribed time interval, usually 1, 3, or 5 min after worker commences resting.

Reflecting Surfaces. Room surfaces from which significant sound reflections occur, or special surfaces used particularly to direct sound throughout the space.

REL. The **NIOSH** REL (Recommended Exposure Limit) is the highest allowable airborne concentration that is not expected to injure a worker. It may be expressed as a ceiling limit or as a time-weighted average (TWA), usually for 10-hr work shifts.

Relative Humidity. The ratio of water vapor present in the ambient air to the water vapor present in saturated air at the same temperature and pressure.

Relative Risk. A ratio between two risks; specifically, the incidence rate among a group of exposed individuals divided by the incidence rate among a group of unexposed individuals.

Repetitive Stress Disorders. Medical conditions of the joints caused by repetitive motion, for example, carpal tunnel syndrome.

Reportable Quantity. See **RQ**.

Reproductive Health Hazard. Any chemical or physical agent that has a harmful effect on the adult male or female reproductive system or on the developing fetus or child.

Residual Noise Level. The noise that exists at a point as a result of the combination of many distant sources, individually indistinguishable.

Resource Conservation and Recovery Act. See **RCRA.**

Respirable Fraction. That portion of an airborne particulate that can be inhaled into the lungs; generally accepted to be limited to particles less than 5 **μm**.

Respirator Protection. A device used to provide a source of clean breathing air to the user by either filtering workroom air or by supplying fresh air from another source.

Respirator Protection Factor. A measure of the efficiency of a particular respirator obtained by simultaneously evaluating the concentration of a contaminant inside and outside of the respirator facepiece. The numerical protection factor reflects the ratio of the concentration of the contaminant outside of the facepiece to the concentration of the contaminant inside of the facepiece. See **Fit Test**, **Qualitative Fit Test** and **Quantitative Fit Testing**.

Respiratory System. The breathing system, including the lungs and air passages (trachea or windpipe, larynx, mouth, and nose), as well as the associated system of nerves and circulatory supply.

Retrospective Study. Also, **Case-Control** or **Case History**. A study for identifying the causal relationship between workplace factors (chemical or physical **agents**, temperature, etc.) and the occurrence of adverse health effects in a population. This is done by evaluating disease incidence with *past* exposures. Contrast with **Prospective Study**.

Reverberation. The persistence of sound in an enclosed space, as a result of multiple reflections after the sound source has stopped.

Reverberation Time (t_{60}). For any given frequency, the time required for the average sound pressure level, originally in a steady state, to decrease 60 dB after the source is stopped.

Reversible Injury. An injury that will cease, or an injury that can be treated, such that the adverse health effect will cease and/or healing can take place.

Rise. The vertical distance between treads.

Risk. The probability of an adverse health effect; risk is often said to be a function of exposure level and toxicity.

Risk Assessment. The evaluation of risk, either qualitatively or quantitatively, that might result from an exposure.

Root-Mean-Square Value (rms). The square root of the arithmetic mean of the squares of a set of values.

Routes of Entry. The method by which a chemical enters the body is called the route of entry. The routes of entry are inhalation, absorption (eye or skin contact), and ingestion.

RQ. Reportable Quantity; used in several pieces of legislation. Briefly, the RQ of a chemical is the amount which, if released into the environment, must be reported to EPA and state agencies (**SARA Title III**).

Run. The distance between risers or the tread distance minus the nosing or overhang.

RTECS. "Registry of Toxic Effects of Chemical Substances," published by **NIOSH**. Presents basic toxicity data on thousands of materials. Its objective is to identify "all known toxic substances" and to reference the original studies.

S

Safety. The practical certainty that an adverse health effect will not occur if an agent is used in a specified quantity and under specified conditions.

Safety Glasses. Spectacle eye protection designed for industrial use with lenses designed for impact (3 mm thick lenses).

Safety Low Voltage. Voltage less than 24V. This voltage when in contact with a person generally has insufficient current to cause fibrillation.

Sample. A subset of the population.

Sampling Pump. In general, a calibrated air pump used to perform **Air Sampling**.

SARA. Superfund Amendments and Reauthorization Act. Also referred to as **SARA Title III** and the **Emergency Planning and Community Right-to-Know Act** of 1986. The disaster in Bhopal, India, in 1987 added impetus to the passage of this law. It is intended to support local and state emergency planning efforts. It provides citizens and local governments with information about potential chemical hazards in industries, etc., in their communities. SARA calls for facilities that store hazardous materials to provide officials and citizens with data on the types (flammables, corrosives, etc.), the amounts on hand (daily, yearly), and their specific locations. Facilities are to prepare and submit inventory lists, **MSDS**s, and **Tier 1** and **Tier 2** inventory forms. EPA SARA Information Hotline (800) 535-0202.

SCBA. Self-Contained Breathing Apparatus. A respirator type in which the user carries a supply of compressed breathing air in a tank on his/her back. Air passes from the tank, through a regulator and is delivered into the facepiece of the respirator. SCBAs offer the highest level of protection of any respirator system and are frequently used for emergency response activities involving highly toxic materials.

Secondary Guard. A guard that requires the operator to use certain devices or to adjust the guard to protect the user. If not used properly the person can put his hands or body in the hazard zone.

Sectional Ladders. Nonself-supporting ladder, made in even-length sections, joined together to obtain a desired height.

Sensitization. A state of immune-response reaction in which further exposure elicits an immune or allergic response. A person previously exposed to a certain material is more sensitive when he experiences further contact with it.

Sensitizer. A material which on first exposure causes little or no reaction but which on repeated exposure may cause a marked response in the individual not necessarily limited to the contact site. Skin sensitization is the most common form. Respiratory sensitization to a few chemicals is also known to occur.

SERC. State Emergency Response Commission. The state organization responsible for implementing and enforcing **SARA Title III**.

Sign. Objective evidence of an adverse health effect. Compare with **Symptoms**.

Silica Gel. A type of **Solid Sorbent Air Sampling** media.

Simple Asphyxiant. See **Asphyxiant**.

Simple Sound Source. A source that radiates sound uniformly in all directions under free-field conditions.

Skin. In this context, "skin" is a notation indicating possible significant contribution to a worker's overall exposure to a material by way of absorption through the skin, mucous membranes, and eyes by either direct or airborne contact.

SKN. Skin effects; e.g., erythema, rash, sensitization of skin.

Slope. The angle of the stair established by measuring the relationship of stair nosings to the horizontal plane.

Slow Response. A selectable mode of operation of a sound level meter of analyzer in which the indicator has high damping and therefore slowly responds to change in sound level. This mode tends to provide an average reading.

Slurry. A pourable mixture of solid and liquid.

Smoke. Dry particles and droplets (carbon, soot, PAHs, etc.) generated by incomplete combustion of an organic material combined with and suspended in the gases from combustion.

Sociocusis. Hearing loss resulting from the accumulative effects of daily nonindustrial noise exposure.

Solid Sorbent. See **Sorbent**.

Soln. Solution. A uniformly dispersed mixture. Solutions are composed of a solvent (water or another fluid, for example) and a dissolved substance, called the solute.

Solubility in Water. A term used to express the percentage of a material, by weight, that will dissolve in water at ambient temperature; useful in determining cleanup methods for spills and fire-extinguishing methods for a material. Solubility, commonly expressed in g/L, is expressed as:

"negligible," less than 0.1%
"slight," 0.1 to 1.0%
"moderate," 1 to 10 1%
"appreciable," more than 10%
"complete," soluble in all proportions

Solution. See **Soln.**

Solvent. A material that can dissolve other materials to form a uniform mixture. Water is the so-called "universal solvent."

Soot. Particles formed by combustion (complete or incomplete) and consisting mostly of carbon. Soot gives smoke its color.

Sorbent. Also Solid Sorbent; a sampling medium (such as charcoal tubes or silica gel) used to collect airborne chemical contaminants. See **Air Sampling**.

Sound. The auditory sensation produced through the organs of hearing usually by vibrations transmitted in a material medium, commonly air.

Sound Absorption. The change of sound energy into some other form, usually hearing, in passing through a medium or on striking a surface. In addition, sound absorption is the property possessed by materials and objects, including air, of absorbing sound energy.

Sound Analyzer. A device for measuring the band-pressure level or pressure-spectrum level of a sound as a function of frequency.

Sound Level Meter. A device to measure sound level or weighted sound level, constructed in accordance with the standard specifications for sound level meters set up by the American National Standards Institute (**ANSI**).

Sound Power Level (w or PWL). The sound power level of a sound source, in decibels, is 10 times the logarithm to the base 10 of the ratio of the sound power radiated by the source to a reference power.

Sound Pressure Level (L_p or SPL). The sound pressure level, in decibels, of a sound is 20 times the logarithm to the base 10 of the ratio of the pressure of this sound to the reference pressure. The reference pressure shall be explicitly stated.

Spray Paint Booth. An enclosure with one open side that is used to contain the overspray generated when spray painting.

Stability. An expression of the ability of a chemical to remain unchanged under reasonable conditions of storage or use.

Stack. The terminal component of a local exhaust system, i.e., the place at which exhaust air is discharged to the atmosphere.

Standard Deviation, SD. The positive square root of the variance of a distribution; a measure of the representativeness of the mean.

Standard Man. A "representative" human being having a body mass of 70 kg (154 lb) and a body surface area of 1.8 m^2 (19.4 ft^2).

Standardized Mortality Ratio (SMR). The ratio of the observed number of deaths in a population to the number of deaths expected in that population if the population had the same age structure as the standard population.

Static Pressure. Air pressure exerted uniformly in all directions, not affected by air movement.

Steady-State Noise. Noises that are continuous or that consist of impulses spaced less than 1 sec apart are considered to be steady-state noises.

Steady-State Vibration. After a vibrating system has been acted upon by a definite force for a sufficient time, it will follow a definite cycle of events described as steady-state vibration.

STEL. Short-Term Exposure Limit. The **TWA** exposure for any 15-min period of time cannot exceed the **PEL-STEL (OSHA)** or **TLV-STEL (ACGIH).**

Step. Basic unit of a stair consisting of a tread and a riser.

Stepladder. Self-supporting portable ladders nonadjustable in length, with a hinged back and flat steps.

STP. Standard Temperature and Pressure. Usually refers to 298 K (25°C) and 760 mmHg (1 atm).

STS. Threshold Shift Significant.

Structure-Borne Sound. Sound energy transmitted through the solid media of the building structure.

Subcutaneous. Beneath the skin.

Sublime. To change from the solid to the vapor phase without passing through the liquid phase, for example, dry ice.

Subpart Z. See **Z list**.

Superfund Amendments and Reauthorization Act. See **SARA**; see also **CERCLA**.

Susceptible. An individual who lacks some or all resistance to a particular disease or environmental **agent**.

Symptom. Subjective evidence of an adverse health effect. Compare with **Sign**.

Synergy. An interaction or potentiation of materials to give a greater result than either material alone. For example, either smoking or exposure to asbestos alone can cause lung cancer. If an individual smokes and is exposed to asbestos, the risk of lung cancer is much greater than the individual risks from smoking of asbestos exposure combined.

Synonyms. Alternative names by which a substance may be known.

SYS. Systemic effects. Effects on the metabolism and excretory functions.

System Ground. The return wire of an AC circuit also known as the common wire and usually has white insulation on it.

T

Target Organ Effects. Effects, resulting from chemical exposure, to specific organs or organ systems (such as the liver, kidneys, nervous system, lungs, skin, and/or eyes).

TC_{Lo}. Toxic Concentration, low. The lowest airborne concentration of a substance to which humans or animals have been exposed and that has produced a toxic effect.

TDL. Toxic dose level.

TD_{Lo}. The lowest dose of a substance introduced by any route other than inhalation over any given period of time and reported to produce any toxic effect in humans or to produce tumorigenic or reproductive effects in animals or humans.

temp. Temperature.

Temperature Regulation. The body's ability to maintain its temperature within a restricted range of temperatures under conditions of different or varying **ambient temperature**, **metabolic rate**, etc.

Temporary Threshold Shift (TTS). The component of threshold shift that shows progressive reduction with the passage of time when the apparent cause has been removed.

TER. See **Teratogen**.

Teratogen. Physical defects in a developing embryo caused by the mother's exposure to a chemical **agent** or material.

Thermal Strain. The sum of an individual's physiological responses to a thermal stress.

Thermal Stress. The sum of the environmental and metabolic heat loads placed on an individual.

Threshold Limit Value. See **TLV**.

Threshold of Hearing of a continuous sound. The minimum value of the sound pressure that excites the sensation of hearing.

Threshold Planning Quantity. The amount of an extremely hazardous substance listed in **SARA Title III** that triggers reporting and emergency planning.

Tier I Report. A hazards inventory report required under SARA Title III that summarizes the amounts of hazardous materials and their storage locations for different hazard categories.

Tier II Report. A chemical inventory report required under SARA Title III that summarizes in detail the amounts of specific hazardous chemicals and their storage locations in a facility.

Time-Weighted Average. Also **TWA**. A way of expressing exposure such that the amount of time spent exposed to each different concentration level is weighted by the amount of time the worker was exposed to that level.

$$TWA = \frac{C_1T_1 + C_2T_2 + \ldots C_nT_n}{T_1 + T_2 + \ldots T_n}$$

where

C_1, C_2, C_n, etc. = airborne concentrations of some chemical (expressed in either ppm or mg/m^3)

T_1, T_2, T_n, etc. = the length of time the worker was exposed to C_1, etc.

Tinnitus. A subjective sense of noises in the head or ringing in the ears for which there is no observable external cause.

TLV. Threshold Limit Value. A term used by **ACGIH** to express the airborne concentration of a material to which "nearly" all workers can be exposed day after day without adverse effects. "Workers" means healthy individuals. The young, old, ill, or naturally susceptible will have lower tolerances and need to take additional precautions. The ACGIH expresses TLVs in three ways:

TLV-TWA, the allowable time-weighted average concentration for a normal 8-hr workday or 40-hr week

TLV-STEL, the short term exposure limit or maximum concentration for a continuous exposure period of 15 min (with a maximum of four such periods per day, with at least 60 min between exposure periods, and provided that the daily TLV-TWA is not exceeded)

TLV-C, the concentration that should not be exceeded at any time

TLV-Skin. See Skin.

Top Cap. The uppermost horizontal piece of a portable stepladder.

Torr. 1 mmHg pressure; see **atm**.

Total Pressure. In a ventilation system, Total Pressure is the algebraic sum of **Static Pressure** and **Velocity Pressure**.

Toxic. In general, toxic describes the ability of a material to injure biological tissue. As specifically defined in the **OSHA Hazard Communication Standard**, there are several criteria for a chemical to be considered toxic, for example, having:

an LD_{50} of 50 to 500 mg/kg when administered orally to albino rats weighing 200 to 300 g each

an LD_{50} of 200 to 1000 mg/kg when administered by continuous contact for 24 hr to the bare skin of albino rabbits weighing 2 to 3 kg each

or an LC_{50} of 200 to 2000 ppm (gas or vapor) or 2 to 20 mg/L (mist, fume, or dust) when administered by continuous inhalation for 1 hr to albino rats weighing 200 to 300 g each

See also **Acute Toxicity**.

Toxicant. Often used as a synonym for **Poison**.

Toxicity. The degree to which an **agent** can cause harmful effects.

Toxicology. The study of the nature, effects, and detection of poisons in living organisms. Also, substances that are otherwise harmless but prove toxic under particular conditions. The basic assumption of toxicology is that there is a relationship among the dose (amount), the concentration at the affected site, and resulting effects.

Toxic Substance. Any chemical or material that (l) has evidence of an acute or chronic health hazard; and (2) is listed in the *NIOSH Registry of Toxic Effects of Chemical Substances (RTECS)*, provided that

The substance causes human toxicity at any dose level

Causes cancer or reproductive effects in animals at any dose level

Has a median lethal dose (LD_{50}) of less than 500 mg/kg of body weight when administered orally to rats

Has a median LD_{50} of less than 1000 mg/kg of body weight when administered by continuous contact to the bare skin of albino rabbits

Has a median lethal concentration (LC_{50}) in air of less than 2000 ppm by volume of gas or vapor, or less than 20 mg/l of mist, fume, or dust when administered to albino rats.

Toxic Substances Control Act. See **TSCA**.

Toxin. Usually, a poison that is manufactured by some life form (as opposed to manufactured chemicals).

TPQ. See Threshold Planning Quantity. As it relates to **SARA Title III**, for any chemical the TPQ is the quantity above which the chemical's presence at a factory, etc., must be reported to the EPA and the state.

Trade Name. The name given to a chemical product by the manufacturer (or supplier).

Trade Secret. Confidential information (such as formula, recipe, process, etc.) that gives the manufacturer an advantage over competitors, Manufacturers can withhold proprietary information from an **MSDS**. **OSHA** permits this provided that the trade secret claim can be substantiated, the MSDS indicates that data are being withheld and the properties and health effects of the withheld chemicals are included.

Transmission Loss. The ratio, expressed in decibels, of the sound energy incident on a structure to the sound energy that is transmitted.

Transport Velocity. The minimum velocity needed in ductwork to prevent particulates carried in the air from settling out.

Tread. The walking surface of the step defined as the distance from the front of the step to the riser wall behind.

Trestle Ladder. Self-supporting portable ladders, nonadjustable in length, consisting of two sections hinged at the top to form angles with the base.

TSCA. Toxic Substances Control Act, Public Law PL 94-469, found in 40 CFR 700-799. Basically, the law says that chemicals are to be evaluated prior to use and can be controlled based on risk. The act provides for a listing of all chemicals that are to be evaluated prior to manufacture or use in the US. EPA, TSCA Office, Office of Pesticides and Toxic Substances, 401 M St. SW, Washington, D.C., 20460, Information (202) 554-1404.

Turbulence. Random movement of air characterized by swirling, changing patterns causing pressure loss and noise.

TWA. See **Time-Weighted Average**.

Two-Hand Trip Device. A guard that uses location of start or initiator buttons to keep the hands a safe distance from the hazard zone. After starting the press, the hands cannot travel fast enough to reach the hazard area before it is safe again.

Tympanic Membrane. The membrane separating the middle ear from the external ear, commonly called the ear drum.

U

UEL. See **Upper Explosive Limit, Lower Explosive Limit**.

UFL. Upper Flammable Limit; identical to **UEL**.

UL. Underwriters Laboratories, Fire Protection Division, 1285 Walt Whitman Rd., Melville, NY, 11747-3081 (516) 271-6200.

Underwriters Laboratories. See **UL**.

Unstable. An unstable chemical will vigorously polymerize, decompose, condense, or become self-reactive under conditions of shock, pressure, or temperature.

Upper Explosive Limit. UEL. The highest concentration of a material in air that will produce an explosion in fire or will ignite when it contacts an ignition source (high heat, electric arc, spark, or flame). A higher concentration of the material in a smaller percentage of concentration of air may be too rich to be ignited. See also **Flammable Limits**.

UV. Ultraviolet (light).

V

Vapor. The gaseous state of a material suspended in air that would be a liquid or solid under ordinary conditions.

Vapor Density. The weight of a vapor or gas as compared to the weight of an equal volume of air; (**MW** of gas)/29 = vapor density, therefore, materials lighter than air have vapor densities of <1.0 and materials heavier than air have vapor densities >1.0. All vapors and gases will mix with air, but the lighter materials will tend to rise and dissipate. Vapors heavier than air are likely to concentrate in low, enclosed places (along or under floors; in sumps, sewers, manholes, trenches, and ditches) creating fire, explosion, or health hazards.

Vapor Pressure. The pressure exerted by a saturated vapor above its own liquid in a closed container. The lower the boiling point of a substance, the higher its vapor pressure. Vapor pressures are useful (with evaporation rates) in estimating how quickly a material becomes airborne within the workplace and thus how quickly a worker can be exposed to it.

Variance. In general, the square of the standard deviation; the mean of the square of the differences between the mean value of an underlying distribution and randomly selected values from the same underlying distribution of a random variable.

Velocity Pressure. Pressure exerted in the direction of airflow, with magnitude increasing with airflow velocity.

Ventilation. The removal and replacement of air to control conditions (e.g., air contaminant concentrations) in the workplace.

Ventilation Volume Flow Rate. The quantity of ventilation air flow measured as air volume per unit of time (e.g., cubic feet per minute, CFM; or cubic meters per second, m^3/s).

Vibration. Vibration is an oscillation wherein the quantity is a parameter that defines the motion of a mechanical system.

Vibration Isolation. Any of several means of preventing transmission of sound vibrations from a vibrating body to the structure in which or on which it is mounted.

Vibration Isolator. A resilient support that tends to isolate a system from steady-state excitation.

Vibration Meter. An apparatus for the measurement of displacement, velocity, or acceleration of a vibrating body.

Vital Statistics. In the context of **epidemiology**, birth death, marriage, divorce, and illness/**morbidity** data.

VOC. Volatile Organic Components, Volatile Organic Compounds, Volatile Organic Chemicals.

Volatility. The measure of a material's tendency to vaporize or evaporate at ambient conditions.

Voltage. Voltage is the potential electrical difference that makes it possible for current to flow in a circuit.

VP. See **Vapor Pressure**.

W

Warning Properties. Warning properties are signs or indications that exposure to a chemical is occurring. Typical warning properties include the presence of an odor and irritation of the mucous membranes, the eyes, the respiratory system, or the skin. Not all chemicals have noticeable warning properties and in general it is not safe to rely upon warning properties as an indicator of exposure.

Water Reactive. Describes a material that reacts with water.

WBC. White blood cell effects.

WEEL. Workplace Environmental Exposure Level. An **AIHA** guideline for exposures to selected chemicals.

Weighting Network. An electrical network designed to be incorporated in a sound level meter such that the latter conforms to a specified weighting curve. The three networks are designated A, B, and C, according to **ANSI** S1.4-1971. See also **Sound Level**.

Wet Bulb Globe Temperature. A commonly used method for anticipating heat stress situations in the work environment. Outdoors, it is a temperature index that is calculated using the **Ambient** or **Dry Bulb Temperature**, the **Globe Temperature** and the **Natural Wet Bulb Temperature**. Indoors, the **Ambient Temperature** is not used.

Wettedness. The amount of an individual's skin that is wetted by sweat.

WHF. Workplace Health Fund, 815 16th Street NW, Washington, D.C., 20006. A nonprofit agency targeting occupational health, (202) 842-7833.

White Noise. A noise whose spectral density (or spectrum level) is substantially independent of frequency over a specified range.

Wipe Testing. The collection of chemicals, etc., by wiping a surface (often with a 100-cm^2 piece of filter paper) and analyzing. This is at best a qualitative method (as opposed to a quantitative method).

Work History. In this context, a worker's history of exposure as a function of task, trade, chemical used, etc. Frequently used for epidemiological studies.

Workplace Environmental Exposure Level. See **WEEL**.

Workplace Health Fund. See **WHF.**

Z

Z List. Also known as Z Tables. The list of **OSHA PELs**; OSHA's Toxic and Hazardous Substances Tables Z-1, Z-2, and Z-3 of air contaminants, found in 29 **CFR** 1910.1000.

INDEX

INDEX

A

D

E

F

G

H

I

N

O

T

U

V